Introduction to Discrete Linear Controls

Theory and Application

OPERATIONS RESEARCH
AND INDUSTRIAL ENGINEERING

Consulting Editor: J. William Schmidt

CBM, Inc., Cleveland, Ohio

Applied Statistical Methods
I. W. Burr

Mathematical Foundations of Management Science
and Systems Analysis
J. William Schmidt

Urban Systems Models
Walter Helly

Introduction to Discrete Linear Controls: Theory and Application
Albert B. Bishop

In preparation:

Integer Programming: Theory, Applications, and Computations
Hamdy A. Taha

Transform Techniques for Probability Modeling
Walter C. Giffin

Analysis of Queueing Systems
J. A. White, J. W. Schmidt, and G. K. Bennett

INTRODUCTION TO
DISCRETE LINEAR CONTROLS
Theory and Application

Albert B. Bishop

DEPARTMENT OF INDUSTRIAL AND SYSTEMS ENGINEERING
THE OHIO STATE UNIVERSITY
COLUMBUS, OHIO

ACADEMIC PRESS   New York   San Francisco   London   1975

A Subsidiary of Harcourt Brace Jovanovich, Publishers

To L. G. Mitten

ACADEMIC PRESS, INC.
111 Fifth Avenue, New York, New York 10003

United Kingdom Edition published by
ACADEMIC PRESS, INC. (LONDON) LTD.
24/28 Oval Road, London NW1

Library of Congress Cataloging in Publication Data

Bishop, Albert Bentley, Date
    Introduction to discrete linear controls.

    (Operations research and industrial engineering
series)
    Bibliography: p.
    Includes index.
    1.   Control theory.   2.   Discrete-time systems.
3.   Feedback control systems.   I.   Title.
QA402.3.B53   629.8'312   74-10203
ISBN 0-12-101650-1

Contents

Chapter VI   Inverse Transformation

Chapter VII   System Performance: Measures and Environmental Effects

Chapter VIII   Parameter Selection in First-Order Systems Considering Sampling and Instrumentation Errors

Preface

This book is an introduction to discrete linear controls. It is written for those engineers, operations researchers, and systems analysts involved with the design, analysis, and operation of discrete-time decision processes. The basic theory is developed directly from the underlying discrete mathematics in an effort to provide the user with an understanding of the nature of discrete controls and equip him in as simple and straightforward a manner as possible with the necessary tools and techniques to deal with such systems.

This approach is somewhat rare in the current control theory literature, in which the theory of discrete controls is developed as an extension of the theory of continuous systems and usually in the context of electromechanical circuits.† For those whose interests lie in the areas of conceptual models of various discrete man–machine systems or the automation of inherently discrete production processes, the presentation here precludes the necessity of devoting time and energy to the learning of classical continuous controls in order to eventually gain the material they need. This is not to say that the continuous theory learned might not be useful, especially if the processes involved do have continuous features in their operation, but for many purposes discrete-systems theory will fully suffice. Furthermore, as computer control of manufacturing processes continues to advance and as quantitative analysis and optimal design of an ever-widening variety of societal and ecological systems involving human decision makers emerges, the appropriateness of discrete models and hence the need for ready access to the methodology of discrete control systems is rapidly increasing.

As in most subjects, the more extensive the background, both analytical and empirical, that the reader brings to his study of the material herein, the greater his potential for not only gaining a basic understanding of the material presented but also extending it in innovative ways. Throughout the study of the text, the reader is encouraged to question and explore, to develop more realistic or effective models, and to devise new approaches to the derivation and manipulation of models and the simplification of calculation procedures. Above all, he should be continually searching his field of experience for areas of application. It has been assumed in preparing the text, however, that the

† See, for example, *Digital and Sampled Data Control Systems* by J. T. Tou (1959, see reference list for detailed information) or *Discrete-Time and Computer Control Systems* by J. A. Cadzow and H. R. Martens (1970). The latter develops discrete control theory with a minimum of dependence on continuous system theory. A recent book devoted completely to discrete theory, and hence an exception to the point being made here is *Discrete-Time Systems* by J. A. Cadzow (1973).

reader will have a background in both differential and integral calculus and be familiar with the basic concepts of classical optimization theory for analytical functions. Although numerous opportunities exist for applying a variety of alternative optimization techniques, these are either mentioned in passing or left entirely to the reader. Sufficient knowledge of probability theory for the reader to be familiar with basic definitions, notions of independence, moments, joint moments, and common distributions is assumed, although much of this material is reviewed briefly in the context of its usage in the text. On the other hand, no particular knowledge of discrete mathematics is assumed. The calculus of finite differences and solution procedures for linear difference equations with constant coefficients is covered in detail in Chapters III–VI. In addition, only a cursory familiarity with the notions of limits is assumed. This is in spite of the fact that the applicability of the z transform depends on the convergence of an infinite sum, the conditions for which we state and then assume hold from there on. No background whatsoever in control theory is assumed.

The book provides a series of building blocks upon which one can formulate models and devise analysis and design exercises which can extend the coverage in the text to best suit the background and interest of the teacher and students. Specifically, Chapter I is a basic introduction to systems analysis, discrete systems, the concept of control, and the role of models in system analysis and design. In Chapter II the development of system difference equations is illustrated with respect to a generalized discrete-process control system, a production–inventory control system, and a simplified flow analysis of the criminal justice system. Chapter III introduces some concepts from the calculus of finite differences useful in the formulation and solution of difference equations. Solution of linear difference equations with constant coefficients by classical means is discussed in Chapter IV. Chapter V introduces the z transform as a more flexible approach to the formulation and solution of linear difference equations, and Chapter VI presents the inverse transformation. In Chapter VII criteria for evaluating system performance are discussed. This is followed by examining the performance of a simplified first-order process control system when perturbed by each of several types of common system disturbances. This performance evaluation is extended in Chapter VIII to include the effects of measurement and sampling errors and a series of examples is presented to illustrate the selection of an optimal value for the control system parameter for each of several types of disturbance given several possible performance criteria. Chapter IX is devoted entirely to system stability and tests to determine the conditions under which a system will operate stably. The properties and performance of several types of second-order system are presented in Chapter X. Emphasis is given to the analysis of the ranges of parameters for stable operation and the interrelationships between these parameter values and the effects of random measurement errors. Chapter XI considers extensions to higher order systems. The signal-flow graph is introduced here as a convenient means of representing and manipulating complex systems. Effects of delay in sensing and feeding back information for decision-making purposes is included among several miscellaneous concluding topics.

The exercises at the end of each chapter are designed to extend the material presented in the text. Each chapter has several drill-type problems to test understanding of each new topic and technique from that chapter. Many of these are presented sequentially so that a course instructor will always have some problems he can assign upon completion of each section or subsection. As new steps in problem solution are covered in the text, exercises are available to apply that step to the results of previously completed steps. Other exercises provide opportunity for additional study of systems or techniques or require verification of expressions presented in the text with only partial or no derivation. Others force attention to new formulations or areas of application. A few might be considered minor topics for research. Because of the building-block nature of these exercises, there are numerous cross references among them. Difference equation models developed in exercises in Chapter II are solved in several stages by classical means in exercises in Chapter IV and by the z transform in Chapters V and VI. System performance under a variety of environmental conditions is evaluated in exercises in Chapters VII, VIII, or X and stability established in exercises in Chapter IX. It is hoped that the familiarity the reader gains with a few specific systems in this way will permit concentration on each new topic as it is introduced without having to feel out a new system structure at every turn.

Acknowledgments

The author is indebted to many people, only a few of whom can be acknowledged here. I was first attracted to the industrial engineering–operations research arena because of an intense fascination with the results one of my instructors could get from mathematical representations of bits and pieces of quality control and production control systems. This man, who later became my graduate adviser, friend, and continuing source of stimulating ideas and encouragement, was Loring G. Mitten, now chairman of the Management Science Division at the University of British Columbia. A powerful member of his profession and the world's best and most unselfish adviser, the true extent of his contributions can be fully understood only by his advisees. It was he who suggested a text of this kind almost twenty years ago, and whose continued encouragement led, at long last, to its completion. Out of respect and appreciation, I dedicate this book to him. I hope it is worthy of his high standards.

I also acknowledge the help provided by David Baker and William Morris, my department chairmen during the lengthy writing process, in making available the resources of The Ohic State University Department of Industrial and Systems Engineering to assist in this undertaking. Numerous students commented on and weeded out errors in several versions of course notes. All such efforts are appreciated, but Seetharama Narasimhan deserves special mention for his assistance with both text and exercises. Salah Elmaghraby performed an outstanding comprehensive review of an early manuscript which led to numerous substansive changes, all of which should result in significant improvements. The patient typists who readied the material for course notes and text manuscript deserve special credit for enduring this impatient author. My thanks to Mrs. Cindy Dickinson, Mrs. Judy Crowl, Mrs. Lois Graber, Mrs. Carol McDonald, and especially to Miss Joan Case who single-handedly typed the entire final manuscript. Finally, my deepest thanks to my wife, Louise, and my children, John, Sue, and Jim, for their love and support which were so essential to bringing this task to completion.

Introduction to Discrete Linear Controls

Theory and Application

| Chapter I | **Systems Theory and Discrete Linear Control Systems** |

The discrete, linear, time-invariant system has in recent years become an object of increasing interest in many areas. Much of this attention has come from operations researchers and systems analysts involved with either human decision makers who tend to make specific, individually identifiable decisions or digital or time-shared system components whose information outputs occur periodically. Furthermore, because of the relative simplicity of mathematical models of such systems, beneficial insights can often be gained by modeling a wide variety of systems as though they were discrete, linear, and time invariant.

In this first chapter we introduce the concept of a system and then define the basic terms of discreteness, etc., in the context of a system. The notion of control and the components and structure of a control system are then examined. The chapter closes with a discussion of models and their role in system analysis and design.

1.1 Systems Theory

The 1950s saw the rise of "operations research" with its emphasis on finding optimal solutions to operational problems (Churchman *et al.*, 1957). An oft-stated feature of OR methodology is a "systems approach," which basically means that the researcher should carefully strive to consider all those factors which are likely to have a reasonably significant effect on the solution of the problem. For example, the routing and scheduling of trucks among terminals of a common carrier cannot be done properly without consideration of the company's truck maintenance program and the materials

1

handling capability at the loading docks. Aircraft instruments and controls have to be designed with both the motor skills and the information processing capacity of the pilot explicitly in mind. Of particular importance to the industrial engineer, the layout of single work stations has had to give way to production line design involving not only production but also materials handling and storage.

With the systems approach came the compiling of lists of "pertinent" factors as one of the first steps in any problem-solving effort. To staff a tool crib one needs to know the types and numbers of tools handled, the frequency of requests for each type by time of day and week, and the service-time distributions. An added frill might be the interference patterns that result from the presence of more than one attendant. Further study usually produces additions and deletions from the list and some understanding of the interrelationships among the items listed. This identification of items to include in the system and the descriptions of their interrelationships is termed *model building*. The description itself is the *model*, which serves both as a source of learning and insight and as a vehicle to optimize system structure and performance.

In order to obtain or even define an optimum solution to a problem, the problem-solver needs a criterion of optimality and a scale on which to evaluate competing solutions. In industrial settings one usually seeks to maximize profit or to minimize cost, although surrogate measures involving product quality, adherence to deadlines, and customer service, all of which contribute to profit in complex ways, are often used. The trucking company may attempt to minimize delivery time or damage to freight. The industrial engineer could attempt to maximize throughput or minimize the bank sizes of his production line. Elsewhere, particularly in the public sector, benefit or effectiveness often share the spotlight or even replace profit and cost as the basis of evaluation. Both cost and performance must be explicitly considered by the designer of an interceptor missile system, where performance could involve maximizing the probability of intercept or minimizing the damage inflicted by an attack force. The aircraft cockpit designer, however, is essentially completely interested in flight safety with equipment cost involved only as a constraint, if at all.

The systems approach of the operations researcher has undergone considerable extension and formalization in recent years resulting in what many refer to today as *systems theory*. The history of this evolution and discussions of the principal current formulations are presented by Klir (1972). Brockett (1970) presents an engineering oriented discussion of linear systems, and Howard provides extensive coverage of dynamic probabilistic systems in his two-volume set divided into Markov (1971a) and semi-Markov and decision processes (1971b). In this book we will be extensively involved with the systems approach of model building, criterion formulation, and optimization of performance of discrete linear decision systems, referred to here as *control systems*. This is the type of system of particular interest to the manager, public official, operations researcher, and design engineer. We will draw heavily on available systems theory, but only to the extent necessary to motivate, derive, and explain the points being developed. The reader is referred to the sources listed above for further discussion of systems theory.

Because of our primary interest in discrete systems, a discussion of what is meant by "discrete," "discrete system," and other terms basic to our exposition is in order at this point. We will then turn our attention to control theory and introduce the concept of a decision or control system. The chapter concludes with a general discussion of models and their role in systems analysis and design.

1.2   Discrete Systems

A discrete event is a specific happening readily distinguishable from other events. Examples include the inauguration of a president, the opening of a supermarket, the dispatching of a bus, or the completion of the manufacture of the ith engine block in a production run. Often, however, the discrete character of an event is a matter of definition. For example, the flow of water through a hydroelectric station is, under normal operating conditions, a continuous phenomenon. Yet one could define as a discrete event the passing of the one-billionth gallon through the station. Similarly, the height of water in a reservoir is a continuous variable. Yet it can be discretized by measuring to the nearest foot only and attaching an integer (discrete) measure to the level. Time is often described in discrete terms such as the number of days to repay a loan. It may also frequently be expressed in units corresponding to the occurrence of a sequence of discrete events. For example, *time i* could be defined as the time at which the ith engine block is completed or as the end of the ith week in a production control plan in which factory schedules are issued weekly. Obviously, to convert to clock or calendar time, i must be multiplied by the time between events and the result added to the time corresponding to the origin of the sequence.

As used herein a *discrete system* is one whose output occurs naturally on a discretized time scale, often referred to as *discrete time*. The engine-block manufacturing line is a good example. It is obviously not meant that the line exists or operates only at those instances at which a block is finished, but that the meaningful descriptors of the operation of the line are, for the most part, the characteristics of the block produced. Since each succeeding set of such characteristics is attached to succeeding engine blocks, it can also be ascribed to the discrete points in time at which the blocks are completed. When described in this way, the engine-block line is a discrete system.

The discreteness of the process output, however, is not the only factor which determines the discreteness of a system. A Fourdrinier machine produces paper in a continuous sheet. The quality characteristics such as density and moisture content are determined, however, by moving a gage across the bed of the machine. At the completion of a scan, the gage signals are analyzed and a discrete-control action initiated to adjust for any noted deviations from standard. Thus the control of this continuous product is accomplished by a discrete-control system. Similarly, a central computer which sequentially monitors a number of processes on a time-shared basis supplies each unit in turn with a discrete-control signal regardless of the nature of the processes or their outputs.

In summary, the term discrete system refers in this book to any system whose operation or output is conveniently described on a discrete time scale; although, in general, the system characteristics, such as the height of an individual engine block, are given continuous measures. Many authors prefer the term "discrete-time system," which is really a more apt description of what is meant. In general, the index i is used to refer to discrete time. As stated previously, i is related to continuous time t by the relationship

$$i = t/T, \qquad i \text{ integer}, \tag{1.2.1}$$

where T is the time between events. Usually, functions of discrete time are written simply in terms of the argument i, e.g., $f(i)$, with T suppressed. However, where real-time considerations are important, conversion from $f(i)$ to $g(t)$, the comparable function in real time, is accomplished simply by substitution of t/T for i in $f(i)$. For example, the function of discrete time

$$f(i) = 3i^2 + 2i$$

can be expressed in terms of continuous time t as

$$g(t) = \frac{3}{T^2} t^2 + \frac{2}{T} t.$$

For $T = 2$,

$$g(t) = 0.75t^2 + t.$$

Conversely, for

$$g(t) = t^3 + 3t$$

and $T = 2$,

$$f(i) = 8i^3 + 6i.$$

Two additional properties which will usually be assumed for the systems discussed herein will now be defined. These are the properties of "linearity" and "time invariance." The reader is referred to the work of Howard (1971a, Chapter 2) for a complete and well presented treatment of the theory of discrete, linear, time-invariant systems.

Linearity

As is well known to engineers, analysts, and operations researchers, a linear relationship between a dependent variable and a group of independent variables is one which can be expressed as a linear surface, i.e., as a line, plane, or hyperplane. If it requires 1.5 minutes to test a circuit board regardless of whether it is the first, seventeenth, or whichever number tested during a production run, the time t required for the test can be expressed as $t = \sum_{k=1}^{n} 1.5 = 1.5n$, where n is the number of circuit boards in the current batch. This equation is, of course, the equation for a straight line passing through the origin, and we say that the total test time is a linear function of the number of items to be tested. Similarly, if the direct cost to manufacture one unit of product of type k is c_k, regardless of how many items of that type have already been produced and what types and how many of other kinds of items are being made, total production cost C can be expressed as

$$C = c_0 + \sum_k c_k n_k, \tag{1.2.2}$$

where n_k is the number of units of product type k manufactured and c_0 represents fixed costs. Equation (1.2.2) is the equation of a hyperplane, a linear surface, with cost–axis intercept c_0.

To extend the notion of linearity to discrete systems, consider Fig. 1.2.1.

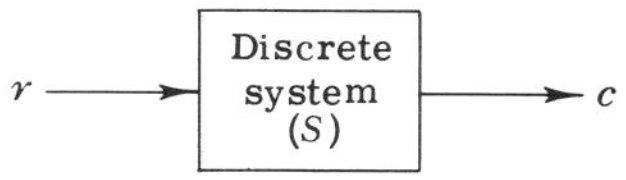

FIG. 1.2.1. Discrete system with input r and output c.

r and c are vectors of discrete-time inputs and outputs, respectively, and are often referred to as input and output *signals*. S represents the transformation performed by the discrete system on the input to produce the output. It is often referred to as the *system operator*. Specifically, using i as the index of discrete time,

$$r = \{r(0), r(1), \ldots, r(i), \ldots\},$$
$$c = \{c(0), c(1), \ldots, c(i), \ldots\},$$

and

$$c = S(r).$$

Thus, $r(i)$ is the input to the system at time i and $c(i)$ is the system output at time i. $c(i)$ is a function of all elements of r up to and including $r(i)$, the exact functional form depending on the nature of the system. Note the assumption that both $r(i)$ and $c(i)$ are not defined for $i < 0$, a valid assumption for physically realizable systems.

Now suppose an input signal r_1 produces an output signal c_1 and that an input signal r_2 produces the output signal c_2, i.e., $c_1 = S(r_1)$ and $c_2 = S(r_2)$. Let us define the composite input r as a linear combination of r_1 and r_2, i.e.,

$$r = ar_1 + br_2,$$

where a and b are constants, and define c as the output signal resulting from input r. Thus,

$$c = S(r) = S(ar_1 + br_2).$$

We then say that *the system is linear if*

$$S(ar_1 + br_2) = aS(r_1) + bS(r_2); \tag{1.2.3}$$

i.e., if the output resulting from a linear combination of input signals is made up of the same linear combination of the individual outputs which result from those inputs alone, we have a linear system. c is then a *linear function* of r, and S is said to be a *linear operator*.

It is apparent that the primary significance of system linearity is the applicability of the principle of superposition, which permits us to investigate the behavior of the system for each of several relatively simple inputs and to predict its behavior under much more complicated linear combinations of these inputs.

Time Invariance

Time invariance refers to the constancy of system performance over time. The 1.5 minutes used above as the time required to test a circuit board may be applicable only so long as an experienced technician runs the tests. A new operator could conceivably take longer. Furthermore, as the new operator begins to get the hang of the test set, his unit operating times may begin to decrease in some complicated pattern toward 1.5 minutes. The time required to test n circuit boards could then no longer be expressed simply as $t = \sum_{k=1}^{n} 1.5 = 1.5n$ as before but by the more complicated expression

$$t = \sum_{k=1}^{n} t_k,$$

where t_k is the time required to test the kth board. Chances of predicting values for t_k are, of course, remote; so planning will undoubtedly be much less precise while the new operator learns his job. Further, suppose we wished to modify the cost equation (1.2.2) to reflect inflation or the enactment of a new labor contract. The c_k's would then have to be made functions of time with a resulting increase in the complexity of the model. Thus time invariance is an important property of systems, since, when it obtains, system modeling and hence the ability to formulate and implement good designs can be greatly simplified.

To express time invariance in the context of discrete systems we define r^k and c^k as r and c, respectively, delayed in time by k units. For example, given

$$r = \{r(0), r(1), \ldots, r(i), \ldots\} = \{1, 2, 3, \ldots, (i+1), \ldots\},$$

then

$$r^3 = \{r^3(0), r^3(1), \ldots, r^3(i), \ldots\} = \{0, 0, 0, 1, 2, 3, \ldots\}$$

or,

$$r^3(i) = \begin{cases} 0 & \text{for} \quad i = 0, 1, 2, \\ i - 2 & \text{for} \quad i \geq 3. \end{cases}$$

Now, given that $c = S(r)$, we say that *the system is time invariant if for all* $k \geq 0$,

$$S(r^k) = c^k;$$

i.e., if the output resulting from a given input is completely unaffected by any delay in the occurrence of the input except for the delay itself, the system is time invariant.

Except for the general background material on control systems and models in the remainder of this chapter, we will deal hereafter exclusively with discrete, linear, time-invariant systems. We now turn our attention to control theory and control systems.

1.3 Control Theory

To control means† "to exercise restraining or directing influence over; to dominate; regulate." Control theory is that set of theories, procedures, and techniques useful in the synthesis and analysis of control systems. Note the explicit inclusion of a decision-making function in this definition. Since decision systems are a very common and important type of system, control theory constitutes a very important subset of general systems theory. In

---

† *Webster's New International Dictionary*, 2nd ed., s.v. "control."

particular, the systems approach discussed earlier is part of the basic modus operandi of control theory. Therefore, we will be involved with modeling control systems and manipulation of the models to achieve satisfactory and, where possible, optimal designs.

Although many examples of effective controllers date back to the late nineteenth century,† control theory as we know it emerged from the complex of electrical, mechanical, and hydraulic controls used by the military in World War II. These controllers, which enjoyed usefulness long before any rigorous associated theory was developed, were generally some sort of dynamic tracking device which kept an antenna or gun aimed automatically at an aircraft or other target. As the theory evolved, it naturally centered around the operation of these existing devices which were usually of a continuous, deterministic nature. Thus the mathematical systems models were almost always in the form of differential equations.

Obviously, to obtain an explicitly closed-form expression for any system variable as a function of time and the various parameters so that design or analysis decisions could be made, it was necessary to solve these differential equation models. Thus, the state of the art of solution methods for differential equations also had an important bearing on the early development of control theory. It is not surprising then that the bulk of the work in the 1940s and early 1950s was directed to deterministic, time-invariant, linear systems, systems which can be represented by linear differential equations with constant coefficients. Fortunately, many actual control systems were of this variety or close enough that these models yielded useful results. Furthermore, and of much greater importance to our purposes here, many common discrete systems also are essentially linear and have parameters whose values do not vary appreciably with time, so that a great deal of progress can be made with models based on linear *difference* equations with constant coefficients. Indeed, the approach followed in this book is based on this premise.

The great potential of control theory and its applications was indicated by Norbert Wiener in *Cybernetics* (1948) and *The Extrapolation, Interpolation, and Smoothing of Stationary Time Series* (1949), in which he pointed out the similarities of all levels of systems and organizations and the important roles that command and control play in their performance. He provided an approach to the handling of random variables as would be involved with sampling and measuring errors and developed the concept of optimum systems and techniques for their realization. The ramifications of sampling are of particular importance in discrete control, especially when discrete controls

---

† Wiener (1948) states that "the first significant paper on feedback mechanisms is an article on governors, which was published by Clerk-Maxwell in 1868." It is interesting to note that the governor itself had been in use since approximately 1775.

are applied to inherently continuous processes or extremely high speed discrete processes such as characterize much of today's manufacturing. Random (or "stochastic") factors in general are significant in almost all manufacturing, management-control, and other man–machine systems and their treatment is an important part of the material which follows. Furthermore, any profit-minded manager or system designer or the tax-paying public appreciates the benefits of optimally or near-optimally performing processes. This is particularly true with society oriented systems such as pollution control and water resources where budgets are likely to be tight and other needed resources limited. Optimization is a recurring theme in the following work, which includes several different approaches to system optimization.

Work in nonlinear systems (Truxal, 1955; Cosgriff, 1958) also began to flourish in the late 1940s followed closely by the emergence of state-space system representations (La Salle and Lefschetz, 1961) and the theory of optimal control (Chang, 1961; Merriam, 1964; Tou, 1963).

Since its early beginnings control theory has developed rapidly until today dozens of books on some phase of controls are published annually and literally hundreds of articles and correspondence items appear each month in technical journals. There is a plethora of available material. Nevertheless, the sources available to the industrial engineer, manager, quality-control engineer, social scientist, and operations researcher for a quick entry to the field of discrete-control theory are almost nonexistent. It is to assist these people to gain a basic understanding and working knowledge of discrete controls that this book is written.

1.4 Control Systems

The fundamental features of control systems are most easily presented in the context of examples. Several such examples are presented in this section to illustrate the concepts of open- and closed-loop control, discrete-system representation, feedback and feedforward, and the great range of applicability of control-system models. After examination of these examples, we will attempt to codify the basic elements essential for control-system operation.

Control-System Examples

Open-Loop Temperature Control

Figure 1.4.1 is a block diagram of a simple home heating system of the type used before the advent of the thermostat. An initial setting of the gas valve admits fuel to the furnace at a rate determined by the valve setting. The

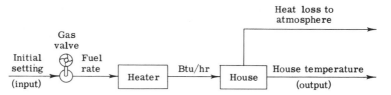

FIG. 1.4.1. Open-loop house temperature control.

heater then, at a corresponding rate, produces heat energy which passes into the rooms of the house. Some of this heat energy escapes as losses to the atmosphere, the amount of loss being dependent on the inside and outside temperatures, the tightness of the house structure, wind conditions, and the frequency of door and window openings and closings. The rest of the heat energy is used to warm the house. For a given valve setting, the temperature reached in the house is a function of the heat losses, which makes it dependent upon a number of factors over which the person who set the valve has no control. For example, a shift in wind direction or a window left open could radically change the house temperature, and the system as shown could do absolutely nothing about it. This is because there is no system element which can automatically detect the temperature shift and reset the gas valve to restore the desired temperature. In other words, there is no mechanism for comparing information about the actual output with the desired output and using the result to adjust the system operation. Such a system is, therefore, called an *open-loop* control.

Closed-Loop Temperature Control

When the occupant of the house becomes uncomfortable and shuts the window or adjusts the valve, he is acting as a sensor, comparitor, and feedback link. If he is considered part of the system, the system would then be a *closed-loop* or *feedback* system. Figure 1.4.2 depicts such as closed-loop temperature control system except a thermostat is shown as the sensor and comparitor providing completely automatic control. Note that from the gas valve through the heat loss and house temperature, this system is the same as

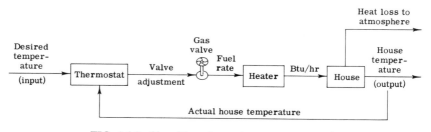

FIG. 1.4.2. Closed-loop house temperature control.

the one in Fig. 1.4.1. The external input to the system from the occupant of the house, however, is the desired temperature setting of the thermostat and not a gas valve setting as in the open-loop system. The thermostat, through some drive mechanism, does the adjusting of the gas valve in accordance with the deviation between the desired and actual temperatures. Note that in this case, the output is dependent not upon the input directly, but on the *difference* between the input and output. If temperature falls for any reason, the thermostat will sense it and send additional fuel to the heater to compensate. A glance at Figs. 1.4.1 and 1.4.2 reveals the appropriateness of the names "open loop" and "closed loop."

Simple Discrete Controller

The temperature controller just discussed is a continuous-time system. For an example of a discrete system, consider the previously mentioned engine-block line. Although many dimensional and other quality characteristics are of importance to the proper functioning of the engine, let us focus our attention on the diameter of the number-one cylinder. Figure 1.4.3 shows those elements of the engine-block line which produce and control this dimension.

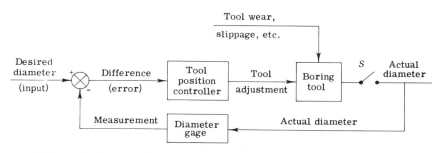

FIG. 1.4.3. Control of diameter of number-one cylinder on engine-block line.

The final diameter is cut by a boring tool. However, the boring operation is not carried out continuously. Instead, it takes place periodically as each new engine block is moved into position. This is represented in the figure by the switch S, which is considered to be closed instantaneously to initiate work on each new engine block. Since the boring tool is subject to wear and perhaps some slippage or even setup error, after boring the cylinder is gaged and the measured diameter compared to the desired value. The difference between the desired and actual diameters, usually referred to as the *error*, is fed to a controller which makes any needed adjustment in the setting of the boring tool. Usually this adjustment will be completed prior to the start of boring the next engine block, i.e., before S is closed again. Note that the symbol $\otimes$ indicates addition or subtraction as designated by the plus and minus signs on the input arrows.

Generalized Automatic Process Controller with Sampling

The system just described could be made more complicated in any of several ways. First, were measurement errors associated with the gage significant, an additional exogenous variable would have to be associated with the gage. In addition, if variations in cylinder diameter from one block to the next were not expected to be very large, the gaging operation might be restricted to every three or four blocks or to a randomly selected subset of blocks. In such cases, sampling errors would also enter the system. A generalized process-control system which embodies both sampling and measurement errors is shown in Fig. 1.4.4. As shown, it is immaterial whether the actual

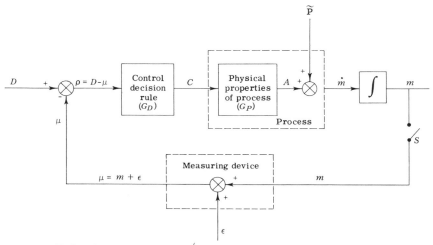

FIG. 1.4.4. Generalized automatic process controller with sampling.

product is discrete or continuous. The drawing of the sample for measurement and adjustment purposes is, however, done periodically as represented by the switch S.

The process output m existing at the instant S is closed is detected by the measuring device, which because of sampling distributions, measurement inaccuracies, or both, adds an error ε to the true value m. The value μ available for comparison with the desired level of operation (often referred to as "process bogie") D is thus the sum of the true value and the measurement error. The error signal ρ is given by

$$\rho = D - \mu = D - m - \varepsilon. \tag{1.4.1}$$

The controller produces a control signal C that is a specified function G_D of ρ and perhaps previous values of ρ if the controller is equipped with memory.

The control signal C reacts with the physical properties of the process being controlled to produce a controlled change A in process output. G_P is used here to represent the functional relationship between A and C. The output is also influenced by the effects of uncontrollable variables $\tilde{P}$, such as raw material imperfections, tool wear, etc. Since A is defined as an instantaneous adjustment and $\tilde{P}$ as an instantaneous perturbation, each affects the rate of change of output $\dot{m}$. The actual output m is the integration of these rates of change over time.

Production Inventory Control System with Feedforward of Customer Orders

Figure 1.4.5 presents a possible configuration for a production inventory control system for a single product or product line. Since production decisions are introduced into the factory only at certain, often periodic, times, the

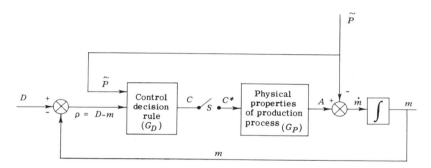

FIG. 1.4.5. Production inventory control system with feedforward of customer orders.

system is shown with switch S between the decision rule and the production process. Each production decision C is based upon an analysis of customer order rate $\tilde{P}$ and inventory level m, or, more specifically, on the difference ρ between the desired level D and the actual level m. When S is closed, the current value of C, here designated C^*, is introduced into the factory, whose physical properties (delays, buildup times, etc.) are represented by G_P. The factory produces at a rate A which, in combination with the depletion rate due to $\tilde{P}$, causes inventory to change at a rate $\dot{m}$. These changes add up (are integrated) over time to give actual inventory level m.

Note that the decision rule G_D is here based on two types of information, the deviation of inventory level from that desired and the pattern of customer orders. This is a *multi-input decision process*. Further note that the feeding of the customer order information to the decision maker is referred to as *feedforward*.

Criminal Justice System Feedback Model

As an example of the wide range of real world situations which can be studied as control systems using control theory methodology, consider the flow through the criminal justice system of persons accused of committing crimes. A simplified model of this flow is shown in Fig. 1.4.6. The model shown is based on a model used by Belkin *et al.* (1973) to assist planners to assess the effects in terms of dynamic flows through the system of possible changes in such factors as policies governing arrests, use of station-house warnings, probation, incarceration, and rehabilitation programs. Blumstein and Larson (1969) had shown as early as 1969 the benefits of viewing the criminal justice system in this way.

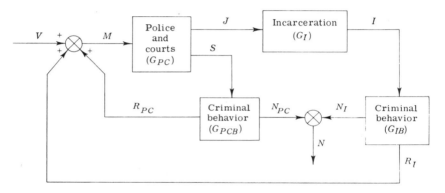

FIG. 1.4.6. Criminal justice system feedback model.

In the model as shown here, V represents virgin arrests, i.e., people arrested for the first time. V is of course a function of time as are all the flows in the system. The independent variable time, which in this case can be either continuous or discrete as best suits the purposes of the researcher, for simplicity is suppressed in the notation used. Total arrests M are the sum of virgin arrests, arrests of recidivists (repeat offenders) whose most recent previous offense did not result in incarceration, R_{PC}, and arrests of recidivists whose most recent previous offense resulted in incarceration, R_I. These arrests result in police and possibly court action, resulting in the incarceration after certain time delays of a portion of those arrested. G_{PC} represents the functional relationship between M and the number sent to jail, J, and the number returned to society without incarceration, S. G_I represents the relationship between those jailed, J, and those released from jail, I, and accounts primarily for time spent until parole or completion of sentence. All those released, either with or without incarceration, return to an environment where interactions with criminal activity is a possibility. Some will eventually succumb and become recidivists. Others will not and thus exit the system. N represents the number leaving the system and is the sum of N_{PC} and N_I, the numbers leaving the system of previously nonincarcerated and incarcerated people, respectively. G_{PCB} represents the relationships among S, R_{PC}, and N_{PC}, and G_{IB} the relationships among I, R_I, and N_I. These relationships involve time lags, the effects of behavior patterns, and the conditions of release on one's propensity to commit further crimes.

Although the reader can immediately see many areas in which the model as shown lacks degrees of reality, such as the common treatment by police and courts of all arrestees regardless of crime type or prior record, the model still enabled Belkin *et al.* to make workable estimates of system loads. They furthermore used the model and historical corrections data to work backward to estimate the pattern of virgin arrests, a normally very illusive item for those studying the criminal justice system to obtain.

Basic Elements of Control Systems

Diagrams similar to those above could be constructed for manpower assignments for work on a production line, harvesting crops, or mobilization of a military force. Buildups over time of mass-transit systems, water storage resources, and sewage disposal plants in the face of forecast and actual demands can also be represented in this general manner. So can the flow of reports for a management information storage and retrieval system. Gordon (1969) has used discrete-control models to study the control of traffic at critical intersections, and Belcher (1971) has applied it to control the headway in car-following situations on highways. In short, many management and operational functions at all levels—hardware, tactical, and strategic—are control functions and as such can be analyzed and synthesized as control systems using the vast technology of control theory.

Control theory provides the base from which quantitative answers can be obtained concerning the characteristics required of the instrumentation used to measure quality characteristics of piece parts and the effects of delays in correcting measured deviations from process specifications. The nature of transient effects resulting from adjustments made in process setting or from uncontrollable factors can be determined. Effects of various classes of inventory-control rules can be investigated, and preferred values of control-system parameters determined based on economic criteria. In some cases, optimal responses to forecast or actual inputs, such as customer orders, can be derived. One can even investigate social phenomena such as policies in setting bail or scheduling hospital patients for radiology.

Control-system formulations and techniques are many and flexible and in general are determined by the characteristics of the system involved and the nature of the control problem. However, as may be seen from Figs. 1.4.1 through 1.4.6 control systems have many features in common. Generally, such systems will consist of:

(1) *An input or reference level* In Fig. 1.4.3 the input is a design bogie for some quality characteristic of the product being produced. In an aircraft tracking system, the aircraft position would be the input. In any case it is the so-called desired level of system operation and may, of course, be a function of time. Sometimes multiple inputs are involved such as when several characteristics of a product are interrelated and must be controlled simultaneously. Although no reference level is shown explicitly in Fig. 1.4.6, prison capacity could have an effect on the court's decision to incarcerate.

(2) *An output or controlled variable* This is the quality characteristic or characteristics being controlled. In Fig. 1.4.3 the output is some characteristic of the product. In Fig. 1.4.5 it is the actual inventory level. The house temperature is the output in Figs. 1.4.1 and 1.4.2, and in an aircraft tracking system it is the direction of the tracking antenna.

(3) *A process or plant* This is the process or device which produces the output. It is the production facility of Figs. 1.4.3 and 1.4.4, the heater in Figs. 1.4.1 and 1.4.2, and would be the antenna drive motor in an aircraft tracking system. In the criminal justice system, what constitutes the plant depends on the functional use of the model.

(4) *A controller or decision-making device* Here is where adjustments to be made in the level of plant operation are calculated or, in the case of man–machine systems, are decided upon. In the open-loop system of Fig. 1.4.1, the controller, namely the gas valve, is set once and for all when the process is started. In the closed-loop system of Fig. 1.4.2, however, the valve is subject to continual adjustment as the heat losses vary.

(5) *Perturbations or noise* These are the results of environmental effects over which no direct control, at least within the control system itself, is available. The system can at best react to the presence of these effects either as they occur or are forecast. Customer orders in Fig. 1.4.5 and heat loss in Figs. 1.4.1 and 1.4.2 are examples of such perturbations.

Perturbations, or " noise " as such disturbances are called in electrical and information systems, are almost always present in control-system applications. Wind loading on antennas, breakage of warehoused items, tool wear in production processes, variations in worker performance as functions of ability, mood, or the activities of the night before, and heat losses in homes and chemical reactions are facts of life with which control-system designers must contend on a continual basis. For this reason, essentially all control systems have one or more feedback loops to permit the system to compensate for these anticipated perturbations. The presence of the feedback loop also permits automatic response to changes in input, an essential feature for a tracking device. Because of this, two more elements common to almost all control systems are:

(6) *A feedback device* This is the device which measures some feature of the output and processes the information for ready comparison with the input. The measuring device of Fig. 1.4.3, perhaps in combination with appropriate recording devices, fulfills this function.

(7) *A comparator or detector* This is the device which actually compares the recorded output with the input. Examples are the mechanical differential, the electrical bridge, and a human observer. The thermostat of the temperature control system acts as a comparator, but is actually the feedback device and comparator combined.

All feedback control systems will contain at least one of each of these types of elements although, as has been stated, the system configurations may vary widely.

1.5 System Models

As previously mentioned, analysts have long realized the usefulness of models in the design and analysis of systems. Control systems are no exception. It seems appropriate, therefore, to review briefly at this point something of the nature, types, and uses of models and the factors to be considered in their selection and development.

As used here, the term model refers simply to an abstraction of the real system. It is an abstraction, moreover, which embodies those features of the actual system which must be treated explicitly in any analysis or synthesis of the system. The block diagrams of Figs. 1.4.1 through 1.4.6 are models of actual systems. They show the basic elements of their respective systems and indicate the interconnections among them. On the other hand, they do *not* show the location of the components in their environment, the physical appearance of any element, or the quantitative relationships existing between, say, the valve setting and the flow rate of the fuel entering the heater in Fig. 1.4.1 or 1.4.2. Additional models would be required to portray these additional aspects of the systems. A layout diagram and pictures or artists sketches would be required to show layouts and physical features. A calibration curve or mathematical equation would be used to display quantitative relationships.

Basically, models can be classified into three generic categories (Churchman *et al.*, 1957):

(1) *Iconic*—models bearing a physical resemblance to the actual item or system they represent. Examples are photographs, blue prints, and aircraft models used in wind tunnel tests.

(2) *Analog*—models in which one physical property is used to represent another. The calibration curve cited above is an analog model in which horizontal distance represents (is an "analog" for) valve setting, and vertical distance represents fuel flow rate. Analog computers are excellent examples of this category of model.

(3) *Symbolic*—models in which a symbol, usually a number or a letter, is used to represent a physical quantity. Mathematical and symbolic logic expressions are of this variety.

Even though there may be some difficulty in placing every conceivable model into one of these three categories (e.g., the block diagrams of Figs. 1.4.1 through 1.4.6 seem to be somewhere between iconic and analog models and a real challenge is posed in classifying a program for a digital computer), this categorization has proven useful. Note that as one moves down the list from iconic towards symbolic the physical resemblance between model and "real world" decreases but with an accompanying increase in generality, flexibility, and manipulability. It is certainly easier to rewrite an expression for the lift on an air foil to account for a change in foil shape than it is to restructure a wind tunnel model of the foil. However, the observation of wind tunnel tests is a lot more direct than interpretation of the solution of a set of flight-dynamics equations.

Other categorization schemes could of course be used. In fact any of the descriptors discussed with respect to systems, such as linear, nonlinear, discrete, continuous, time-invariant, deterministic, and stochastic, can also be applied to models of systems. It should be made clear, moreover, that the model chosen to represent a given system need not necessarily fall into all of the same categories that the actual system does. A high speed production process making light bulbs is inherently discrete because the output consists of clearly identifiable individual objects. Yet models of such systems may, for reasons of convenience or the preference of the analyst, treat the system as continuous in time. Conversely, a chemical plant producing liquid petroleum products may well be modeled as a discrete system by treating the product in units of a day's production or the quantity held in a given container. In the former case, the time between the emergence of individual light bulbs is considered short enough so as to be continuous for study purposes. In the latter case, the differences in operating conditions (or at least their effects on process output) which could occur in the period of production taken as the discrete unit of time were deemed unimportant.

Intentional deviations from reality are almost always present in models. There are of course the deviations caused by selection of appropriate system features for inclusion in the model and the rejection of other features thought not to be significant for the purposes intended. Additional deviations are also built in, however, for a number of reasons. Included are:

(1) *Ease in representation* Occasionally certain relationships are so complex that they cannot be represented in a reasonable way, so simplified approximations are used instead. This is almost always done when tabular data are to be used to structure a mathematical model. In such cases curve fitting, usually with some low order polynomial, is performed.

(2) *Enhancement of understanding* Since much of the success of a design or analysis effort depends on the insights gained through construction and manipulation of the model, it is often more efficient to work, at least at first, with a relatively simple model. Not only can construction and manipulation be facilitated, but interpretation of results and hence the gaining of understanding of the system enhanced. Additional advantages may also accrue when results and conclusions are transmitted to responsible decision makers. Managers, like all people, tend to distrust and therefore discard that which they do not understand. Successful implementation may thus depend on the simplicity and straightforwardness of the model.

(3) *Ease in manipulation and calculation* The more complex and complicated a model is made, the more trouble and expense is likely to be incurred in utilizing the model to obtain insights or quantitative results. A compromise, therefore, must often be sought between the realism and manipulability of a model. The preponderance of linear models of systems for which optimum behavior in the face of constraints is sought is due almost entirely to the superiority of the simplex algorithm and other linear programming techniques over their nonlinear counterparts. Many a nonlinear system has been "optimized" by linear programming. Similarly, many varieties of queuing systems have been analyzed as though the arrivals were Poisson distributed and service times exponential. Because of the advanced nature of methods for analyzing and solving linear differential and difference equations, many control systems are treated as being linear whether or not they really are linear.

As long as the model builder is aware of the deviations from reality built into a model, the effects of these deviations can usually be compensated for, at least subjectively. Even though model revision is sometimes called for before final implementation of results is appropriate, there is no need for such intentional deviations to produce improper final conclusions. Unintentional departures from reality, however, caused by basic unfamiliarity with the system involved or by oversight, can unknowingly lead to wrong conclusions. Fortunately, the ridiculousness of some such conclusions alert the analyst that something is amiss, so that corrections can be made before implementation is begun.

It is with these points in mind that the approach followed in this book was adopted. The need for a direct approach to the development of discrete controls has already been mentioned with regard to the increasing prevalence of inherently discrete systems currently requiring design and analysis. An additional point should be added, however. The human decision-making process is itself essentially discrete. Ideas and conclusions about phenomena external to the person himself are for the most part clearly distinguishable. They do not come pouring out in a continuous, amorphous stream. Therefore, any system involving human observers, decision makers, or feedback links must be considered as discrete. For the same reason, managers, public officials, potential customers, and others to whom the benefits of a particular system must be sold, are much more likely to become convinced if they can understand, at least conceptually, how the system operates. In general, a periodic sampling which results in a specific machine adjustment is much more meaningful to both the process engineer and the production manager than a hypothetical continuous extraction of information from the process output and never-ceasing decisions concerning changes in process operating conditions. Interpretation of the results of discrete-control adjustments is also much more straightforward than those of continuous adjustments.

Our emphasis on linear time-invariant systems is basically for the same reasons linear continuous systems have received the major emphasis in the past. Many real world systems are linear or sufficiently so that linear models may be used to obtain meaningful insights regarding their operation. In addition, the mathematics of discrete linear time-invariant systems, namely linear difference equations with constant coefficients and linear optimizing techniques, are much better developed than nonlinear ones.

The inclusion of both deterministic and stochastic factors is simply a recognition of the fact that the components of man–machine systems do not operate in a strictly deterministic way. Many of the more significant determinants of system behavior are indeed such random factors as sampling errors, instrumentation errors, and worker and raw-material variability.

As stated earlier in our discussion of systems theory, the two basic functions of models are (1) to describe a system and (2) to be used as a vehicle to optimize system structure or performance. The same model could of course serve both functions. It is important to emphasize that the *descriptive* role must be done well before the optimizing or *prescriptive* role is played.

Exercises

1.1 Discuss *one* of the following topics basing your presentation on material found in the operations research or systems analysis literature:

    a. The importance of a "systems approach" in solving design and operational problems.

    b. The steps involved in systems analysis and design.

    c. Instances in which operations research or systems analysis has failed.

1.2 Give three examples, other than those cited in the text, of discrete-time systems. For each example state whether the system is

    a. linear   **b.** time invariant

1.3 Describe an open-loop and a closed-loop system of potential interest to an I.E. or OR analyst. For each system list inputs, essential elements of the system itself, the basic structure of the system, and the outputs. Block diagrams should be useful in describing these systems.

1.4 For the systems shown in each of the following figures, identify each of the seven basic control-system elements listed on pages 16 and 17:

    a. Fig. 1.4.4   **b.** Fig. 1.4.5   **c.** Fig. 1.4.6

1.5 The ideal operating temperature for a chemical reactor is 300°F. A continuous record of this temperature is made on a pen recorder which is read periodically and adjustments made to each of the three raw material input valves accordingly. The physical properties of the reactor and each raw material have been carefully determined so that the relationships between the valve settings, fuel flow, and the resulting reactor temperature are well known. A scale showing degrees of valve adjustment to effect a given temperature change is mounted over the valves. Draw a block diagram of the reactor and the control system showing or referring to each of the basic control system elements listed on pages 16 and 17. State all assumptions made in deriving your diagram.

1.6 A simplified description of the manufacture of paper on a Fourdrinier machine is as follows. Fiber and water are mixed in controllable portions in a stock mixer. The resulting mixture flows into the headbox at a rate controlled by the stock valve. The exit from the headbox is an opening whose horizontal dimension is essentially the width of the final sheet of paper and whose height, which is relatively small, is controlled by a series of adjustable plates called slice gates. The stock flows onto a moving wire table where a combination of gravity and suction from a vacuum device remove much of the water present. The sheet then enters the dryer section consisting of several hundred yards of steam-filled rollers where additional water is removed to produce the paper. The travel time for the sheet through the dryer section can be on the order of minutes. Just before the wind-up reel a sensing head containing, among other devices, a basis-weight gage and a moisture gage scans across the sheet. The basis-weight gage measures essentially the sum of the weights of the fiber and water per unit area of sheet. The moisture gage measures the fractional moisture, i.e., the weight of water per unit area divided by the basis weight. A discrete value for each measurement representing an average over the width of the sheet is calculated after each scan and used to control the consistency (the fraction of fiber in the original stock), the stock valve, the slice gates, the vacuum pressure on the wire, machine speed, and steam pressure in the dryer rolls. Draw a block diagram representing these interrelationships. Identify each of the seven basic control-system elements. State specifically how the weight of fiber per unit area is determined.

Chapter II | Discrete Control-System Models

As discussed in Chapter I, the first step in the systems approach is a listing of the essential components of the system and the establishment of the inter-actions among them. The result of this effort is the first cut at a systems model, usually a relatively qualitative abstraction in flow chart or similar form. Figures 1.4.1 through 1.4.6 are examples of this sort of model. For quantitative analysis of systems and for the formulation of preferred or optimal designs, however, all relationships must be made explicit, hopefully to the point where they are understood precisely enough to be expressed in mathematical terms.

Development of mathematical models of control systems is the subject of this chapter. Several of the systems previously discussed are used as examples to illustrate this development. Since the major interest in the study of control systems is usually in the system output and perhaps other variables as func-tions of time, these models are of necessity dynamic. Thus, for continuous-time systems the system models are generally in the form of differential equations; for discrete-time systems, difference equation models result.

Specifically, we will use the engine-block cylinder-diameter controller, the generalized discrete process controller, and simple versions of the production inventory control system and the criminal justice system feedback models introduced in Chapter I to illustrate the development of system difference equation models. For the discrete-product engine-block system, the difference equation is derived directly. For the generalized process controller, however, a more detailed and much more complicated sampled-data continuous-system model is first developed to serve as a point of departure for the derivation of several difference equation models, with particular attention being paid to the assumptions underlying the derivations. The forms of the production inventory and the criminal justice system configurations presented were chosen to facilitate model formulation, but are nevertheless reasonably representative of real world systems.

2.1   Difference Equations

Since difference equations provide the basic structure of almost all our models of discrete-time control systems, it is important at this juncture to define the terms we will use in discussing difference equations. Specifically:

(1) *Independent variable*   The variable with respect to which all other system variables and parameters are functionally expressed. In discrete-time control-system equations, the independent variable is almost always discrete time or an index related to discrete time.

(2) *Dependent variables*   The variable whose explicit functional relationship with the independent variable is sought as the solution of the difference equation. This will often be a process output level, a production rate, or an inventory level which we wish to express as a function of sampling period or planning period. In the engine-block example it is the cylinder diameter.

(3) *Forcing function*   A known or assumed function of the independent variable which influences the functional relationship between the dependent and independent variables. In control-system work, forcing functions can include such factors as system inputs and external perturbations.

(4) *Difference*   The difference between the values of a function at two successive instants of discrete time. For example, the difference of process output at time i, represented by $\Delta m(i)$, is given by

$$\Delta m(i) = m(i + 1) - m(i). \tag{2.1.1}$$

Higher differences are obtained by taking differences of differences. An extensive discussion of differences and their properties is presented in Chapter III.

(5) *Difference equation*   An equation containing one or more irreducible orders of difference of the dependent variable with respect to the independent variable. By irreducible is meant that summation of the equation will not eliminate the presence of the differences without simultaneously creating a summation.

(6) *Linear difference equation*   A difference equation in the form of a linear combination of the dependent variable and whatever of its differences are present.

(7) *Linear difference equation with constant coefficients*   A linear difference equation in which the combinatorial coefficients of the various differences

are constant real numbers, i.e., are independent of the independent and dependent variables.

(8) *Order* The difference between the degree of the largest and smallest differences present (counting the dependent variable itself as the zeroth difference).

The equation

$$a_0 \Delta^n x(i) + a_1 \Delta^{n-1} x(i) + \cdots + a_{n-1} \Delta x(i) + a_n x(i) = f(i), \quad (2.1.2)$$

where i is the independent variable, the index, $x(i)$ the dependent variable, $\Delta^k x(i)$ the kth difference of $x(i)$, $f(i)$ the forcing function, and a_k a constant real number, the coefficient of the $(n - k)$th difference of $x(i)$, $k = 0, \ldots, n$, is a linear difference equation with constant coefficients. It is in what is called "difference-operator" form since the operator Δ appears throughout the equation. Substitution for each such difference of its definition in terms of $x(i + 1)$, $x(i)$, etc., yields, after expanding and grouping terms in like arguments of x,

$$b_0 x(i + n) + b_1 x(i + n - 1) + \cdots + b_{n-1} x(i + 1) + b_n x(i) = f(i), \quad (2.1.3)$$

which is the linear difference equation in what is called "expanded form." Note that the order of the equation is readily apparent in this form as the difference between the largest and smallest arguments of the dependent variable, i.e., $x(i + n)$ and $x(i)$, subtraction of which gives n.

It can be shown, based on material to be presented in Chapter III, that

$$b_k = \sum_{p=0}^{k} (-1)^{k-p} C_{k-p}^{n-p} a_p, \quad (2.1.4)$$

where C_k^n represents the number of combinations of n items taken k at a time. Further, it can be shown that

$$a_k = \sum_{p=0}^{k} C_{k-p}^{n-p} b_p. \quad (2.1.5)$$

Thus conversion from the difference-operator form to expanded form and vice versa is straightforward, which allows the analyst some choice in the exact form his equations take. As will be observed in the following sections, mixed forms involving both difference operators and varying values of the independent variable sometimes result during the evolutionary stages of model development. Usually it is advantageous to convert to one or the other of the "pure" forms given by (2.1.2) or (2.1.3) for solution purposes. In this text, the expanded form is almost always used. Solution techniques for linear difference equations with constant coefficients are presented in Chapters IV, V, and VI.

2.2 Control of Cylinder Diameter

Consider the system for producing and controlling the diameter of the number-one cylinder on the engine block discussed in Chapter I and shown in Fig. 1.4.3. For ease of illustration, we make the following assumptions: (1) all blocks are gaged so there is no sampling error; (2) gaging errors are not significant; (3) the perturbations imposed on the boring-tool setting (the tool wear, slippage, etc.) add to the adjustments originated by the position controller; (4) control adjustments are always complete before work is begun on the next block; and (5) the control rule used is to adjust the tool a fixed portion of the deviation between the desired and actual diameters.

Now let i designate the engine block currently involved, where i is measured relative to process start-up or other appropriate time origin such as the start of a production period or shift. We may now conveniently define the following: $m(i)$ is the actual diameter of cylinder number one on engine-block i; D the desired cylinder diameter, assumed independent of i; $\rho(i)$ the difference between D and $m(i)$; K the proportionality constant used by the controller to multiply $\rho(i)$ to determine the amount of adjustment for the boring tool; $a(i)$ the amount of tool adjustment resulting from the gaging of block i; and $p(i)$ the net effect of all perturbations (wear, etc.) affecting the tool during the period between boring block i and block $i + 1$.

The system difference equation is now derived as follows. By definition

$$\rho(i) = D - m(i), \tag{2.2.1}$$

so that, using the proportional-control rule,

$$a(i) = K\rho(i) = KD - Km(i). \tag{2.2.2}$$

Since this adjustment is complete when boring begins on block $i + 1$ and since the perturbations are additive, the difference between $m(i + 1)$ and $m(i)$ must be the sum of $a(i)$ and $p(i)$. Thus,

$$m(i + 1) - m(i) = a(i) + p(i) \tag{2.2.3}$$

or

$$m(i + 1) = m(i) + a(i) + p(i). \tag{2.2.4}$$

Equation (2.2.4) is a simple continuity relationship which states that the tool setting at the time block $i + 1$ is bored equals the setting in effect when the previous block was bored plus the adjustment effected by the controller plus the net effect of all perturbations imposed on the tool in the interim.

Substitution of (2.2.2) into (2.2.3) and rearranging yields

$$m(i + 1) - (1 - K)m(i) = KD + p(i) \tag{2.2.5}$$

which is the desired system difference equation. Note that it is with constant coefficients and forcing function $KD + p(i)$. The equation could now be solved by methods to be discussed later to obtain $m(i)$ as an explicit function of K, D, and the sequence of perturbations $p(0)$ through $p(i - 1)$.

The procedure just explained can be used to derive system equations for any system. However, the difficulty of the task and the form of the resulting equation or set of equations depend on the nature and complexity of the system being modeled. The generalized discrete controller given in Fig. 1.4.4 will now be used to illustrate some of the many possible situations which can arise when attempting to model even a relatively simple, straightforward system.

2.3 Generalized Discrete Process Controller with Sampling

Sampled-Data Formulation

In dealing with a generalized system where the product could be inherently either discrete or continuous and possibilities for output sampling must be maintained, it is convenient to consider the index i as related to the number of times the switch S is closed relative to the appropriate time origin. If we now think of the time between two successive closings of S as a sampling period, then we can explicitly define the ith sampling period as the time between the ith and $(i + 1)$st closing of the sampling switch. If the samples are equally spaced in time by the increment T, then the ith sampling period corresponds to time t in the range $iT \leq t \leq (i + 1)T$.

We can now make the time relationships among the variables in Fig. 1.4.4 explicit by including an argument designating the sample from which they resulted. Thus, $m(i)$ is the process output at the instant S is closed for the ith time, and $\varepsilon(i)$ the error associated with the measurement of $m(i)$. Therefore

$$\mu(i) = m(i) + \varepsilon(i). \tag{2.3.1}$$

$D(i)$ is the input or desired level of process output in effect when the ith sample is measured and

$$\rho(i) \doteq D(i) - \mu(i) = D(i) - m(i) - \varepsilon(i). \tag{2.3.2}$$

Since the desired levels of process operation are usually constant for any given brand or model, in what follows we will consider $D(i)$ as a constant D, although a constant input does not necessarily apply in all cases. Thus

$$\rho(i) = D - m(i) - \varepsilon(i). \tag{2.3.3}$$

As shown in Fig. 1.4.4, the control signal C, the process adjustment A, and the external perturbations $\tilde{P}$ are continuous phenomena, so must be represented as $C(t)$, $A(t)$, and $\tilde{P}(t)$, respectively. In the ith sampling interval, $C(t)$ may depend on any or all of the measured deviations observed up to and including instant i and the time elapsed since the start of the interval, $t - Ti$. The functional relationship is represented by the symbol G_D. Defining $R(i)$ as the set of all measured deviations from sampling instant one through instant i, i.e.,

$$R(i) = \{\rho(1), \rho(2), \ldots, \rho(i)\}, \tag{2.3.4}$$

permits this relationship to be written explicitly as

$$C(t) = G_D[R(i), t - Ti], \qquad Ti \leq t \leq T(i + 1). \tag{2.3.5}$$

In many practical cases where both currency of information and limitations of information storage are important, only $\rho(i)$ or possibly the most recent two or three values of measured deviation would actually be represented in G_D. In other cases, some function such as the average of all measured deviations observed to date might be used. Several specific formulations will be discussed later in this section. As written, however, (2.3.5) represents the most general form of the equation possible.

Similarly, $A(t)$ depends upon $C(t)$ and the physical properties of the process, represented functionally by G_P. This may be expressed as

$$A(t) = G_P[C(t)] = G_P\{G_D[R(i), t - Ti]\}, \qquad Ti \leq t \leq T(i + 1). \tag{2.3.6}$$

Note that the validity of (2.3.6) depends on the vanishing of the control signal $C(t)$ by the end of the current interval. Were this not the case, $A(t)$ would depend not only on $C(t)$ for t values in the current interval but also on control signals from past intervals. Although such systems could certainly exist, especially in production planning where plans may be formulated two or more periods ahead and then updated at the end of each period, proper modeling techniques can almost always avoid such carryovers. Equation (2.3.6) is thus quite appropriate except in very rare cases which we choose to ignore here.

Changes in process output also occur continuously. As pictured, the rate of change of process output at any time t is the sum of $A(t)$ and $\tilde{P}(t)$, i.e.,

$$\dot{m}(t) = \frac{dm(t)}{dt} = A(t) + \tilde{P}(t). \tag{2.3.7}$$

During the ith sampling period, this rate of change may be expressed in terms of the current measured deviation from the desired output by substituting (2.3.6) for $A(t)$ in (2.3.7) which yields:

$$\dot{m}(t) = G_P\{G_D[R(i), t - iT]\} + \tilde{P}(t), \qquad iT \leq t \leq (i + 1)T. \tag{2.3.8}$$

Equation (2.3.8) is a piecewise-continuous differential equation which can be solved for $m(t)$ once specific functional forms for G_P and G_D are inserted. It is typical of the equations which result when an inherently continuous system is sampled periodically and a sequence of discrete control actions follows. This type of system is generally called a *sampled-data system* and the equation a *sampled-data model*. Note that the system behavior is described continuously at all points in time. For this reason, the system might be more properly designated a "sampled-data continuous system" since sampling can also be applied to truly discrete systems for which continuous representation between sampling instants would be inappropriate. Detailed coverage of sampled-data systems and models is provided in many standard control theory texts [see, for example, Tou (1959), Cadzow and Martens (1970), and Truxal (1955)].

Discrete Formulation

A truly discrete model describes a system only at the sampling instants, i.e., it includes only those values of the system variables which exist at the instants the samples are taken. Such models are inherently much easier to derive and to manipulate than the sampled-data models. They are, therefore, preferred for use in those situations where they apply. Two such situations are:

(1) When the system under study is truly discrete such as the engine-block line wherein the process output is not defined between the completion times of the items made.

(2) When, whether or not the process output can be defined between the sampling instants, the system behavior in these intervals can be ignored in any evaluation of system performance or effectiveness. This would apply to systems which are relatively certain to be well-behaved between sampling instants, i.e., no wild transients are likely to result from the kinds of control actions or external perturbations anticipated which would significantly affect the process output or the cost or efficiency of operation.

When either of these conditions is met (or assumed to be met by the analyst or designer) difference equation models can be formulated as follows. The difference in the process output between two successive sampling intervals, say i and $i + 1$, is $m(i + 1) - m(i)$. As mentioned previously, this is represented by the symbol $\Delta m(i)$, i.e.,

$$\Delta m(i) = m(i + 1) - m(i), \tag{2.3.9}$$

which is formally defined in the calculus of finite differences as the "first forward difference" of $m(i)$. For truly discrete systems, (2.3.9) derives directly from the situation involved as in the case of Eq. (2.2.3) for the cylinder-diameter control system. In the case of the inherently continuous system for which a discrete model is being developed, $\Delta m(i)$ is the result of the continuous accumulation of the instantaneous changes throughout the sampling interval. This accumulation is of course the integration of $m(t)$ over the interval so that

$$\Delta m(i) = \int_{t=Ti}^{T(i+1)} \dot{m}(t)\,dt = \int_{t=Ti}^{T(i+1)} A(t)\,dt + \int_{T=Ti}^{T(i+1)} \tilde{P}(t)\,dt. \quad (2.3.10)$$

Under the assumption that each control signal vanishes prior to the end of the sampling interval, we may further define

$$c(i) = \int_{t=Ti}^{T(i+1)} C(t)\,dt = \int_{t=Ti}^{T(i+1)} G_D[R(i),\, t - Ti]\,dt = g_D\,[R(i)] \quad (2.3.11)$$

where g_D represents the discrete functional relationship between the integrated control signal $c(i)$ and the sequence of measured deviations from process bogie. Note that $c(i)$ contains no information regarding the instantaneous rate of control within the sampling interval, just the total effect of the entire signal by the end of the interval. This is analogous to the production planning situation with a weekly planning period in which, instead of specifying the production levels continuously throughout the week, only the desired total production for the week is stated. This would be done of course only for production processes for which the instantaneous rates had little effect on the over-all cost or efficiency of production or where predetermined rules had been established to compute instantaneous schedules for the week for any given production order $c(i)$.

Given a discrete control signal $c(i)$, the physical properties G_P of the process, and some predetermined way of converting $c(i)$ into a continuous schedule for the period, the total adjustment effected during the period is given by

$$a(i) = \int_{t=Ti}^{T(i+1)} A(t)\,dt = \int_{t=Ti}^{T(i+1)} G_P[C(t)]\,dt = g_P[c(i)]. \quad (2.3.12)$$

Here g_P represents the relationship between the total adjustment which takes place during the period $a(i)$ and the desired adjustment $c(i)$, for the corresponding instantaneous pattern for $C(t)$ and the physical characteristics of the process given by G_P.

We further define

$$p(i) = \int_{t=Ti}^{T(i+1)} \tilde{P}(t)\,dt \quad (2.3.13)$$

as the net effect of all external perturbations entering the process during the ith sampling interval.

Substitution of (2.3.12) and (2.3.13) into (2.3.10) yields

$$\Delta m(i) = a(i) + p(i) = g_P[c(i)] + p(i). \tag{2.3.14}$$

Further substitution of (2.3.11) and (2.3.3) results in

$$\begin{aligned}\Delta m(i) &= g_P\{g_D[R(i)]\} + p(i) \\ &= g_P\{g_D[D - m(0) - \varepsilon(0), \ldots, D - m(i) - \varepsilon(i)]\} + p(i) \end{aligned} \tag{2.3.15}$$

which is a true difference equation with the independent variable the discrete-time index i. (Note that whether a measurement is actually made at the time defined by $i = 0$ depends on the time base for the model. In general, such an initial measurement is part of the system; although, as in the cylinder-diameter control model, it is not. Cylinder number one, corresponding to $i = 1$, is the first item measured.)

To solve such equations, we must of course specify the functions g_D and g_P, or derive them from their continuous equivalents G_D and G_P, respectively. We omit further discussion of the relationships between the discrete and continuous functions here. Instead, we illustrate the derivation of specific system difference equations by specifying sets of relationships to be incorporated into (2.3.15).

First, however, a word regarding the construction of block diagrams representing discrete systems or discrete models of systems is in order. It is convenient to omit specific representation of sampling switches and instead to display the discrete nature of the system by labeling the components and signals involved with symbols defined in discrete terms and by substituting summation and difference operations for integrations and differentiations, respectively. This format is illustrated in Fig. 2.3.1, a block diagram representing the discrete formulation of the generalized process controller described by Eq. (2.3.15). All symbols are as defined in the development of Eq. (2.3.15).

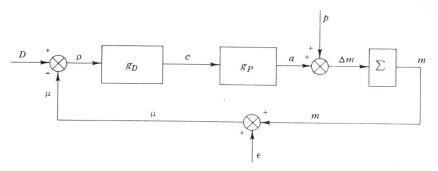

FIG. 2.3.1. Generalized automatic process controller with sampling—discrete representation.

Rapid-Response Controller

In the very common situation in which the entire desired adjustment $c(i)$ is effectively completed within the sampling period,

$$a(i) = c(i), \qquad (2.3.16)$$

i.e., g_P is simply an identity function. Then (2.3.15) may be written

$$\Delta m(i) = g_D[D - m(0) - \varepsilon(0), \ldots, D - m(i) - \varepsilon(i)] + p(i). \qquad (2.3.17)$$

Controllers for which (2.3.16) holds are sometimes referred to as "rapid-response controllers." Figure 2.3.2 is the diagram of the rapid-response version of the generalized automatic process controller with sampling. Note that it is identical to Fig. 2.3.1 except for the omission of the box containing the symbol g_P (the physical properties of the process being controlled). Usually specific reference to c is also omitted. In all further discussion of process controllers the rapid-response feature will be assumed unless it is explicitly stated to the contrary.

We will now examine several specific forms for the control decision rule g_D.

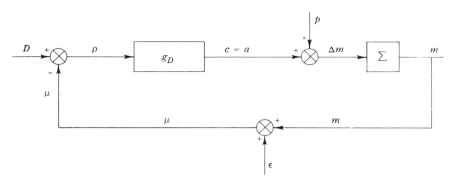

FIG. 2.3.2. Rapid-response discrete automatic process controller with sampling.

Proportional Control

The simplest control decision rule is to base $c(i)$ solely on the most recent measured deviation $\rho(i)$. That is, we define g_D such that

$$c(i) = K\rho(i) = KD - Km(i) - K\varepsilon(i), \qquad (2.3.18)$$

where K is a constant. Since the control signal is a constant portion of the measured deviation $\rho(i)$, this type of control is known as a "proportional controller." In combination with a rapid-response capability, as it is here, it is called a "simple proportional controller" and is the type represented by

Eq. (2.2.2) in the previous discussion of the engine-block line. Substitution of (2.3.18) into (2.3.17) results in the expression

$$\Delta m(i) = KD - Km(i) - K\varepsilon(i) + p(i). \qquad (2.3.19)$$

Rearranging to locate all terms involving the dependent variable $m(i)$ on the left-hand side of the equation, we obtain

$$\Delta m(i) + Km(i) = KD - K\varepsilon(i) + p(i), \qquad (2.3.20)$$

a linear difference equation in difference-operator form.

Note that substitution of (2.3.9) for $\Delta m(i)$ and grouping terms yields

$$m(i + 1) - (1 - K)m(i) = KD - K\varepsilon(i) + p(i), \qquad (2.3.21)$$

which is identical to the expanded-form system equation (2.2.5) for the engine-block line derived earlier except for the addition here of the measurement error $\varepsilon(i)$. Thus, the same continuity relationships pointed out earlier apply here also. That is, the process output at instant $T(i + 1)$ is equal to the sum of the process output at the previous sampling instant Ti, the adjustment just made, and the net effect of all perturbations caused by factors external to the system which occurred during the period $Ti \leq t \leq T(i + 1)$.

Difference Control

As the name implies difference control is based on the difference between the two most recent observed deviations, i.e., g_D is defined such that

$$
\begin{aligned}
c(i) &= L[\rho(i) - \rho(i - 1)] \\
&= L\{[D - m(i) - \varepsilon(i)] - [D - m(i - 1) - \varepsilon(i - 1)]\} \\
&= -L[m(i) - m(i - 1)] - L[\varepsilon(i) - \varepsilon(i - 1)], \qquad (2.3.22)
\end{aligned}
$$

where L is a constant. In difference-operator form this can be expressed as

$$c(i) = L\,\Delta\rho(i - 1) = -L\,\Delta m(i - 1) - L\,\Delta\varepsilon(i - 1). \qquad (2.3.23)$$

Substitution of (2.3.23) into (2.3.17) and rearranging yields

$$\Delta m(i) + L\,\Delta m(i - 1) = -L\,\Delta\varepsilon(i - 1) + p(i). \qquad (2.3.24)$$

Replacement of the difference operators by actual differences and grouping of terms gives

$$m(i + 1) - (1 - L)m(i) - Lm(i - 1) = -L\varepsilon(i) + L\varepsilon(i - 1) + p(i), \qquad (2.3.25)$$

a second-order linear difference equation with constant coefficients. Note that some memory capacity must be available for difference control to be implemented.

Control based on differences of differences (called "second differences") and other more complicated difference-type functions of observed deviations can obviously be effected. Explicit discussion of such systems, however, must wait until we have discussed differences in more detail in Chapter III.

Summation Control

Summation control is based on the sum of all previous observed deviations, i.e., g_D is defined such that

$$c(i) = N \sum_{n=0}^{i} \rho(n) = N \sum_{n=0}^{i} [D - m(n) - \varepsilon(n)]$$

$$= ND(i + 1) - N \sum_{n=0}^{i} m(n) - N \sum_{n=0}^{i} \varepsilon(n), \qquad (2.3.26)$$

where N is a constant. Substitution of (2.3.26) into (2.3.17) yields, after rearranging terms,

$$\Delta m(i) + N \sum_{n=0}^{i} m(n) = ND(i + 1) - N \sum_{n=0}^{i} \varepsilon(n) + p(i) \qquad (2.3.27)$$

which may more conveniently be written as

$$m(i + 1) - (1 - N)m(i) + N \sum_{n=0}^{i-1} m(n) = ND(i + 1) - N \sum_{n=0}^{i} \varepsilon(n) + p(i).$$
$$(2.3.28)$$

At first glance the order of (2.3.28) may appear to be undefined in that the values of process output at all points in time from zero to $i + 1$ are present. However, after discussing summations in more detail in Chapter III and techniques for solving linear difference equations in Chapters IV, V, and VI, we will see that (2.3.28) can be solved in straightforward fashion as a second-order linear difference equation.

Combinations

Proportional, difference, and summation control are often used together in the same control system in various combinations. The reasons for the selection of any particular combination and the responses of the resulting system to various perturbation patterns are discussed in Chapter X. For now, let us develop the difference equation representing the combination of all three. g_D is thus defined such that

$$c(i) = K\rho(i) + L \Delta\rho(i - 1) + N \sum_{n=0}^{i} \rho(n). \qquad (2.3.29)$$

Substitution of (2.3.29) into (2.3.17) and appropriate grouping of terms yields

$$m(i + 1) - (1 - K - L - N)m(i) - (L - N)m(i - 1) + N\sum_{n=0}^{i-2} m(n)$$

$$= [K + N(i + 1)]D - (K + L + N)\varepsilon(i) + (L - N)\varepsilon(i - 1)$$

$$- N\sum_{n=0}^{i-2} \varepsilon(n) + p(i), \tag{2.3.30}$$

which is only one of many forms in which this result could be expressed.

With (2.3.30) we have the basis for formulating the difference equation for any two-way combination or single-type control. For example, to find the difference equation for a proportional-plus-difference controller, one can simply set N in Eq. (2.3.30) to zero. The equation for a simple proportional controller can be obtained by setting L and N to zero, etc. It will be shown subsequently that (2.3.30) as given can be solved as a third-order linear difference equation, as is summation-plus-difference control ($K = 0$). Proportional-plus-difference ($N = 0$) and proportional-plus-summation ($L = 0$) control equations can be solved as second-order linear difference equations.

General nth-Order Controller

Obviously difference and summation control and combinations thereof involving proportional control represent special cases of more general types of controllers whose difference equations are of second or third order. For example, a general second-order controller results when g_D is defined such that

$$c(i) = K_0 \rho(i) + K_1 \rho(i - 1), \tag{2.3.31}$$

where K_0 and K_1 are constants. Thus control is based on the two most recent observed deviations. Substitution of (2.3.31) into (2.3.17) and grouping of terms yields

$$m(i + 1) - (1 - K_0)m(i) + K_1 m(i - 1)$$

$$= (K_0 + K_1)D - K_0 \varepsilon(i) - K_1 \varepsilon(i - 1) + p(i). \tag{2.3.32}$$

Difference control results when $K_0 = -K_1 = L$ and proportional-plus-difference when $K_0 = K + L$ and $K_1 = -L$. Relationships involving summation control must await study of summations and difference equation solution techniques.

General nth-order control is based on the n most recent observed deviations, i.e., by defining g_D such that

$$c(i) = \sum_{k=0}^{n-1} K_k \rho(i - k), \tag{2.3.33}$$

where K_0 through K_{n-1} are constants. Substitution into (2.3.17) yields

$$m(i+1) - (1 - K_0)m(i) + \sum_{k=1}^{n-1} K_k m(i-k) = D \sum_{k=0}^{n-1} K_k - \sum_{k=0}^{n-1} K_k \varepsilon(i-k) + p(i),$$

$$(2.3.34)$$

an nth-order linear difference equation with constant coefficients. Any specific control rules, such as those involving higher order differences and summations, can be expressed as special cases of the generalized equation of the same order. Thus study of generalized equations can be extremely useful.

On the other hand, specific controllers such as the types listed above and various forecasting smoothing functions, to be discussed in connection with production inventory control systems, are of particular interest because of certain inherent properties often found useful in controller operation. These properties will be discussed in later chapters. Another reason for interest in difference- and summation-type controls is the relative simplicity of differencing and summing elements in computer circuits.

2.4   Production–Inventory Control System

The system difference equation for the discrete model of the production-inventory control system of Fig. 1.4.5 will be developed directly from basic definitions instead of from modifications of piecewise-continuous differential equations as was done for the generalized process controller. Direct development of discrete models is almost always easier and more straightforward, but involves more risks of inadvertently omitting or misrepresenting important aspects of the system operation. Naturally, care is required regardless of the approach taken.

We again use i as the index of discrete time. In actuality, it corresponds to the ith production period relative to some (perhaps arbitrary) time origin. Other definitions are the following: $m(i)$ is the actual inventory level at the beginning of period i; $D(i)$ the desired inventory level at the beginning of period i (calculated perhaps from some optimizing procedure external to the control system);

$$\rho(i) = D(i) - m(i) \qquad (2.4.1)$$

is the deviation between desired and actual inventory at the beginning of period i; $p(i)$ the total customer orders arriving during period i; $c(i)$ the production order covering period i based on some function g_D of current and perhaps past deviations from desired inventory levels (potentially $\rho(1)$ through $\rho(i)$), the history of customer orders (potentially $p(1)$ through $p(i-1)$, since $p(i)$ has not yet occurred at the beginning of interval i), and possibly forecasts of $p(i)$ and other future orders; and $a(i)$ the total amount produced during period i which is some function g_P of the production order $c(i)$ and perhaps past production orders which could have some residual effect on the condition of the manufacturing facility at the beginning of period i.

As in the case of the process controller, we define $R(i)$ as the set of all measured deviations $\rho(i)$ from period 1 through period i. Let us further represent by $P(i)$ the set of all customer orders arriving during period 1 through i, by $p(i, k)$ the forecast of customer orders for period $i + k$ made at the beginning of period i, and by $P(i, k)$ the set of customer-order forecasts made at the beginning of period i for periods i through $i + k$. Thus

$$P(i, k) = \{p(i, 0), p(i, 1), \ldots, p(i, k)\}. \tag{2.4.2}$$

Note that a value of $k = 0$ refers to a forecast for the current period i. In terms of these definitions, the production order $c(i)$ can be expressed as

$$c(i) = g_D[R(i), P(i - 1), P(i, k)], \tag{2.4.3}$$

where the value of k depends on the forecasting technique used. In cases where forecasts are based on smoothing and extrapolation of historical data, $P(i, k)$ need not appear explicitly in the expression for $c(i)$ since each $p(i, k)$ can be expressed in terms of the elements of the sets $R(i)$ and $P(i - 1)$.

For ease in illustration, consider the situation in which previous production orders have no significant effect on the production for the current period (or that unfilled portions of past orders are accounted for in making up the current order), so that $a(i)$ may be expressed simply as

$$a(i) = g_P[c(i)]. \tag{2.4.4}$$

If all items produced during period i are added to inventory before the end of the period, and if all customer orders received during period i are filled or backordered before the end of the period (a backordered item is considered a negative item of inventory), then

$$\Delta m(i) = a(i) - p(i). \tag{2.4.5}$$

This is a materials-balance equation similar to (2.3.14) developed for the discrete simple proportional controller. Note that here in (2.4.5), however, the sign of $p(i)$ is negative because of the definition in terms of customer orders which act to deplete the inventory. Substitution of $m(i + 1) - m(i)$ for $\Delta m(i)$ and rearranging (2.4.5) yields the basic continuity equation

$$m(i + 1) = m(i) + a(i) - p(i). \tag{2.4.6}$$

This states that the inventory level at the beginning of period $i + 1$ is equal to the inventory level at the beginning of period i plus the amount produced during period i minus the amount ordered by customers during period i.

Returning to the development of the system difference equation, we utilize (2.4.1) through (2.4.5) to obtain:

$$\Delta m(i) = g_P[c(i)] - p(i)$$
$$= g_P\{g_D[R(i), P(i-1), P(i,k)]\} - p(i)$$
$$= g_P\{g_D[D(1) - m(1), \ldots, D(i) - m(i), p(0), \ldots,$$
$$p(i-1), p(i,0), \ldots, p(i,k)]\} - p(i). \qquad (2.4.7)$$

This can be solved for $m(i)$ for any assumed patterns of $D(i)$, $p(i)$, and $p(i,k)$, and functional relationships g_D and g_P. A block diagram representing this discrete version of the production–inventory control system is given in Fig. 2.4.1. All symbols are as defined above.

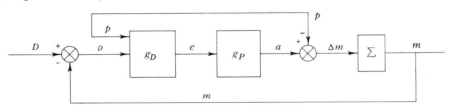

FIG. 2.4.1. Discrete production–inventory control system with feedforward of customer orders.

Rapid-Response Production Capability

Thus far nothing has been said about the physical properties of the production process. One possible situation is that sufficient production capacity is always available to permit completion of the entire production order $c(i)$ within period i. This is equivalent to the rapid-response controller and corresponds to g_P being an identity function such that

$$a(i) = c(i). \qquad (2.4.8)$$

This greatly simplifies the system model, allowing (2.4.7) to be written

$$\Delta m(i) = g_D[R(i), P(i-1), P(i,k)] - p(i), \qquad (2.4.9)$$

which may be a great help for the analyst but a real cause for concern for the management and owners who would prefer greater utilization of capacity.

Let us consider two specific types of control scheme for the rapid-response production situation.

Memoryless Order Rule

One type of control decision rule of practical interest because of its simplicity is to make the production order $c(i)$ a linear combination of the most recent values of observed inventory deviation and customer orders. That is, g_D is defined such that

$$c(i) = K_e p(i) + K_P p(i-1) = K_e D(i) - K_e m(i) + K_P p(i-1), \quad (2.4.10)$$

where K_e and K_P are constants. The resulting system difference equation in expanded form is

$$m(i+1) - (1 - K_e)m(i) = K_e D(i) - p(i) + K_P p(i-1), \quad (2.4.11)$$

a first-order equation whose left-hand side is identical to (2.3.21), the system equation for the rapid-response proportional process controller. Obviously, all of the control decision rules discussed in the context of the generalized process controller can be considered for use in production inventory control.

Forecasting

In a sense, the term $K_P p(i-1)$ in (2.4.10), the order rule for memoryless control, is a zeroth-order forecast of the customer orders for period i, $p(i)$. Certainly, if there is any period-to-period consistency in the pattern of orders, the terms $p(i)$ and $K_P p(i-1)$ in (2.4.11) will tend to offset each other for $K_P > 0$; and, although the effects of this cancellation can better be evaluated after the equation is solved, intuitively one can see the benefits of anticipating even a portion of incoming orders.

In general, what we seek is as accurate a forecast as possible of $p(i)$ regardless of what it may be based on. Thus we would tend to use an ordering rule of the form

$$c(i) = g_{De}[R(i)] + K_F p(i, 0), \quad (2.4.12)$$

where K_F is a constant and $g_{De}[R(i)]$ any appropriate rule for determining the effects of inventory patterns on the current order. The forecast $p(i, 0)$ could be completely exogenous, being based on market surveys, external economic indicators, or simply management whim. Conversely, it could be based completely on an extrapolation of the pattern of past orders, perhaps utilizing some of the well conceived and tested procedures of Brown (1963) and Box and Jenkins (1970). Sometimes combinations of exogenous and historical system data are used.

Where a forecast based on exogenous factors is used, one simply inserts a number for $p(i, 0)$. In the case of a forecast involving extrapolation of the pattern of customer orders, however, it is first necessary to solve the forecasting equation to get $p(i, 0)$ in terms of $p(0)$ through $p(i-1)$ which then becomes part of the forcing function of the system equation. For example, if Brown's single exponential smoothing is used to produce the forecast

$$p(i, 0) = (1 - a)p(i-1, 0) + ap(i-1), \quad (2.4.13)$$

where a $(0 \leq a \leq 1)$ is the so-called smoothing constant, $p(i - 1, 0)$ is the forecast of orders for period $i - 1$ made at the beginning of period $i - 1$, and $p(i - 1)$ is of course the actual orders received during period $i - 1$. Equation (2.4.13) can be arranged to obtain

$$p(i, 0) - (1 - a)p(i - 1, 0) = ap(i - 1) \qquad (2.4.14)$$

which is a first-order linear difference equation in expanded form with forcing function $ap(i - 1)$. Solving (2.4.14) by the methods to be discussed in Chapters IV, V, and VI yields

$$p(i, 0) = \sum_{k=0}^{i-1} p(k)(1 - a)^{i-1-k} + p(0, 0)(1 - a)^i, \qquad (2.4.15)$$

where $p(0, 0)$ is any initial estimate, made at $i = 0$, of orders to be received between $i = 0$ and $i = 1$. Equation (2.4.15) must now be substituted in (2.4.12) along with whatever form is being investigated for $g_{De}[R(i)]$ and the results substituted into the system equation

$$\Delta m(i) = a(i) - p(i) = c(i) - p(i). \qquad (2.4.16)$$

Expression of the results of these steps as a linear difference equation in expanded form is left as an exercise for the reader.

Fixed-Delay Production Capability

A simplifying assumption useful in many analyses is to consider the production facility simply as a fixed delay. Thus, a production order received at the beginning of period i will not begin to reach inventory until period $i + \tau$, where τ is a positive integer representing the number of periods of delay in accomplishing production. Obviously in any plant the amount of time required to get an order into production depends on many factors including order size, total production capacity, and the presence of other orders. Nevertheless, the fixed-delay assumption can apply reasonably well in some cases and provide useful insights in many more.

With a fixed production delay,

$$a(i) = c(i - \tau), \qquad (2.4.17)$$

which yields the following system difference equation:

$$\Delta m(i) = g_D[R(i - \tau), P(i - 1 - \tau), P(i - \tau, k)] - p(i). \qquad (2.4.18)$$

Note that in a practical situation involving production lags, the inventory decision rule would almost always take into account those production orders which had already been issued but whose resulting production has not yet emerged from the plant. Thus, in general, (2.4.18) would be supplemented by reference to $c(i-1)$ through $c(i-\tau+1)$. The case in which τ is a random variable is covered in Chapter V using z transform techniques.

2.5 Criminal Justice System Feedback Model

There are a myriad of possible structures that models of the criminal justice system could assume depending on the intended use and level of application. For illustrative purposes we will show here the fairly simple, composite form used by Belkin *et al.* (1973). Although it is conceptually a specific representation of the general system shown in Fig. 1.4.6, it will be developed here directly. Figure 2.5.1 depicts the resulting system diagram in

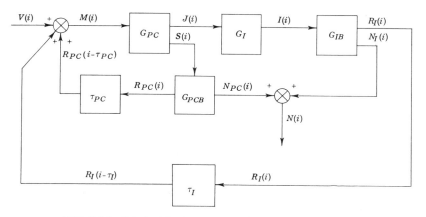

FIG. 2.5.1. Criminal justice system switching and delay model.

which the basic decisions are manifest in the portion of people arrested who are incarcerated and the portions of both incarcerated and nonincarcerated who become recidivists. The major physical property involved is the mean time between arrests for those who choose to recidivate.

Specifically, the symbols in the figure represent the following:

$V(i)$ the number of virgin arrests during time period i,
$M(i)$ the total number of arrests during period i,
$J(i)$ the number incarcerated during period i,

$S(i)$ the number arrested but not incarcerated during i (i.e., those whose cases were dismissed, probationers, fines, those found innocent, etc.),

$I(i)$ the number released from jail during i,

$R_I(i)$ the number released from jail during i who later recidivate,

$N_I(i)$ the number released from jail during i who go straight,

$R_{PC}(i)$ the number released by the police or courts during i who later recidivate,

$N_{PC}(i)$ the number released by the police or courts during i who go straight,

$N(i)$ the total number released by police, courts, and jail during i who go straight,

τ_I the mean time between arrest and rearrest for persons incarcerated,

τ_{PC} the mean time between arrest and rearrest for persons not incarcerated.

Note that with this set of definitions time spent in police involvement, courts, and jail are included in the calculation of mean time between arrests. There are obviously other more precise ways to model the system but this simplified form was apparently felt to be adequate for the purposes of the researchers. Therefore, the following relationships exist:

$$M(i) = V(i) + R_I(i - \tau_I) + R_{PC}(i - \tau_{PC}),$$
$$I(i) = J(i) = aM(i), \qquad S(i) = (1 - a)M(i),$$
$$R_I(i) = b_I I(i), \qquad N_I(i) = (1 - b_I)I(i),$$
$$R_{PC}(i) = b_{PC} S(i), \qquad N_{PC}(i) = (1 - b_{PC})S(i),$$
$$N(i) = N_I(i) + N_{PC}(i),$$

where a is the probability an arrested person is incarcerated, $0 \le a \le 1$, b_I the probability a person released from jail is rearrested, $0 \le b_I \le 1$, and b_{PC} the probability a person released by the police or courts is rearrested, $0 \le b_{PC} \le 1$.

The next step is of course the combining of the above individual relationships to form the system model. First, however, the question must be answered as to what the model is to be used for. Do we want to know the jail population as a function of discrete time, or the total number of arrests, or the police-court releases who are rearrested? Whatever we decide upon as our dependent variable, it is apparent that the virgin arrests $V(i)$ is our forcing function and τ_I, τ_{PC}, a, b_I, and b_{PC} are the basic system parameters. Any of the remaining symbols can be considered as the dependent variable; in fact, for a complete systems analysis we might wish to treat each variable in turn as the dependent variable and formulate and solve each of the resulting difference equations. Since each variable carries associated costs, both financial and societal, only in this way can a complete analysis of the implications of various sets of parameter values be evaluated on a system-wide basis.

We will demonstrate the formulation of the system equation with $M(i)$ taken as the dependent variable. From the above equations we get

$$
\begin{aligned}
M(i) &= V(i) + R_I(i - \tau_I) + R_{PC}(i - \tau_{PC}) \\
&= V(i) + b_I J(i - \tau_I) + b_{PC} S(i - \tau_{PC}) \\
&= V(i) + b_I a M(i - \tau_I) + b_{PC}(1 - a)M(i - \tau_{PC}),
\end{aligned} \tag{2.5.1}
$$

which may be rearranged to obtain

$$
M(i) - b_{PC}(1 - a)M(i - \tau_{PC}) - b_I a M(i - \tau_I) = V(i). \tag{2.5.2}
$$

Equation (2.5.2) is a linear difference equation in expanded form with dependent variable $M(i)$ and forcing function $V(i)$. Its order is the maximum of τ_{PC} and τ_I. Note that if the time period used is one month, with mean time between arrests measured in years, (2.5.2) could be of extremely high order, a factor which we will see in Chapters IV, V, and VI imposes some severe problems, both analytical and computational.

2.6 Conclusion

In this chapter we have been introduced to the linear difference equation with constant coefficients and have observed how models of a wide variety of systems can be formulated in linear-difference equation form. It therefore behooves us to become familiar with the available techniques for manipulation and solution of this type of equation in order to predict the performance of these systems, to derive preferred system structures and decision rules, and to find optimal values for system parameters. Such solution techniques are presented in Chapters IV, V, and VI, Chapter IV being devoted to what might be called "classical methods" and V and VI to the z transform approach.

In addition, the finite mathematics with which difference equations are involved, e.g., the properties of differences and summations, may not be as familiar to many readers as the continuous mathematics traditionally at the core of undergraduate engineering and science programs. For this reason, a brief introduction to the calculus of finite differences is presented in Chapter III.

Any reader already familiar with any of the topics in Chapters III through VI could skip the sections involved.

Exercises

2.1   Give the order of each of the following linear difference equations with constant coefficients:

   a. $\Delta^3 x(i) + 3\,\Delta^2 x(i) - 2\,\Delta x(i) - x(i) = f(i)$

   b. $2x(i + 4) - 5x(i + 3) + x(i + 2) + 3x(i + 1) + 4x(i) = f(i)$

   c. $x(i + 4) - 2x(i + 3) - x(i + 2) + 3x(i + 1) = f(i)$

   d. $3\,\Delta^4 x(i) - 2x(i) = f(i)$

   e. $x(i + 3) - 2x(i + 2) + x(i) = f(i)$

   f. $\Delta^5 x(i) - 2\,\Delta^3 x(i) + 3\,\Delta x(i) = f(i)$

2.2   A controller of the type discussed in Section 2.2 is to be used to control a turret lathe which has been set up to turn the outside diameter of bushings to a target value of 2.0000 inches.

   a.   If a setup error causes $m(1)$ to be 1.9000 inches and no other perturbations occur ($p(i) = 0$ for all i), use the system difference equation to determine the diameters of the first five bushings turned (i.e., those at 1, 2, 3, 4, 5):

   (1)   in terms of the proportionality constant K;

   (2)   for the specific case of $K = 0.5$.

   b.   Given that the diameter of the first bushing turned measures exactly 2.0000 inches and that tool wear is such that in the absence of control, the diameter of each successive bushing would increase by 0.0001 over that of the previous one, find the diameters of bushings 2 through 5:

   (1)   in terms of the proportionality constant K;

   (2)   for the specific case of $K = 0.5$.

2.3   The discrete-time controller shown below has a fixed delay of τ_1 periods in obtaining a measurement (i.e., the output existing at time i gets recorded at time $i + \tau_1$), and an additional delay of τ_2 in feeding this measurement back for comparison with the desired operating level D (a constant). The "physical properties of the process," g_P, are such that $a(i) = c(i)$.

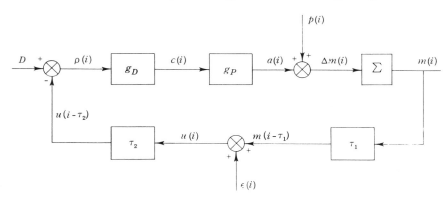

Derive the system difference equation which relates system output $m(i)$ to the desired level D, the perturbations $p(i)$, and the measurement error $\varepsilon(i)$, for each of the following cases (any form will do):

a. $\tau_1 = 0$, $\tau_2 = 2$, and g_D such that $c(i) = K\rho(i)$, i.e., a proportional controller.

b. $\tau_1 = 0$, $\tau_2 = 0$, and g_D such that $c(i) = K\rho(i) + L\,\Delta\rho(i-1)$, i.e., a proportional-plus-difference controller.

c. $\tau_1 = 0$, $\tau_2 = 0$, and g_D such that $c(i) = K\rho(i) + N\sum_{n=0}^{i-1}\rho(n)$, i.e., proportional-plus-summation control.

d. $\tau_1 = 2$, $\tau_2 = 1$, and $c(i) = K\rho(i) + N\sum_{n=0}^{i-1}\rho(n)$.

e. General τ_1 and τ_2 and $c(i) = K\rho(i) + L\,\Delta\rho(i-1)$.

f. General τ_1 and τ_2 and $c(i) = K\rho(i) + L\,\Delta\rho(i-1) + N\sum_{n=0}^{i-1}\rho(n)$.

g. General τ_1 and τ_2 and $c(i) = \sum_{k=0}^{n-1} K_k\,\rho(i-k)$, i.e., a generalized nth-order controller.

2.4 (a) through (g). Repeat the corresponding part of 2.3 for $a(i) = A \cdot c(i - \tau)$, i.e., there is an attenuation of the control signal by a factor A and a delay of τ periods.

2.5 The discrete production inventory control system shown below has a fixed delay of τ_I periods in feeding back inventory-level information and a delay of τ_L periods in feeding in customer-order information for decision-making purposes.

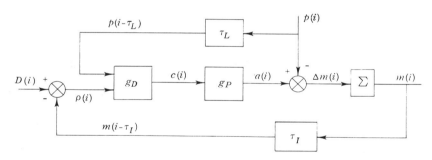

Derive the system difference equation which relates inventory level $m(i)$ to the desired level $D(i)$ and the customer orders $p(i)$, for each of the following cases (any form will do):

a. $\tau_I = \tau_L = 1$, g_P such that $a(i) = c(i)$, and g_D such that $c(i) = K_e\,p(i) + K_P\,p(i-1)$, where K_e and K_P are constants.

b. General τ_I and τ_L, and $a(i) = c(i) = K_e\,p(i) + K_P\,p(i-\tau_L)$.

c. $\tau_I = 0$, $a(i) = c(i)$, and g_D such that $c(i)$ is the sum of $K_e\,p(i)$ and a forecast of customer orders for period i based on the average of the orders received for the last n periods.

   d. The same as part (c) except instead of the rapid response capability g_P is represented by a delay of τ periods.

   e. $\tau_I = 0$, $a(i) = c(i)$, and g_D such that $c(i)$ is the sum of $K_e \rho(i)$ and a forecast of customer orders based on single exponential smoothing with smoothing constant a and $p(0,0) = 0$ (see Eq. (2.4.15)).

   f. The same as part (e) except instead of the rapid response capability g_P is represented by a delay of τ periods.

2.6   The block diagram of the inventory-control system shown below is due to Sargent (1966). It allows for consideration of orders placed but not yet filled in deciding on new orders.

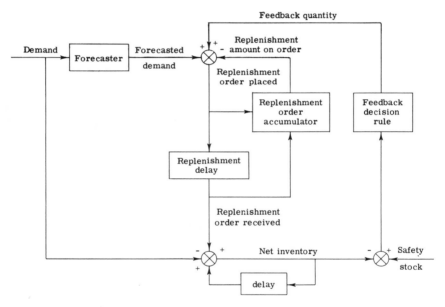

   a.   Specify a set of functions for the boxes shown in the diagram, attempting to be as realistic as possible.

   b.   Using discrete time as the independent variable, define a set of symbols to represent the functions and variables in the system.

   c.   Derive the difference equation relating net inventory (as dependent variable) to demand and safety stock (as forcing functions).

   d.   Derive the difference equation relating replenishment order placed (as dependent variable) to demand and safety stock.

   e.   Derive the difference equation relating replenishment amount on order (as dependent variable) to demand and safety stock.

2.7 Given the simplified switching and delay model of the criminal justice system presented in Fig. 2.5.1, consider each of the following as the dependent variable and formulate the difference equation which relates it to the forcing function $V(i)$, the number of virgin arrests during period i. Assume $\tau_I \geq \tau_{PC}$.

    a. $J(i)$ **b.** $S(i)$ **c.** $R_I(i)$ **d.** $N_I(i)$
    e. $R_{PC}(i)$ **f.** $N_{PC}(i)$ **g.** $N(i)$

2.8 Acknowledging that the model of the criminal justice system in Fig. 2.5.1 may fail to properly represent the real world in a number of ways, revise the model to include at least one feature which improves the realism of the model. For the revised model, write the difference equations which treat $M(i)$ and $J(i)$ as the dependent variable, with $V(i)$ as the forcing function in each case.

2.9 A city's water supply system consists of an intake reservoir of constant cross-sectional area A_1 square feet which receives river water at a controllable hourly rate of $r_1(i)$ gallons. Uncontrollable leakage from the reservoir is L_1 gallons per hour. Water can be pumped from this intake reservoir into a filter at a controllable rate of $r_2(i)$ gallons per hour. The flow characteristics of the filter can be modeled by $\Delta f(i) = \alpha_1 f(i) + r_2(i)$, where $f(i)$ is the flow rate in gallons per hour from the filter into a pure-water reservoir with constant cross section A_2 square feet and leakage L_2 gallons per hour. The heights of water in each reservoir, $h_1(i)$ and $h_2(i)$, can be measured each hour as can the demand in gallons for the previous hour, $d(i-1)$. Derive a difference equation relating the intake from the river (as dependent variable) to demand and the leakages (as forcing functions).

2.10 Write the system difference equation for the closed-loop system presented in the answer to Exercise 1.3.

Chapter III | **The Calculus of Finite Differences**

In mathematics a *calculus* is a method of computation or any process of reasoning using symbols.† The calculus of finite differences is a calculus which applies to functions which are defined only at certain discrete values of the independent variable. In the majority of discrete control problems, the points of definition of such functions are evenly spaced points of the independent variable time. Thus a discrete function $g(t)$ would be defined only for $t = iT$, where i is an integer and T the length of the time interval.

In the material which follows, many similarities with the traditionally more familiar differential and integral calculus will be obvious.

3.1 Differences

For the function $g(t)$ defined only for t an integer multiple of T, the expression

$$[g(t + T) - g(t)]/T \tag{3.1.1}$$

is defined as the *difference quotient* of $g(t)$. Note that were $g(t)$ defined over continuous time the limit of this quotient as T approaches zero would be the derivative of $g(t)$ with respect to t. In the case of functions of discrete time, not only is this limiting process unnecessary, it is not possible.

It is convenient in developing and utilizing the difference calculus to define the time scale used so that it will be measured in units of numbers of periods rather than in units of time itself (see Miller, 1960, Chapter 1). To do this, we

† Webster's New International Dictionary, 2nd ed., s.v. "calculus."

define the normalized independent variable $i = t/T$ and the function

$$f(i) = f(t/T) = (1/T)g(t). \tag{3.1.2}$$

Then,

$$(1/T)g(t + T) = f((t + T)/T) = f((t/T) + 1) = f(i + 1). \tag{3.1.3}$$

Using (3.1.2) and (3.1.3), the difference quotient becomes

$$[g(t + T) - g(t)]/T = (1/T)g(t + T) - (1/T)g(t) = f(i + 1) - f(i) = \Delta f(i), \tag{3.1.4}$$

where, as previously stated, $\Delta f(i)$ is defined as the first forward difference of $f(i)$. Note that use of the symbol Δ implies a time unit corresponding to the length of one period.

Other forms of differences can be defined,† including a backward difference (sometimes called a rearward difference) which occasionally is referred to in the literature involving the optimization of discrete functions. The first backward difference of a function $f(i)$ is

$$\nabla f(i) = f(i) - f(i - 1). \tag{3.1.5}$$

Since $\nabla f(i)$ can be easily expressed in terms of $\Delta f(i)$ as

$$\nabla f(i) = \Delta f(i - 1), \tag{3.1.6}$$

we will confine our discussion of the properties of differences to those of the much more widely used forward differences.

Properties and Special Formulas

Differencing is a linear operation. Thus the usual associative, commutative, and distributive laws may be applied. To illustrate, let $f(i)$ and $g(i)$ be any two functions of discrete time i. Then for any constants a and b,

$$\begin{aligned} \Delta[af(i) + bg(i)] &= [af(i + 1) + bg(i + 1)] - [af(i) + bg(i)] \\ &= a[f(i + 1) - f(i)] - b[g(i + 1) - g(i)] \\ &= a\,\Delta f(i) + b\,\Delta g(i). \end{aligned} \tag{3.1.7}$$

The following special formulas apply for differencing of products and quotients, respectively:

$$\Delta[f(i)g(i)] = f(i + 1)\,\Delta g(i) + g(i)\,\Delta f(i) \tag{3.1.8}$$

† For their definitions and uses consult a text on numerical analysis such as Nielsen (1956).

and

$$\Delta[f(i)/g(i)] = [g(i)\,\Delta f(i) - f(i)\,\Delta g(i)]/g(i)g(i+1). \tag{3.1.9}$$

The similarity to the comparable differentiation formulas is apparent. The multiplication formula is derived as follows. By definition,

$$\Delta[f(i)g(i)] = f(i+1)g(i+1) - f(i)g(i). \tag{3.1.10}$$

We now add and subtract $f(i+1)g(i)$ and group terms to obtain

$$\begin{aligned}
\Delta[f(i)g(i)] &= f(i+1)g(i+1) - f(i+1)g(i) + f(i+1)g(i) - f(i)g(i) \\
&= f(i+1)[g(i+1) - g(i)] + g(i)[f(i+1) - f(i)] \\
&= f(i+1)\,\Delta g(i) + g(i)\,\Delta f(i). \tag{3.1.11}
\end{aligned}$$

Obviously $f(i)g(i+1)$ could have been added and subtracted instead to obtain the alternative expression

$$\Delta[f(i)g(i)] = g(i+1)\,\Delta f(i) + f(i)\,\Delta g(i). \tag{3.1.12}$$

An additional form for the multiplication formula is

$$\Delta[f(i)g(i)] = g(i)\,\Delta f(i) + f(i)\,\Delta g(i) + \Delta f(i)\,\Delta g(i). \tag{3.1.13}$$

The division formula is based on the following derivation:

$$\begin{aligned}
\Delta\left[\frac{f(i)}{g(i)}\right] &= \frac{f(i+1)}{g(i+1)} - \frac{f(i)}{g(i)} \\
&= \frac{f(i+1)g(i) - f(i)g(i+1)}{g(i+1)g(i)} \\
&= \frac{f(i+1)g(i) - f(i)g(i) - f(i)g(i+1) + f(i)g(i)}{g(i+1)g(i)} \\
&= \frac{g(i)[f(i+1) - f(i)] - f(i)[g(i+1) - g(i)]}{g(i+1)g(i)} \\
&= \frac{g(i)\,\Delta f(i) - f(i)\,\Delta g(i)}{g(i+1)g(i)}. \tag{3.1.14}
\end{aligned}$$

These results and the formulas for the differences of several common functions of discrete time are listed in Table 3.1.1. Items 1 through 4 in the table correspond to the above formulas. The remaining items are derived by straightforward application of the definitional formulas for the first forward difference,

$$\Delta f(i) = f(i+1) - f(i). \tag{3.1.15}$$

Item 5, where $f(i)$ is constant for all i, simply verifies the fact that constants do not change with time. The linear function in item 6 has, as is well

Table 3.1.1

Formulas for Common First Forward Differences

Item	Function	First forward difference
1	$af(i)$, a constant	$a\,\Delta f(i)$
2	$af(i) + bg(i)$, a, b constants	$a\,\Delta f(i) + b\,\Delta g(i)$
3a	$f(i)g(i)$	$f(i+1)\,\Delta g(i) + g(i)\,\Delta f(i)$
3b	$f(i)g(i)$	$g(i+1)\,\Delta f(i) + f(i)\,\Delta g(i)$
3c	$f(i)g(i)$	$f(i)\,\Delta g(i) + g(i)\,\Delta f(i) + \Delta f(i)\,\Delta g(i)$
4	$\dfrac{f(i)}{g(i)}$	$\dfrac{g(i)\,\Delta f(i) - f(i)\,\Delta g(i)}{g(i+1)g(i)}$
5	$f(i) = a$, a constant	0
6	$f(i) = ai$, a constant	a
7	$f(i) = a^i$, a constant	$(a-1)a^i$
8	$f(i) = \log_a i$	$\log_a\left(\dfrac{i+1}{i}\right)$
9a	$f(i) = \sin ai$	$2\sin(a/2)\cos a(i + 1/2)$
9b	$f(i) = \sin ai$	$(\cos a - 1)\sin ai + \sin a \cos ai$
10a	$f(i) = \cos ai$	$-2\sin(a/2)\sin a(i + 1/2)$
10b	$f(i) = \cos ai$	$-\sin a \sin ai + (\cos a - 1)\cos ai$

known, a constant change from one period to the next. Note that if the constant a in item 7 is 2, then $\Delta f(i) = f(i) = 2^i$. The result listed for the logarithmic function in item 8 is based on the relationship $\log_a b - \log_a c = \log_a(b/c)$. The forms listed for the sine and cosine terms in 9 and 10 result from the use of trigonometric identities involving differences of two sine functions, differences of two cosine functions and, alternatively, the sine of the sum of two angles and the cosine of the sum of two angles. In general in the work that follows, 9b and 10b can be used more directly than 9a and 10a.

Higher Order Differences

Formulas for second and higher order differences can be derived by successive application of (3.1.15) to the next lower order difference. Thus, the second forward difference of $f(i)$, $\Delta^2 f(i)$, is

$$
\begin{aligned}
\Delta^2 f(i) = \Delta[\Delta f(i)] &= \Delta f(i+1) - \Delta f(i) \\
&= [f(i+2) - f(i+1)] - [f(i+1) - f(i)] \\
&= f(i+2) - 2f(i+1) + f(i).
\end{aligned}
\tag{3.1.16}
$$

The third forward difference of $f(i)$, $\Delta^3 f(i)$, is found in a similar manner by taking the first forward difference of the second forward difference given in

(3.1.16). This yields

$$\Delta^3 f(i) = f(i+3) - 3f(i+2) + 3f(i+1) - f(i). \tag{3.1.17}$$

A pattern appears to be emerging, which, if it holds, would permit a general expression for the nth forward difference, $\Delta^n f(i)$, to be written directly without having to build on the intermediate lower order differences. The hypothesized form is that $\Delta^n f(i)$ contains $n+1$ terms involving $f(i+n)$ through $f(i)$, which alternate in sign beginning with $f(i+n)$ positive, and that the term involving $f(i+n-j)$ is multiplied by the binomial coefficient

$$C_j{}^n = \frac{n!}{(n-j)!j!}. \tag{3.1.18}$$

This form can be expressed as

$$\Delta^n f(i) = \sum_{j=0}^{n} (-1)^j C_j{}^n f(i+n-j). \tag{3.1.19}$$

This will be tested by induction. Comparisons with Eqs. (3.1.15) through (3.1.18) show (3.1.19) holds for n of 1, 2, and 3. We now must show that if it holds for a general n, it will hold for $n+1$. We do this by taking the first forward difference of $\Delta^n f(i)$ as follows:

$$\Delta^{n+1} f(i) = \Delta[\Delta^n f(i)] = \Delta\left[\sum_{j=0}^{n} (-1)^j C_j{}^n f(i+n-j)\right]$$

$$= \sum_{j=0}^{n} (-1)^j C_j{}^n f(i+1+n-j) - \sum_{j=0}^{n} (-1)^j C_j{}^n f(i+n-j). \tag{3.1.20}$$

Now a term involving $f(i+n+1)$ is obtained when $j=0$ in the first summation. No such term exists in the second summation. Therefore, the coefficient of $f(i+n+1)$ in $\Delta^{n+1} f(i)$ is

$$(-1)^0 C_0{}^n - 0 = 1. \tag{3.1.21}$$

$f(i+n)$ appears in both summations. It corresponds to $j=1$ in the first and $j=0$ in the second. Its coefficient in $\Delta^{n+1} f(i)$ is therefore

$$(-1)^1 C_1{}^n - (-1)^0 C_0{}^n = -n - 1 = -(n+1). \tag{3.1.22}$$

Similarly, for $j=2$ in the first summation and $j=1$ in the second, the coefficient of $f(i+n-1)$ is found to be

$$(-1)^2 C_2{}^n - (-1)^1 C_1{}^n = \frac{n(n-1)}{2} + n = \frac{n^2+n}{2} = \frac{(n+1)n}{2}. \tag{3.1.23}$$

In general, the coefficient of $f(i + n + 1 - j)$ is the difference of $(-1)^j C_j^n$ from the first summation and $(-1)^{j-1} C_{j-1}^n$ from the second for $j = 1$ through n. These coefficients may be expressed as follows:

$$(-1)^j C_j^n - (-1)^{j-1} C_{j-1}^n$$

$$= (-1)^j \frac{n!}{(n-j)! j!} - (-1)(-1)^j \frac{n!}{(n-j+1)!(j-1)!}$$

$$= (-1)^j \frac{n!}{(n-j)!(j-1)!} \left[\frac{1}{j} + \frac{1}{(n-j+1)} \right]$$

$$= (-1)^j \frac{n!}{(n-j)!(j-1)!} \left[\frac{(n-j+1)+j}{j(n-j+1)} \right]$$

$$= (-1)^j \frac{n!}{(n-j+1)! j!} [n+1]$$

$$= (-1)^j \frac{(n+1)!}{(n+1-j)! j!} = (-1)^j C_j^{n+1}. \qquad (3.1.24)$$

Finally, the term $f(i)$ does not appear in the first summation so its coefficient is obtained only from the second summation as

$$0 - (-1)^n C_n^n = -(-1)^n = (-1)^{n+1}. \qquad (3.1.25)$$

Note that (3.1.21), (3.1.22), (3.1.23), and (3.1.25) all fit the form developed in (3.1.24).† Thus, the $(n + 1)$st forward difference of $f(i)$ may be written

$$\Delta^{n+1} f(i) = \sum_{j=0}^{n+1} (-1)^j C_j^{n+1} f(i + n + 1 - j), \qquad (3.1.26)$$

which is the form desired since $n + 1$ replaces n in $\Delta^n f(i)$ as given in (3.1.19). Thus, we have proved by induction that (3.1.19) is the nth forward difference of the function of discrete time $f(i)$.

3.2 Factorial Polynomials

A frequently occurring function in many discrete-time systems is $f(i) = i^k$, where k is usually but not necessarily an integer. Finding differences of such

---

† Actually the two special cases of $f(i + n + 1)$ and $f(i)$ could also have been derived in the manner of (3.1.24), since the factorial of a negative integer is infinite. Thus, for $f(i + n + 1)$, which corresponds to $j = 0$ in the format $f(i + n + 1 - j)$, the $(j - 1)!$ is infinite and the second term in the first line of (3.1.24) vanishes. Similarly, for $j = n + 1$ the $(n - j)!$ causes the first term in the first line of (3.1.24) to vanish.

functions, though straightforward, is cumbersome. If, for example, k is a positive integer,

$$\Delta i^k = (i + 1)^k - i^k$$
$$= i^k + k i^{k-1} + C_2^k i^{k-2} + \cdots + C_{k-1}^k i + 1 - i^k$$
$$= k i^{k-1} + C_2^k i^{k-2} + \cdots + C_{k-1}^k i + 1. \tag{3.2.1}$$

Polynomials of powers of i would add further complication. An alternative approach to expressing differences of functions of this kind is through use of a particular function of i called the "factorial function." Depending on available software and other system characteristics, this alternative could prove helpful.

Factorial Functions

For k a positive integer, define $i^{(k)}$ as

$$i^{(k)} = i(i - 1)(i - 2) \cdots (i - k + 1). \tag{3.2.2}$$

The symbol $i^{(k)}$ is read "i, k factorial" (Miller, 1960, p. 6). Note that

$$i^{(k)} = \frac{i!}{(i - k)!} \tag{3.2.3}$$

and that

$$i^{(i)} = i!. \tag{3.2.4}$$

For $k > i$, $(i - k)!$ is infinite and $i^{(k)}$ vanishes. Furthermore, for j also a positive integer,

$$i^{(k)}(i - k)^{(j)} = \frac{i!}{(i - k)!} \cdot \frac{(i - k)!}{(i - k - j)!} = \frac{i!}{(i - k - j)!} = i^{(k+j)}. \tag{3.2.5}$$

Now, if (3.2.5) is to hold for $k = 0$ and j positive, i.e.,

$$i^{(0)}(i - 0)^{(j)} = i^{(0+j)} = i^{(j)}, \tag{3.2.6}$$

$i^{(0)}$ must be defined as one. Furthermore, in order for (3.2.5) to hold for negative factorial exponents in general, it must hold in the following case for $j = k > 0$:

$$i^{(-k)}(i + k)^{(k)} = i^{(-k+k)} = i^{(0)} = 1. \tag{3.2.7}$$

Therefore,

$$i^{(-k)} = \frac{1}{(i + k)^{(k)}} = \frac{1}{(i + k)(i + k - 1) \cdots (i + 1)}. \tag{3.2.8}$$

Thus, for k negative,

$$i^{(k)} = \frac{1}{(i + 1)(i + 2) \cdots (i - k - 1)(i - k)}. \tag{3.2.9}$$

We now investigate the first forward difference of $i^{(k)}$. First, for k positive,

$$\Delta i^{(k)} = (i + 1)^{(k)} - i^{(k)}$$
$$= [(i + 1)(i)(i - 1) \cdots (i - k + 2)] - [(i)(i - 1) \cdots (i - k + 2)(i - k + 1)]$$
$$= [(i + 1) - (i - k + 1)][(i)(i - 1) \cdots (i - k + 2)]$$
$$= k i^{(k-1)}, \qquad k > 0. \tag{3.2.10}$$

For k negative,

$$\Delta i^{(k)}$$

$$= \frac{1}{(i + 2)(i + 3) \cdots (i - k)(i - k + 1)} - \frac{1}{(i + 1)(i + 2) \cdots (i - k - 1)(i - k)}$$

$$= \frac{(i + 1) - (i - k + 1)}{(i + 1)(i + 2) \cdots (i - k)(i - k + 1)}$$

$$= k i^{(k-1)}, \qquad k < 0. \tag{3.2.11}$$

Finally, for $k = 0$,

$$\Delta i^{(0)} = 1 - 1 = 0 \tag{3.2.12}$$

which also conforms to the general statement that

$$\Delta i^{(k)} = k i^{(k-1)} \tag{3.2.13}$$

for all integer k. Note the similarity to the formula for the derivative of a continuous variable raised to a power.

Factorial functions can also be defined for noninteger values of k. Since the need for such functions is marginal for our purposes here, this extension will not be pursued.

The Factorial Polynomial

We now define a *factorial polynomial* as a polynomial in factorial functions. A factorial polynomial of order n, where n is a positive integer, is represented by

$$\phi(i) = a_0 \, i^{(n)} + a_1 i^{(n-1)} + \cdots + a_{n-1} i^{(1)} + a_n. \tag{3.2.14}$$

The a's are constants and $a_0 \neq 0$. This expression can of course be written as an ordinary polynomial in i by multiplying out the various factorial functions and grouping terms in like powers of i. Conversely, any ordinary polynomial in i can be put into the form of a factorial polynomial of the same order if it is beneficial to do so. The details of converting from ordinary to factorial polynomials and vice versa is discussed in the next section and illustrated by an example.

One interesting property of factorial polynomials which is not a property of ordinary polynomials is that if

$$\phi(i) = 0, \tag{3.2.15}$$

for all i, then

$$a_0 = a_1 = \cdots = a_{n-1} = a_n = 0. \tag{3.2.16}$$

This may be proven as follows. For (3.2.15) to hold for $i = 0$, certainly $a_n = 0$. If we now difference both sides of (3.2.15), we obtain, using (3.2.13),

$$\Delta\phi(i) = na_0 i^{(n-1)} + (n-1)a_1 i^{(n-2)} + \cdots + a_{n-1} = 0. \tag{3.2.17}$$

For the equality to hold at $i = 0$, a_{n-1} must equal zero. By continued differencing and evaluating the resulting expressions at $i = 0$, we can show each coefficient in turn must be zero.

It may also be shown that factorial polynomials are unique. Thus, if $\phi(i)$ is expressed as in (2.3.14) and in some other fashion such as by

$$\phi(i) = b_{-k} i^{(n+k)} + \cdots + b_{-1} i^{(n+1)} + b_0 i^{(n)} + b_1 i^{(n-1)} + \cdots + b_n, \tag{3.2.18}$$

then

$$b_{-k} = \cdots = b_{-1} = 0 \tag{3.2.19}$$

and

$$b_j = a_j \qquad \text{for } j = 0, \ldots, n. \tag{3.2.20}$$

This is demonstrated by subtracting (3.2.14) from (3.2.18) to obtain

$$b_{-k} i^{(n+k)} + \cdots + b_{-1} i^{(n+1)} + (b_0 - a_0)i^{(n)} + \cdots + (b_n - a_n). \tag{3.2.21}$$

Since (3.2.21) resulted from the subtraction of two expressions for the same polynomial, it must be identically zero. Therefore, each coefficient of the factorial functions of i must be zero by our previous discussion and (3.2.19) and (3.2.20) thus hold. Therefore, $\phi(i)$ is unique.

Finally, for any function of discrete time $f(i)$, we define for positive k,

$$f(i)^{(k)} = f(i)f(i-1)\cdots f(i-k+1), \tag{3.2.22}$$

and, for negative k,

$$f(i)^{(k)} = [f(i+1)f(i+2)\cdots f(i-k)]^{-1}. \tag{3.2.23}$$

$f(i)^{(0)}$ is defined as one, which, though seemingly arbitrary, is consistent with the properties demonstrated below.

In the particular case where $f(i)$ is a linear function of i, specifically,

$$f(i) = a + bi, \tag{3.2.24}$$

where a and b are constants, and for k positive,

$$(a + bi)^{(k)} = [a + bi][a + b(i - 1)] \cdots [a + b(i - k + 1)]. \tag{3.2.25}$$

The first forward difference of $(a + bi)^{(k)}$ is

$$\begin{aligned}
\Delta(a + bi)^{(k)} &= [a + b(i + 1)][a + bi] \cdots [a + b(i - k + 2)] \\
&\quad - [a + bi] \cdots [a + b(i - k + 2)][a + b(i - k + 1)] \\
&= \{[a + b(i + 1)] - [a + b(i - k + 1)]\}\{[a + bi] \cdots \\
&\quad \times [a + b(i - k + 2)]\} \\
&= bk(a + bi)^{(k-1)}. \tag{3.2.26}
\end{aligned}$$

Conversion

Before leaving our discussion of factorial polynomials, we illustrate with a simple third-order polynomial how one converts an ordinary polynomial to a factorial polynomial. Consider the ordinary third-order polynomial in i,

$$P(i) = a_0 i^3 + a_1 i^2 + a_2 i + a_3. \tag{3.2.27}$$

The corresponding factorial polynomial will have to be of third order also to accommodate the i^3 term. By definition,

$$i^{(3)} = i(i - 1)(i - 2) = i^3 - 3i^2 + 2i, \tag{3.2.28}$$

so that

$$i^3 = i^{(3)} + 3i^2 - 2i. \tag{3.2.29}$$

Substitution of (3.2.29) for i^3 in (3.2.27) yields

$$P(i) = a_0 i^{(3)} + (3a_0 + a_1)i^2 + (-2a_0 + a_2)i + a_3. \tag{3.2.30}$$

Similarly,

$$i^{(2)} = i(i - 1) = i^2 - i, \tag{3.2.31}$$

from which

$$i^2 = i^{(2)} + i. \tag{3.2.32}$$

Substitution into (3.2.30) results in

$$P(i) = a_0 i^{(3)} + (3a_0 + a_1)i^{(2)} + (a_0 + a_1 + a_2)i + a_3, \tag{3.2.33}$$

which, after substituting $i^{(1)}$ for i, gives the factorial polynomial

$$\phi(i) = a_0\, i^{(3)} + (3a_0 + a_1)i^{(2)} + (a_0 + a_1 + a_2)i^{(1)} + a_3\, i^{(0)}. \quad (3.2.34)$$

Note that the coefficient of $i^{(3)}$ is the same as that of i^3 in the original polynomial, and that the constant terms in both polynomials are also equal. It should be apparent that these equalities must exist for any order polynomial.

The procedure just illustrated can be followed to express any order regular polynomial as a factorial polynomial of like order. Furthermore, a comparable procedure could be devised for the conversion of a factorial polynomial to a regular polynomial. Conversions of this sort can often be facilitated, however, by making use of certain recurrence relationships among the coefficients involved. When the expression for a factorial function $i^{(k)}$ is expanded and written as a linear combination of powers of i from 1 through k as follows,

$$i^{(k)} = i(i - 1) \cdots (i - k + 1)$$

$$= S_k{}^k i^k + S_{k-1}^k\, i^{k-1} + \cdots + S_1{}^k i$$

$$= \sum_{j=1}^{k} S_j{}^k i^j, \qquad (3.2.35)$$

the combinatorial coefficients $S_1{}^k$ through $S_k{}^k$ are known as *Stirling numbers of the first kind*. When a power of i is expressed as a linear combination of factorial functions such as

$$i^k = \hat{S}_k{}^k i^{(k)} + \hat{S}_{k-1}^k\, i^{(k-1)} + \cdots + \hat{S}_1{}^k i^{(1)} = \sum_{j=1}^{k} \hat{S}_j{}^k i^{(j)}, \qquad (3.2.36)$$

the combinatorial coefficients $\hat{S}_1{}^k$ through $\hat{S}_k{}^k$ are called *Stirling numbers of the second kind*. It can be shown that the recurrence relationship

$$S_j^{k+1} = S_{j-1}^k + j S_j{}^k \qquad (3.2.37)$$

holds for Stirling numbers of both the first and second kinds. It can also be shown that each Stirling number of the first kind can be calculated from

$$S_j{}^k = \frac{1}{j!} \cdot \frac{d^j i^{(k)}}{di^j}\bigg|_{i=0} \qquad (3.2.38)$$

and that each Stirling number of the second kind can be calculated from

$$\hat{S}_j{}^k = \frac{1}{j!}\, \Delta^j i^k\bigg|_{i=0}. \qquad (3.2.39)$$

The interested reader will find the derivations of all of these relationships in Miller (1960, pp. 11–21).

Equation (3.2.38) would be extremely cumbersome to actually use as a computational formula; however, (3.2.39) could be quite useful in converting regular polynomials to factorial ones. Since in practice this is usually the conversion we seek, we are in a bit of luck. We illustrate its use in converting the third-order polynomial of (3.2.27) as follows. The calculations are presented in Table 3.2.1. The number in the row corresponding to i^k and the column corresponding to $i^{(j)}$ is $\hat{S}_j^{\,k}$, for $1 \le j \le k$, calculated from (3.2.39).

Table 3.2.1

Conversion of Regular Polynomial of Third Order to Factorial Polynomial by Use of Stirling Numbers of the Second Kind

Row	Coefficient of i^k	Column			
		$i^{(3)}$	$i^{(2)}$	$i^{(1)}$	$i^{(0)}$
i^0	a_3	—	—	—	1
i^1	a_2	—	—	1	0
i^2	a_1	—	1	1	0
i^3	a_0	1	3	1	0
$i^{(k)}$		a_0	$3a_0 + a_1$	$a_0 + a_1 + a_2$	a_3

For example, in row i^3, column $i^{(2)}$,

$$\hat{S}_2^{\,3} = \frac{1}{2!} \Delta^2 i^3 \Big|_{i=0} = \frac{1}{2!} [(i+2)^3 - 2(i+1)^3 + i^3]_{i=0}$$

$$= \frac{1}{2!} [2^3 - 2 \cdot 1^3 + 0] = \frac{1}{2} [8 - 2] = 3. \tag{3.2.40}$$

A column headed $i^{(0)}$ is added to make provision for the constant term a_3. A column is also provided for the coefficients of i^k in the original polynomial. The coefficient of $i^{(k)}$ in the resulting factorial polynomial is obtained by multiplying each entry in column j by the corresponding entry in the column of coefficients of i^k and summing the products. Note that the results agree with (3.2.34).

The use of factorial polynomials for finding differences of ordinary polynomials may seem at first glance to be a somewhat roundabout approach because of the conversions from ordinary to factorial and back again. However, armed with tables of Stirling numbers (which can easily be compiled using either the definitional formulas or the recurrence relationships given above), one can reduce the conversions to almost instantaneous tasks, thus realizing some benefit from the procedure.

For example, consider the general third-order polynomial given in (3.2.27). The Stirling numbers of the second kind for $k = 1, 2, 3$ are shown in Table 3.2.2. Using the table, the conversion of (3.2.27) to a factorial polynomial proceeds quickly as follows:

$$
\begin{aligned}
P(i) &= a_0 i^3 + a_1 i^2 + a_2 i + a_3 \\
&= a_0[i^{(3)} + 3i^{(2)} + i^{(1)}] + a_1[i^{(2)} + i^{(1)}] + a_2[i^{(1)}] + a_3 \\
&= a_0 i^{(3)} + (3a_0 + a_1)i^{(2)} + (a_0 + a_1 + a_2)i^{(1)} + a_3 \\
&= \phi(i).
\end{aligned}
$$

Table 3.2.2

Stirling Numbers of the Second Kind

k	j		
	1	2	3
1	1		
2	1	1	
3	1	3	1

From (3.2.17), $\Delta\phi(i)$ is found to be

$$
\Delta\phi(i) = 3a_0 i^{(2)} + 2(3a_0 + a_1)i^{(1)} + (a_0 + a_1 + a_2).
$$

Now using the Stirling numbers of the first kind given in Table 3.2.3 for $k = 1, 2$, this difference is easily converted to ordinary polynomial form as follows:

$$
\begin{aligned}
\Delta P(i) &= 3a_0[i^2 - i] + (6a_0 + 2a_1)[i] + (a_0 + a_1 + a_2) \\
&= 3a_0 i^2 + (3a_0 + 2a_1)i + (a_0 + a_1 + a_2).
\end{aligned}
$$

Table 3.2.3

Stirling Numbers of the First Kind

k	j	
	1	2
1	1	
2	-1	1

The doubting reader can compare the above with straightforward differencing of (3.2.27) and if not convinced may run through both methods on a seventh-order polynomial (*after* compiling the appropriate tables of Stirling numbers).

3.3 Summations

Let $F(i)$ be a function defined for all i. From the definition of the first forward difference,

$$F(a + 1) = F(a) + \Delta F(a), \tag{3.3.1}$$

where a is an integer, and

$$F(a + 2) = F(a + 1) + \Delta F(a + 1) = F(a) + \Delta F(a) + \Delta F(a + 1). \tag{3.3.2}$$

Similarly the value of F at any point in time $b > a$, b integer, can be expressed in terms of the sum of the value of F at a and the differences $\Delta F(a)$ through $\Delta F(b - 1)$. This relationship is expressed symbolically as

$$F(b) = \sum_{i=a}^{b-1} \Delta F(i) + F(a). \tag{3.3.3}$$

Normally we like to think in terms of the summation operation being the inverse of the differencing operation, i.e.,

$$f(i) = \Delta F(i). \tag{3.3.4}$$

Now combining (3.3.3) and (3.3.4), we get

$$F(b) = \sum_{i=a}^{b-1} f(i) + F(a). \tag{3.3.5}$$

It is apparent that even though the function $F(i)$ has a unique difference $\Delta F(i)$, the inverse relationship is not unique because of the presence of the constant $F(a)$. Thus, for a given $\Delta F(i)$, an infinite number of values of $F(i)$ can be obtained depending upon the constant involved.

In terms of the indefinite sum, we write

$$F(b) = \sum f(i) + K(b), \tag{3.3.6}$$

where $K(b)$ is defined as a *periodic constant*. To understand why $K(b)$ carries this definition, consider a second function $G(b)$ such that

$$G(b) = \sum f(i). \tag{3.3.7}$$

Now, from (3.3.7) and (3.3.4),

$$\Delta G(b) = f(b) = \Delta F(b). \tag{3.3.8}$$

Furthermore, if we subtract (3.3.7) from (3.3.6), we get

$$F(b) - G(b) = K(b),   \tag{3.3.9}$$

which means that the first forward difference of $K(b)$ is

$$\Delta K(b) = \Delta [F(b) - G(b)] = \Delta F(b) - \Delta G(b) = 0.   \tag{3.3.10}$$

Therefore,

$$K(b + 1) = K(b) = K   \tag{3.3.11}$$

for all i. Thus, should one wish to think of a function of continuous time, $K(t/T)$ must be defined such that for all $t = iT$, $K(t/T) = K$. The periodic constant plays the same role in summation as the constant of integration plays in integration.

Properties and Special Formulas

Obviously, summation is a linear operation so that given two functions of $i, f(i)$ and $g(i)$, and the constants a and b,

$$\sum [af(i) + bg(i)] = a \sum f(i) + b \sum g(i).   \tag{3.3.12}$$

As is customary in expressing both indefinite integrals and indefinite sums, we omitted the periodic constants in (3.3.12) and will continue to do so throughout this work except where specifically required to include them.

An important relationship in summation calculus is the formula for summation by parts,

$$\sum f(i) \, \Delta g(i) = f(i)g(i) - \sum g(i + 1) \, \Delta f(i).   \tag{3.3.13}$$

This particular form is obtained by summing both sides of (3.1.12), one of our formulas for the difference of the product of two functions, and then rearranging terms. A similar formula in which the roles of $f(i)$ and $g(i)$ are reversed could be obtained in like manner from (3.1.11).

In addition, the formulas for the summations of several common functions are of interest. These and the above results are listed in Table 3.3.1. Items 1, 2, and 3 in the table correspond to the above formulas. Items 4 through 9 derive from the corresponding difference formulas. Item 4 comes from summing item 7 in Table 3.1.1 and dividing through by $(a - 1)$. Item 5 is derived by summing item 8 in Table 3.1.1 to obtain

$$\log_a i = \sum \log_a \frac{i + 1}{i} = \sum [\log_a(i + 1) - \log_a i]$$

$$= \sum \log_a(i + 1) - \sum \log_a i.   \tag{3.3.14}$$

Table 3.3.1

Formulas for Summations of Common Functions

Item	Function	Summation
1	$af(i)$, a constant	$a\sum f(i)$
2	$af(i) + bg(i)$, a, b constants	$a\sum f(i) + b\sum g(i)$
3	$f(i)\,\Delta g(i)$	$f(i)g(i) - \sum g(i+1)\,\Delta f(i)$
4	$f(i) = a^i$, $a \neq 1$ constant	$a^i/(a-1)$
5	$f(i) = \log_a i$	$\log_a(i-1)!$
6a	$f(i) = \sin ai$, $a \neq 2m\pi$	$-\dfrac{\cos a(i - \frac{1}{2})}{2\sin(a/2)}$
6b	$f(i) = \sin ai$, $a \neq 2m\pi$	$-(\frac{1}{2})\cotan(a/2)\cos ai - \frac{1}{2}\sin ai$
7a	$f(i) = \cos ai$, $a \neq 2m\pi$	$\dfrac{\sin a(i - \frac{1}{2})}{2\sin(a/2)}$
7b	$f(i) = \cos ai$, $a \neq 2m\pi$	$\frac{1}{2}\cotan(a/2)\sin ai - \frac{1}{2}\cos ai$
8	$f(i) = i^{(k)}$, $k \neq -1$	$\dfrac{i^{(k+1)}}{k+1}$
9	$f(i) = (a + bi)^{(k)}$, $k \neq -1$	$\dfrac{(a + bi)^{(k+1)}}{b(k+1)}$

Now each of the terms on the right-hand side of (3.3.14) is a sum of the logarithms of successive integers which can also be expressed as the log of the products of these integers. Thus the form $\log_a i!$ suggests itself. If indeed $\log_a i!$ is substituted for the first term, then $\log_a(i - 1)!$ must be substituted for the second and we have

$$\log_a i! - \log_a(i - 1)! = \log_a \frac{i!}{(i-1)!} = \log_a i \qquad (3.3.15)$$

which is the form desired since it satisfies (3.3.14). Thus

$$\sum \log_a i = \log_a(i - 1)!. \qquad (3.3.16)$$

The derivations of items 6 and 7 stem jointly from items 9 and 10 in Table 3.1.1. Summing 9b and 10b yields, respectively,

$$\sin ai = \sum [(\cos a - 1)\sin ai + \sin a\cos ai]$$
$$= (\cos a - 1)\sum \sin ai + \sin a \sum \cos ai \qquad (3.3.17)$$

and

$$\cos ai = \sum [-\sin a \sin ai + (\cos a - 1)\cos ai]$$
$$= -\sin a \sum \sin ai + (\cos a - 1)\sum \cos ai. \qquad (3.3.18)$$

Simultaneous solution of (3.3.17) and (3.3.18) for $\sum \sin ai$ and $\sum \cos ai$ followed by use of certain trigonometric identities result in items 6b and 7b in Table 3.3.1. The subsequent obtaining of 6a and 7a involves further use of well-known identities. Finally, items 8 and 9 are derived in straightforward fashion by summing (3.2.13), the expression for $\Delta i^{(k)}$, and (3.2.26), which is $\Delta(a + bi)^{(k)}$, respectively, and rearranging terms.

Definite Sums

In the material which follows, most of our involvement with summations will be with definite sums, i.e., where specified upper and lower limits are imposed on the values of the summation index. In deriving Eq. (3.3.3) we noted that the value of a function $F(i)$ for $i = b \geq a$ was the sum of the value of the function at $i = a$ and the first forward differences of $F(i)$ for $i = a$ through $b - 1$. That is, for $f(i) = \Delta F(i)$,

$$F(b) = \sum_{i=a}^{b-1} f(i) + F(a). \tag{3.3.3}$$

Rearranging terms yields

$$\sum_{i=a}^{b-1} f(i) = F(b) - F(a) \tag{3.3.19}$$

which also may be written in either of the following forms:

$$\sum_{i=a}^{b-1} f(i) = \sum f(i) \Big|_a^b = F(i) \Big|_a^b. \tag{3.3.20}$$

Regardless of the form adopted, the relationship represents the fundamental theorem of the summation calculus.

As an illustration of the use of this formula, consider the problem of finding the sums of the cubes of the first n positive integers. We first express i^3 as a linear combination of the factorial functions $i^{(3)}$, $i^{(2)}$, and $i^{(1)}$ using (3.2.36). The combinatorial coefficients (Stirling numbers of the second kind) are evaluated using (3.2.39) with $k = 3$ or an appropriate table to obtain

$$i^3 = i^{(3)} + 3i^{(2)} + i^{(1)}. \tag{3.3.21}$$

(See row 3 of Table 3.2.2 to compare results.) Now

$$\sum_{i=0}^{n} i^3 = \sum_{i=0}^{n} [i^{(3)} + 3i^{(2)} + i^{(1)}]$$

$$= \left[\frac{i^{(4)}}{4} + 3 \cdot \frac{i^{(3)}}{3} + \frac{i^{(2)}}{2} \right]_0^{n+1}$$

$$= \left[\frac{i(i-1)(i-2)(i-3)}{4} + i(i-1)(i-2) + \frac{i(i-1)}{2}\right]_0^{n+1}$$

$$= \tfrac{1}{4}[(n+1)n(n-1)(n-2) + 4(n+1)n(n-1) + 2(n+1)n]$$

$$= \tfrac{1}{4}(n+1)n[(n^2 - 3n + 2) + (4n - 4) + 2]$$

$$= \tfrac{1}{4}(n+1)n(n^2 + n) = \frac{(n+1)^2 n^2}{4}. \tag{3.3.22}$$

This method can of course be used to find sums of any power of integers except $k = -1$.

As a second example, let us find the definite sum of ia^i, where a is a constant, with summation limits from 0 to n. First note that

$$ia^i = aia^{i-1} = a\, da^i/da. \tag{3.3.23}$$

Therefore,

$$\sum_{i=0}^n ia^i = a \sum_{i=0}^n \frac{da^i}{da} = a \frac{d}{da}\left[\sum_{i=0}^n a^i\right]$$

$$= a \frac{d}{da}\left[\frac{a^i}{a-1}\Big|_0^{n+1}\right] = a \frac{d}{da}\left[\frac{a^{n+1} - 1}{a-1}\right]$$

$$= a \frac{(a-1)(n+1)a^n - (a^{n+1} - 1)}{(a-1)^2} = \frac{(a-1)(n+1)a^{n+1} - a^{n+2} + a}{(a-1)^2}.$$

$$\tag{3.3.24}$$

This procedure can be extended to find the sum of $i^k a^i$ for any nonnegative integer value of k.

The definite sums of a few fairly common functions are listed in Table 3.3.2. Extensive tables of sums may be found in the work of Jolley (1961) and Mangulis (1965). Note that all sums listed in Table 3.3.2 are from 0 to n. Should the sum from, say, m to n, where $m \le n$, be desired, it may be found from

$$\sum_{i=m}^n f(i) = \sum_{i=0}^n f(i) - \sum_{i=0}^{m-1} f(i). \tag{3.3.25}$$

In other words, the appropriate tabulated sum can be evaluated for upper bounds of n and $m - 1$ and the two results subtracted to obtain the sum from m to n.

Finally, to close this section on summations, we illustrate summation by parts. We will use the same example as above, namely the sum of ia^i from 0 to n. In the formula for summation by parts, (3.3.13), let $f(i) = i$ and

Table 3.3.2

Sums of Some Common Functions

Item	Summation	Definite sum
1	$\sum_{i=0}^{n} i^{(k)}, \quad k \neq -1$	$\dfrac{(n+1)^{(k+1)}}{k+1}$
2	$\sum_{i=0}^{n} (a+bi)^{(k)}, \quad k \neq -1$	$\dfrac{[a+b(n+1)]^{(k+1)} - [a-b(k)]^{(k+1)}}{b(k+1)}$
3	$\sum_{i=0}^{n} i$	$\dfrac{n(n+1)}{2}$
4	$\sum_{i=0}^{n} i^2$	$\dfrac{n(n+1)(2n+1)}{6}$
5	$\sum_{i=0}^{n} i^3$	$\dfrac{n^2(n+1)^2}{4}$
6	$\sum_{i=0}^{n} a^i, \quad a \neq 1$	$\dfrac{a^{n+1}-1}{a-1}$
7	$\sum_{i=0}^{n} ia^i, \quad a \neq 1$	$\dfrac{(a-1)(n+1)a^{n+1} - a^{n+2} + a}{(a-1)^2}$
8	$\sum_{i=0}^{n} i^2 a^i, \quad a \neq 1$	$\dfrac{(a-1)^2(n+1)^2 a^{n+1} - 2(a-1)(n+1)a^{n+2} + a^{n+3} - a^2 + a^{n+2} - a}{(a-1)^3}$
9	$\sum_{i=0}^{n} \sin ai, \quad a \neq 2m\pi$	$\dfrac{\cos a((n+1)/2)\cos(an/2)}{\sin(a/2)}$
10	$\sum_{i=0}^{n} \cos ai, \quad a \neq 2m\pi$	$\dfrac{\sin a((n+1)/2)\cos(an/2)}{\sin(a/2)}$
11	$\sum_{i=0}^{n} a$	$a(n+1)$

$\Delta g(i) = a^i$. Using the appropriate difference formula for $f(i)$ and the appropriate summation formula for $\Delta g(i)$, we obtain $\Delta f(i) = 1$ and $g(i) = a^i/(a-1)$. Substitution into (3.3.13) yields

$$\sum_{i=0}^{n} ia^i = \left[(i)\frac{a^i}{a-1} - \sum \frac{a^{i+1}}{a-1}(1) \right]_0^{n+1}$$

$$= \left[\frac{ia^i}{a-1} - \frac{1}{a-1}\left(\frac{a^{i+1}}{a-1}\right) \right]_0^{n+1} = \left[\frac{(a-1)ia^i - a^{i+1}}{(a-1)^2} \right]_0^{n+1}$$

$$= \frac{(a-1)(n+1)a^{n+1} - a^{n+2} + a}{(a-1)^2}, \tag{3.3.26}$$

which agrees with our previous result. Thus in many cases alternative approaches to the determination of sums are available. One can often, therefore, pick an approach which best suits him and the tools at his disposal.

3.4   The Calculus of Finite Differences and Linear Difference Equations

The topics covered in the first three sections of this chapter form the mathematical base for the formulation and solution of linear difference equations. The very name "difference equation" indicates the fundamental role that differences play in the formulation of difference equation models. The solution of a difference equation is basically a summing process. This will become evident in Chapters IV and VI where the material just presented on summations is used extensively in both the classical and z transform solution procedures.

Furthermore, several of the discrete-system models developed in Chapter II involved difference and summation operations in their control decision rules as in the process controller or due to the effects of the physical properties of the process such as the historically based forecasts in the production inventory control system. As will become apparent in succeeding chapters, it is important for solution purposes that these differences and summations, of whatever order, be manipulated so that the difference equation is expressed in expanded form. Note also that Eqs. (2.1.4) and (2.1.5), for conversion of a linear difference equation from difference-operator to expanded form and vice versa, respectively, are based on (3.1.19), the definitional formula for $\Delta^n f(i)$.

We now turn our attention to the solution of linear difference equations with constant coefficients.

Exercises

3.1   Given $f(i) = a^i$ and $g(i) = i^2$, find each of the following:

    a. $\Delta[3f(i) + 2g(i)]$ **b.** $\Delta[f(i)g(i)]$ **c.** $\Delta[f(i)/g(i)]$
    d. $\Delta[g(i)/f(i)]$ **e.** $\Delta^2 g(i)$ **f.** $\Delta^k f(i)$ **g.** $\Delta^2[f(i)g(i)]$

3.2   Find each of the following differences:

    a. $\Delta(2i + 3)$ **b.** $\Delta i^2 2^i$ **c.** $\Delta[\sin ai \,/\cos bi]$
    d. $\Delta^4 f(i)$, in terms of $f(i), f(i + 1)$, etc.   **e.** $\Delta^4 5^i$
    f. $\Delta^4 i^3$ **g.** $\Delta^2 \cos ai$ in terms of functions with argument i
    h. $\Delta^2 \log_a i$

3.3   Write each of the following difference equations in expanded form:

    a. $2 \Delta^4 x(i) + 7 \Delta^3 x(i) - 5 \Delta^2 x(i) + 3x(i) = i^5$

b. $3 \Delta^4 x(i) - 5 \Delta^3 x(i) + \Delta x(i) + 7x(i) = 10^i$
c. $\Delta^5 x(i) + 2 \Delta^3 x(i) + 3 \Delta x(i) + 4x(i) = f(i)$
d. $3 \Delta^4 x(i) + 10 \Delta^3 x(i) + 11 \Delta^2 x(i) + 9 \Delta x(i) + 7x(i) = \sin 4i$

3.4 Write each of the following difference equations in difference-operator form:

a. $4x(i + 3) - 3x(i + 2) + 2x(i + 1) - x(i) = f(i)$
b. $2x(i + 3) + 3x(i + 2) - x(i + 1) + 4x(i) = \cos 3i + i^2$
c. $3x(i + 4) - 2x(i + 3) - x(i + 2) + 5x(i + 1) + 2x(i) = \sin 2i$
d. $2x(i + 4) + 3x(i + 3) - 4x(i + 2) + 2x(i + 1) - x(i) = i^3$

3.5 Write the following difference equations derived in the Exercises in Chapter II in both expanded and difference-operator form: 3-a through 3-g, 4-a through 4-g, 5-a through 5-f, 6-c, 6-d, 6-e, 7-a through 7-g.

3.6 Find the numerical values of each of the following factorial functions $i^{(k)}$:

a. $i = 4, k = 2$ **b.** $i = 5, k = -3$ **c.** $i = 7, k = 7$
d. $i = 10, k = 12$ **e.** $i = 4, k = -6$

3.7 (a) through (e). Find the numerical value of the first forward differences of each factorial function in Exercise 3.6.

3.8 Find the numerical values of each of the following linear factorial functions $(3 + 2i)^{(k)}$:

a. $i = 4, k = 3$ **b.** $i = 5, k = -3$ **c.** $i = 3, k = 3$
d. $i = 3, k = 5$

3.9 (a) through (d). Find the numerical value of the first forward differences of each linear factorial function in Exercise 3.8.

3.10 (a) through (d). Find the numerical value of the second forward differences of each linear factorial function in Exercise 3.8.

3.11 Convert each of the following regular polynomials to a factorial polynomial:

a. $P(i) = i^3 + 2i^2 - i + 4$ **b.** $P(i) = 2i^4 + 3i^3 + i^2 + 2$
c. $P(i) = i^4 - 2i^3 + 4i^2 + 3i - 7$

3.12 Convert each of the following factorial polynomials to a regular polynomial:

a. $\phi(i) = i^{(3)} + 2i^{(2)} - i^{(1)} + 4$ **b.** $\phi(i) = 3i^{(4)} - 2i^{(3)} + i^{(1)} - 5$
c. $\phi(i) = i^{(4)} + 3i^{(3)} - 2i^{(2)} - 4i^{(1)} + 2$

3.13 Find expressions for the following indefinite sums:

a. $\sum (2i + 3)$ **b.** $\sum [3(\cos 4i) - 2 \log_{10} i]$ **c.** $\sum (2^i + 3i^{(2)})$

d. $\sum i^2 2^i$ **e.** $\sum \Delta^4 5^i$ **f.** $\sum (7 + 6i)^{(4)}$

g. $\sum (7 + 6i)^{(-4)}$ **h.** $\sum i \sin 2i$ **i.** $\sum i^{(3)} \cdot 2^i$

3.14 Find the following definite sums:

a. $\sum_{i=0}^{n} i^4$ **b.** $\sum_{i=0}^{n} i \sin i$ **c.** $\sum_{i=0}^{n} 2^i \sin 2i$

d. $\sum_{i=0}^{n} i^{(3)}$ **e.** $\sum_{i=0}^{n} (3 + 2i)^{(2)}$ **f.** $\sum_{i=0}^{n} (3 + 2i)^{(-2)}$

g. $2 \cdot 3 \cdot 4 + 4 \cdot 5 \cdot 6 + 6 \cdot 7 \cdot 8 + 8 \cdot 9 \cdot 10 + 10 \cdot 11 \cdot 12$ **h.** $\sum_{i=0}^{\infty} i^2 4^i$

i. $\sum_{i=1}^{\infty} \frac{1}{(2i + 1)(2i + 3)}$ **j.** $\sum_{i=1}^{10} \ln i$ **k.** $\sum_{i=3}^{6} (2 + 3i)^{(2)} \cdot 4^i$

3.15 Find each of the following:

a. $m(17)$ given $m(0) = 5$ and $\Delta m(i) = i$ for $i \geq 0$

b. $x(15)$ given $x(5) = 0$ and $\Delta x(i) = 2^{i-1}$ for $i \geq 5$

c. $y(10)$ given $y(3) = 20$ and $\Delta y(i) = 2^i \cos 3i$ for $i \geq 3$

| **Classical Solution of Linear Difference Equations with Constant Coefficients**

In this chapter we present the classical methods for solving linear difference equations with constant coefficients. After a discussion of the general nature of the solutions of difference equations, we present procedures for solving homogeneous equations and then particular equations. For the latter, both the methods of undetermined coefficients and variation of parameters are discussed. The similarities between the material presented here and the solution procedures for linear differential equations with constant coefficients will be apparent to readers familiar with the theory of differential equations.

4.1 The Nature of Solutions

For ease in presentation, the expanded form of the equation will be used. Thus we have

$$b_0 x(i+n) + b_1 x(i+n-1) + \cdots + b_{n-1} x(i+1) + b_n x(i) = f(i), \quad (4.1.1)$$

where, as previously defined, i is the independent variable, $x(i)$ the dependent variable, $f(i)$ the forcing function, and b_k the constant coefficient of $x(i+n-k)$, $k = 0, \ldots, n$. Furthermore, $b_0 \neq 0$. In a manner analogous to the solving of a linear differential equation with constant coefficients, we further define $x(i)^{(H)}$ as the solution to the homogeneous equation

$$b_0 x(i+n) + b_1 x(i+n-1) + \cdots + b_{n-1} x(i+1) + b_n x(i) = 0, \quad (4.1.2)$$

and $x(i)^{(P)}$ as the solution to the original equation with the particular forcing function $f(i)$. $x(i)^{(H)}$ and $x(i)^{(P)}$ are referred to as the *homogeneous solution* and *particular solution*, respectively.

Now any $x(i)$ which satisfies the homogeneous equation will not affect the satisfying of the particular equation if it is added to the solution $x(i)^{(P)}$. This is shown as follows. We insert the sum $x(i)^{(P)} + x(i)^{(H)}$ as the dependent variable in (4.1.1) to obtain

$$f(i) = \sum_{k=0}^{n} b_k[x(i + n - k)^{(P)} + x(i + n - k)^{(H)}]$$

$$= \sum_{k=0}^{n} b_k x(i + n - k)^{(P)} + \sum_{k=0}^{n} b_k x(i + n - k)^{(H)}. \tag{4.1.3}$$

The second summation, however, vanishes by definition of $x(i)^{(H)}$, leaving

$$f(i) = \sum_{k=0}^{n} b_k x(i + n - k)^{(P)} \tag{4.1.4}$$

which must hold because of the definition of $x(i)^{(P)}$. Thus the total solution to the difference equation, also called the *complete solution* since it includes all possible $x(i)$ which satisfy (4.1.1), is

$$x(i) = x(i)^{(P)} + x(i)^{(H)}, \tag{4.1.5}$$

where $x(i)^{(H)}$ can be a linear combination of all those $x(i)$ which satisfy the homogeneous equation (4.1.2).

The first step in the solution procedure is to determine the homogeneous solution since, as will be seen later, it will have a bearing on the form of the particular solution. See Miller (1960) and Hildebrand (1952) for detailed discussions of the solution procedures. We now turn our attention to finding $x(i)^{(H)}$.

4.2   The Homogeneous Solution

If we assume a solution to (4.1.2) of the form

$$x(i) = h^i, \tag{4.2.1}$$

then it follows that

$$x(i + n - k) = h^{i+n-k} \tag{4.2.2}$$

for all k, $0 \leq k \leq n$. Substitution into (4.1.2) yields

$$b_0 h^{i+n} + b_1 h^{i+n-1} + \cdots + b_{n-1} h^{i+1} + b_n h^i = 0, \tag{4.2.3}$$

which may be written

$$h^i[b_0 h^n + b_1 h^{n-1} + \cdots + b_{n-1} h + b_n] = 0. \tag{4.2.4}$$

For (4.2.4) to have a nontrivial solution (i.e., $h \neq 0$) for all i, the expression in brackets must equal zero, that is,

$$b_0 h^n + b_1 h^{n-1} + \cdots + b_{n-1} h + b_n = 0. \qquad (4.2.5)$$

Equation (4.2.5) is variously known as the "characteristic equation" (as is its counterpart in the theory of linear differential equations), the "determinantal equation," or the "indicial equation" and is seen to be an ordinary polynomial equation. Since it is an nth-order polynomial equation, it must have exactly n roots. Thus there will be n functions $x(i)$ which satisfy the homogeneous equation, although the occurrence of multiple roots requires the use of some special procedures. In any case, $x(i)^{(H)}$ will be a linear combination of n functions of i, where the combinatorial coefficients are evaluated from so-called boundary conditions (initial conditions, etc., to be discussed in Section 4.4 of this chapter).

Roots of polynomial equations can of course be real or complex, and single or multiple. It facilitates discussion to consider these roots according to the four categories resulting from the combinations of these two characteristics.

Single Real Roots

Each single real root produces a single term in $x(i)^{(H)}$ of the form ch^i, where c is the combinatorial coefficient to be evaluated from boundary conditions.

Single Pairs of Complex or Imaginary Roots

Complex roots always occur in conjugate pairs. As will be demonstrated, it is desirable to treat both roots together. Therefore, we let

$$h_1 = a + jb \qquad (4.2.6)$$

and

$$h_2 = a - jb, \qquad (4.2.7)$$

where a is the real part of the root, b the imaginary part of the root, and $j = (-1)^{1/2}$. Should $a = 0$, the roots are of course purely imaginary, although they are treated as complex.

The contribution of this pair of roots to $x(i)^{(H)}$ is

$$c_1(a + jb)^i + c_2(a - jb)^i, \qquad (4.2.8)$$

where c_1 and c_2 are combinatorial coefficients. This is a very unhandy form to work with or, of even more importance to systems designers and analysts, to interpret. We, therefore, seek a more convenient form for representation of the pair of roots. Figure 4.2.1 shows h_1 and h_2 in the complex h plane. R is the distance from the origin to either root, and θ is the angle between the positive real axis and the line extending from the origin through h_1. Similarly, $-\theta$ is the angle between the positive real axis and the line extending through h_2. It is apparent that

$$R = (a^2 + b^2)^{1/2} \tag{4.2.9}$$

and

$$\theta = \tan^{-1} b/a. \tag{4.2.10}$$

θ is normally expressed in radians. From the well-known identities

$$\cos \theta = \tfrac{1}{2}(e^{j\theta} + e^{-j\theta}) \tag{4.2.11}$$

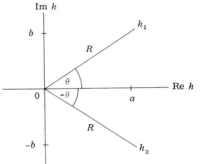

FIG. 4.2.1. h_1 and h_2 in the complex h plane.

and

$$\sin \theta = \frac{1}{2j}(e^{j\theta} - e^{-j\theta}), \tag{4.2.12}$$

one obtains

$$e^{j\theta} = \cos \theta + j \sin \theta \tag{4.2.13}$$

and

$$e^{-j\theta} = \cos \theta - j \sin \theta. \tag{4.2.14}$$

Thus h_1 and h_2 can be expressed as

$$h_1 = R(\cos \theta + j \sin \theta) = Re^{j\theta} \tag{4.2.15}$$

and

$$h_2 = R(\cos \theta - j \sin \theta) = Re^{-j\theta}, \tag{4.2.16}$$

which are read "h_1 is of magnitude R at an angle θ" and "h_2 is of magnitude R at an angle $-\theta$," respectively.

Expressions for $h_1{}^i$ and $h_2{}^i$ are found as follows:

$$h_1{}^i = (Re^{j\theta})^i = R^i e^{j\theta i} = R^i(\cos \theta i + j \sin \theta i) \qquad (4.2.17)$$

and, similarly,

$$h_2{}^i = R^i(\cos \theta i - j \sin \theta i). \qquad (4.2.18)$$

Thus raising a function in this form to a power i raises the magnitude to the power i and multiplies the angle by i. Therefore, the contribution of the pair of roots h_1 and h_2 to the homogeneous solution can be conveniently expressed as

$$
\begin{aligned}
c_1 h_1{}^i + c_2 h_2{}^i &= c_1(R^i \cos \theta i + jR^i \sin \theta i) + c_2(R^i \cos \theta i - jR^i \sin \theta i) \\
&= (c_1 + c_2)R^i \cos \theta i + j(c_1 - c_2)R^i \sin \theta i \\
&= c_3 R^i \cos \theta i + c_4 R^i \sin \theta i, \qquad (4.2.19)
\end{aligned}
$$

where

$$c_3 = c_1 + c_2 \qquad (4.2.20)$$

and

$$c_4 = j(c_1 - c_2). \qquad (4.2.21)$$

We assert without proof that for boundary conditions expressed as real quantities c_1 and c_2 will be complex conjugates so that c_3 and c_4 will always be real numbers. Thus the pair of roots contributes two sinusoidal but real terms to $x(i)^{(H)}$ which are easily interpreted in terms of system performance.

To illustrate, suppose a characteristic equation had a pair of roots $-0.3 \pm j(0.4)$. Then

$$R = (0.09 + 0.16)^{1/2} = (0.25)^{1/2} = 0.5$$

and

$$\theta = \tan^{-1}(0.4/-0.3) = \tan^{-1}(-1.33).$$

Note that the real part of the root h_1 is negative and the imaginary part is positive so that the angle is in the second quadrant. Therefore, $\theta = 180° - 53° = 127°$. Had the roots been $0.3 \mp j(0.4)$, the argument of $\tan^{-1}$ would still be (-1.33) but the angle would have been in the fourth quadrant. Specifically θ would have been $-53°$ or $307°$. For the case cited, the contribution of the pair of roots to $x(i)^{(H)}$ is

$$c_3[0.5^i \cos(127\pi/180)i] + c_4[0.5^i \sin(127\pi/180)i].$$

The $\pi/180$ in the cosine and sine is to convert the arguments to radians. This result is more easily written as $c_3(0.5^i \cos 2.217i) + c_4(0.5^i \sin 2.217i)$.

Multiple Real Roots

If exactly p of the roots of a polynomial are identical, the polynomial is said to have a pth-order root or a root of multiplicity p. Each such root must contribute exactly p terms to $x(i)^{(H)}$.

If, for a pth-order root of value h, we simply used $c_1 h + c_2 h + \cdots + c_p h$ in $x(i)^{(H)}$, the p terms could not be distinguished. They could be collapsed into one term of the form ch^i, where $c = c_1 + c_2 + \cdots + c_p$. Therefore, some modification is necessary to produce p clearly distinguishable terms in $x(i)^{(H)}$. We will illustrate the development of this procedure with a second-order equation with a double root, i.e., a root h of multiplicity two.

Let us assume a solution to the homogeneous equation of the form

$$x(i) = v(i)h^i, \tag{4.2.22}$$

where $v(i)$ is an unknown function of i to be determined. If

$$v(i) = c, \tag{4.2.23}$$

where c is a constant, then

$$x(i) = ch^i \tag{4.2.24}$$

is obviously a solution to the characteristic equation so the assumed form (4.2.22) is guaranteed to yield at least this one solution.

Now, the homogeneous equation corresponding to a second-order linear difference equation is

$$b_0 x(i + 2) + b_1 x(i + 1) + b_2 x(i) = 0, \tag{4.2.25}$$

which, assuming neither b_0 nor b_2 equals zero, yields the characteristic equation

$$b_0 h^2 + b_1 h + b_2 = 0. \tag{4.2.26}$$

The roots of this equation are given by the quadratic formula as

$$h = \frac{-b_1 \pm (b_1{}^2 - 4b_0 b_2)^{1/2}}{2b_0}. \tag{4.2.27}$$

For a double root to occur, the radical must vanish. This requires, for the case at hand, that

$$b_1{}^2 = 4b_0 b_2. \tag{4.2.28}$$

The resulting double root is

$$h = -b_1/2b_0. \tag{4.2.29}$$

Therefore, (4.2.22) may be written

$$x(i) = v(i)(-b_1/2b_0)^i. \tag{4.2.30}$$

Furthermore,

$$x(i + 1) = v(i + 1)h^{i+1} = v(i + 1)(-b_1/2b_0)^{i+1} \tag{4.2.31}$$

and

$$x(i + 2) = v(i + 2)h^{i+2} = v(i + 2)(-b_1/2b_0)^{i+2}. \tag{4.2.32}$$

Substitution into the homogeneous equation results in

$$b_0 v(i + 2)(-b_1/2b_0)^{i+2} + b_1 v(i + 1)(-b_1/2b_0)^{i+1} + b_2 v(i)(-b_1/2b_0)^i = 0. \tag{4.2.33}$$

Dividing out $(-b_1/2b_0)^i$ leaves

$$b_0(-b_1/2b_0)^2 v(i + 2) + b_1(-b_1/2b_0)v(i + 1) + b_2 v(i) = 0 \tag{4.2.34}$$

or

$$(b_1^2/4b_0)v(i + 2) - (b_1^2/2b_0)v(i + 1) + b_2 v(i) = 0. \tag{4.2.35}$$

However, from (4.2.28), $4b_0 b_2$ can be substituted for b_1^2 in (4.2.35) to yield

$$b_2 v(i + 2) - 2b_2 v(i + 1) + b_2 v(i) = 0, \tag{4.2.36}$$

or, after dividing out b_2,

$$v(i + 2) - 2v(i + 1) + v(i) = 0, \tag{4.2.37}$$

the left-hand side of which is precisely the formula for the second forward difference of $v(i)$. Thus,

$$\Delta^2 v(i) = 0. \tag{4.2.38}$$

Summing yields

$$\Delta v(i) = \sum \Delta^2 v(i) + c_1 = c_1, \tag{4.2.39}$$

where c_1 is the periodic constant, and, finally,

$$v(i) = \sum \Delta v(i) + c_2 = \sum c_1 + c_2 = c_1 \sum i^{(0)} + c_2$$
$$= c_1 i^{(1)} + c_2 = c_1 i + c_2. \tag{4.2.40}$$

Thus,

$$x(i) = (c_1 i + c_2)h^i = c_1 i h^i + c_2 h^i. \tag{4.2.41}$$

In this manner the double root produces two clearly distinguishable terms in $x(i)^{(H)}$ for the second-order equation. The above method, moreover, will produce the same result for a double root for any order of difference equation.

Furthermore, this procedure can be extended to show that for a root h of any multiplicity p, regardless of how many other roots may be present, p terms will be produced in $x(i)^{(H)}$ of the form

$$(c_1 i^{p-1} + c_2 i^{p-2} + \cdots + c_{p-1}i + c_p)h^i. \qquad (4.2.42)$$

The proof is left to the interested reader.

To illustrate, the roots of the characteristic equation of the fifth-order difference equation

$$x(i + 5) - 1.5x(i + 4) + 0.5x(i + 3) + 0.25x(i + 2)$$
$$- 0.1875x(i + 1) + 0.03125x(i) = f(i) \quad (4.2.43)$$

consist of a single root at $h = -0.5$ and a fourth-order root at 0.5. The homogenous solution is, therefore,

$$x(i)^{(H)} = c_0(-0.5)^i + (c_1 i^3 + c_2 i^2 + c_3 i + c_4)(0.5)^i. \qquad (4.2.44)$$

Multiple Pairs of Complex or Imaginary Roots

Multiple pairs of complex roots combine the characteristics of complex roots and of multiple real roots discussed in the previous two sections. Therefore, combining the results of these two sections, we find that each pair of complex or imaginary roots of multiplicity p results in $2p$ terms in $x(i)^{(H)}$ of the form

$$R^i[(c_{c1} i^{p-1} + c_{c2} i^{p-2} + \cdots + c_{cp}) \cos \theta i$$
$$+ (c_{s1} i^{p-1} + c_{s2} i^{p-2} + \cdots + c_{sp}) \sin \theta i]. \quad (4.2.45)$$

For example, consider the fourth-order difference equation

$$x(i + 4) + 0.12x(i + 3) + 0.86x(i + 2) + 0.30x(i + 1) + 0.0625x(i) = f(i).$$
$$(4.2.46)$$

The characteristic equation has the double complex roots $-0.3 \pm j(0.4)$, for which it has been shown that $R = 0.5$ and $\theta = 127° = 2.217$ rad. Therefore, the homogeneous solution is

$$x(i)^{(H)} = 0.5^i[(c_{c1}i + c_{c2}) \cos 2.217i + (c_{s1}i + c_{s2}) \sin 2.217i]. \quad (4.2.47)$$

Combinations

In general, the roots of the characteristic equation of any linear difference equation could have one or more roots in each of the four categories just examined. Regardless, the homogeneous solution still consists of a linear combination of the terms formed in the ways discussed for each category of root.

For illustrative purposes only (we would hate to confront such a monster in actual practice), suppose a 15-order equation had the following roots:

Single real: $-0.5, 1, 1.5$
Single pair complex: $0.05 \pm j(0.12)$ $(R = 0.13, \theta = 1.186 \text{ rad})$
Multiple real: double at -1, fourth-order at 0.5
Multiple pair complex: double at $-0.3 \pm j(0.4)$ $(R = 0.5, \theta = 2.217 \text{ rad})$

The homogeneous solution would then be

$$x(i)^{(H)} = c_1(-0.5)^i + c_2 + c_3 1.5^i + 0.13^i[c_4 \cos 1.186 + c_5 \sin 1.186]$$
$$+ (c_6 i + c_7)(-1)^i + (c_8 i^3 + c_9 i^2 + c_{10} i + c_{11})0.5^i$$
$$+ 0.5^i[(c_{12} i + c_{13}) \cos 2.217i + (c_{14} i + c_{15}) \sin 2.217i], \quad (4.2.48)$$

a linear combination of 15 clearly distinct and identifiable terms.

Characteristic Equation Formulation

Before moving on to the particular solution, let us discuss the formulation of the characteristic equation for a specific system difference equation. We will use the rapid-response discrete automatic process controller diagrammed in Fig. 2.3.2.

Proportional-Plus-Difference Process Controller

First consider the proportional-plus-difference controller whose system difference equation, obtained by setting $N = 0$ in (2.3.30), is

$$m(i + 1) - (1 - K - L)m(i) - Lm(i - 1)$$
$$= KD - (K + L)\varepsilon(i) + L\varepsilon(i - 1) + p(i). \quad (4.2.49)$$

Note the presence of terms with arguments less than i, specifically the terms in $m(i - 1)$ and $\varepsilon(i - 1)$. Although such terms were not explicitly included in the previous discussion of the homogeneous equation, they are treated like all other terms in deriving the characteristic equation. Proceeding in the manner of (4.2.1) through (4.2.5), we assume a solution to the homogeneous equation

$$m(i + 1) - (1 - K - L)m(i) - Lm(i - 1) = 0 \quad (4.2.50)$$

of the form

$$m(i) = h^i. \quad (4.2.51)$$

The terms $m(i + 1)$ and $m(i - 1)$ of course become h^{i+1} and h^{i-1}, respectively. Substitution into (4.2.50) yields

$$h^{i+1} - (1 - K - L)h^i - Lh^{i-1} = 0. \tag{4.2.52}$$

Instead of factoring h^i as was done in (4.2.4), we instead factor h^{i-1} to obtain

$$h^{i-1}[h^2 - (1 - K - L)h - L] = 0 \tag{4.2.53}$$

from which the characteristic equation is

$$h^2 - (1 - K - L)h - L = 0. \tag{4.2.54}$$

In general we factor the lowest power of h present so that the characteristic equation will be a polynomial in nonnegative powers of the h and will include a constant term (h^0). This form is obviously the easiest to work with in finding the roots.

An alternative procedure to handle arguments less than i is to increase all arguments of all terms in the system difference equation by whatever amount is needed to bring the smallest argument up to i. Since in (4.2.49) the smallest argument is $i - 1$, the argument in every term is increased by one to yield

$$m(i + 2) - (1 - K - L)m(i + 1) - Lm(i)$$
$$= KD - (K + L)\varepsilon(i + 1) + L\varepsilon(i) + p(i + 1). \tag{4.2.55}$$

Although either method results in the same characteristic equation, the procedure of incrementing the discrete-time arguments for the entire system equation is generally recommended because, as will be discussed in Section 4.3, it also facilitates determination of the particular solution.

Proportional-Plus-Difference-Plus-Summation Process Controller

The system equation (2.3.30) for the process controller involving proportional, difference, and summation control, as noted earlier, is complicated by the presence of summation terms which bring all past process outputs and measurement errors into the equation. The equation must, therefore, be modified in order to proceed. Modification in this case involves taking the first forward difference of the equation to obtain

$$\Delta m(i + 1) - (1 - K - L - N)\Delta m(i) - (L - N)\Delta m(i - 1) + N\sum_{n=0}^{i-2}\Delta m(n)$$
$$= ND - (K + L + N)\Delta\varepsilon(i) + (L - N)\Delta\varepsilon(i - 1) - N\sum_{n=0}^{i-2}\Delta\varepsilon(n) + \Delta p(i). \tag{4.2.56}$$

From (3.3.3), the basic summation relationship,

$$\sum_{n=0}^{i-2}\Delta m(n) = m(i - 1) - m(0) \tag{4.2.57}$$

and

$$\sum_{n=0}^{i-2} \Delta \varepsilon(n) = \varepsilon(i-1) - \varepsilon(0). \tag{4.2.58}$$

We now substitute (4.2.57) and (4.2.58) into (4.2.56), replace all remaining difference terms with their expanded-form equivalents, move the term involving $m(0)$ to the right-hand side, and group terms. This yields

$$m(i+2) - (2-K-L-N)m(i+1) + (1-K-2L)m(i) + Lm(i-1)$$

$$= ND - (K+L+N)\varepsilon(i+1) + (K+2L)\varepsilon(i) - L\varepsilon(i-1)$$

$$+ N\varepsilon(0) + p(i+1) - p(i) + Nm(0), \tag{4.2.59}$$

which is a third-order equation since the difference between the highest and lowest arguments of the dependent variable is $(i+2) - (i-1) = 3$. Derivation of the characteristic equation from this modification of the system equation is trivial.

Note that one could have modified only the left-hand side of (2.3.30) to derive the characteristic equation. It must be remembered, however, that the characteristic equation obtained is for a modification of the system equation (in this case, the first difference), and, as will be seen in Section 4.3, it must be used with the corresponding right-hand side to obtain the particular solution.

Note also that since (4.2.59) is of third order, the characteristic equation will be cubic. Thus its roots could fall into any of the following patterns depending on the relative values of K, L, and N: (1) three single real roots, (2) one single and one double real root, (3) one triple real root, and (4) one single real root and one pair of conjugate complex roots.

Summary

In summary, an nth order linear difference equation must contain exactly n terms in its homogeneous solution. For complex pairs of roots, it is convenient to convert the pair to a cosine and a sine term thus eliminating the imaginary aspects of the roots. Roots of multiplicity p contribute exactly p terms which are mutually distinguishable because of their multiplication by successive powers of i from 0 to $p - 1$. Finally, before proceeding to solve the system equation, we put it in closed form with the lowest value of the argument of the dependent variable equal to i. We now turn our attention to the particular solution $x(i)^{(P)}$.

4.3 The Particular Solution

Two procedures are presented for finding the particular solution to linear difference equations with constant coefficients. Both require the prior determination of the homogeneous solution $x(i)^{(H)}$. The method of undetermined coefficients is restricted to equations with a limited but fairly common set of forcing functions. When applicable, this procedure is generally much easier to apply than the method of variation of parameters, which may be used for any forcing function including perfectly general ones simply in the form $f(i)$.

Undetermined Coefficients

The method of undetermined coefficients is applicable when the forcing function $f(i)$ is a member of what is appropriately called a "finite family," a linear combination of members of such families, or, in most cases, the product of members of these families. A finite family occurs when shifts or changes in the index of a function result in only a finite number of basically different functional forms. For example, if $f(i) = a^i$, where a is a constant, then $f(i + 1) = a^{i+1} = a \cdot a^i$. Likewise, $f(i + 2) = a^{i+2} = a^2 a^i$ and in general $f(i + n) = a^{i+n} = a^n a^i$. Thus for all shifts of n periods, n positive or negative, the resulting function is still basically a^i. The presence of the constant multiplier a^n has no influence in defining family members.

Another common finite family includes the sine and cosine. For $f(i) = \sin ai$, $f(i + 1) = \sin a(i + 1) = \sin a \cos ai + \cos a \sin ai$. Similarly, for $f(i) = \cos ai$, $f(i + 1) = \cos a \cos ai - \sin a \sin ai$. In both cases, the shifted function includes both members of the family, each multiplied by a constant in the form of $\sin a$ or $\cos a$. It is apparent that this type of result would be obtained for shifts of any size since $\sin a(i + n) = \sin an \cos ai + \cos an \sin ai$ and $\cos a(i + n) = \cos an \cos ai - \sin an \sin ai$. Again, the presence of $\sin an$ and $\cos an$ does not affect the designation of family members.

Now consider $f(i) = i^a$, where a is integer. $f(i + 1) = (i + 1)^a = i^a + ai^{a-1} + \cdots + ai + 1 = \sum_{k=0}^{a} C_k^a i^{a-k}$ and, in general,

$$f(i + n) = (i + n)^a = i^a + ani^{a-1} + \cdots + an^{a-1}i + n^a = \sum_{k=0}^{a} C_k^a n^k i^{a-k}.$$

Therefore, in the case of i raised to an integer power, the family must include all integer powers of i from i^a through i^0. Note that since $i^0 = 1$, a constant term is a member of the family i^a.

Examples of functions which are not members of finite families are $\log ai$ and $\tan ai$. $\log a(i + n)$ cannot be manipulated in any meaningful way, so that a new function of i is introduced for each value of n. The same is true for tan $a(i + n)$ which equals $(\tan an + \tan ai)/(1 - \tan an \tan ai)$ but can be reduced no further.

A few of the common functions which are members of finite families and the members of their families are listed in Table 4.3.1.

Table 4.3.1

Common Finite Families

Item	$f(i)$	Members of finite family
1	a^i	a^i
2	e^{ai}	e^{ai}
3	$\sin ai$	$\sin ai, \cos ai$
4	$\cos ai$	$\sin ai, \cos ai$
5	i^a, a integer	$i^a, i^{a-1}, \ldots, i, 1$ or $\{i^k; 0 \leq k \leq a\}$

As a final example of the formulation of a finite family, consider $f(i) = i^a \sin bi$, a integer, which is the product of two functions we have already shown are members of finite families. For a shift of n periods,

$$f(i + n) = (i + n)^a \sin b(i + n)$$
$$= (i^a + ani^{a-1} + \cdots + an^{a-1}i + n^a)(\sin nb \cos bi + \cos nb \sin bi).$$

Thus the family consists of the $2(a + 1)$ members $i^k \sin bi$ and $i^k \cos bi$ for $0 \leq k \leq a$. Thus the members of the family involved represent all combinations of all members of the families of both of the functions being multiplied to form $f(i)$. This result will always occur in product-type functions. That is, whenever i raised to an integer power appears as a factor in a more complex function, the resulting family members will each contain a factor of i raised to an integer power and there will furthermore be a member involving each power of i from the zeroth through the power of i in the original function. Similarly, when either $\sin ai$ or $\cos ai$ is a factor in a function, the associated finite family will contain members having $\sin ai$ as a factor and those with $\cos ai$ as a factor.

The Method

We will now outline the steps involved in the method of undetermined coefficients and then illustrate its use in two examples. The method proceeds as follows:

1. Find the homogeneous solution $x(i)^{(H)}$.

2. Construct the entire finite family for every term in the forcing function $f(i)$.

For example, if $f(i) = c_1 a^i + c_2 i^b$, where a, b, c_1, and c_2 are constants and b is integer, we list the family members for a^i, which is of course a^i only, and the members for i^b, which is i^b, i^{b-1}, ..., i, and 1 (or simply "constant").

3. Check the homogeneous solution to see if any member of any finite family has the same functional form as any term in $x(i)^{(H)}$. If so, every member of the family whose member corresponds to a term in $x(i)^{(H)}$ is multiplied by the smallest integer power of i such that no member of the family will then have the same functional form as any term in $x(i)^{(H)}$.

To illustrate, consider the $f(i)$ in (2) above and assume $x(i)^{(H)} = c_3 i + c_4$, which it could if its characteristic equation had a double root of one. Two members of the family of i^b, namely, i and the constant, have the same functional form as a term in $x(i)^{(H)}$. Thus every member of the family of i^b must be multiplied by some power of i. Simple multiplication by i would not be sufficient since the constant member of the family would then become i which still corresponds to the i term in $x(i)^{(H)}$. Therefore, multiplication by i^2 is required to fully remove the correspondence between the family and all terms in $x(i)^{(H)}$. The resulting family terms therefore become a^i, which remains unchanged since it is in a different family, and i^{b+2}, i^{b+1}, ..., i^3, and i^2. No i or constant term remains.

It should be well noted that failure to multiply a family by a sufficiently high power of i to clear all correspondence with $x(i)^{(H)}$ will cause a breakdown in step 5, which follows, such that it will be impossible to evaluate the coefficients of all the family members. Thus any error of this type will be quickly recognized. If, however, too high a power of i is used, incorrect answers will be obtained which of course would not necessarily be recognizable at that point. In any case, care should be taken to determine the proper multipliers for each finite family involved.

4. Construct $x(i)^{(P)}$ as a linear combination of all family members, modified where appropriate by multiplication by powers of i in step 3, for the families corresponding to all terms in the forcing function $f(i)$.

For the situation illustrated above in which the terms resulting after clearing of correspondences with $x(i)^{(H)}$ were a^i, i^{b+2}, i^{b+1}, ..., i^3, and i^2, the assumed form for $x(i)^{(P)}$ would be

$$x(i)^{(P)} = A \cdot a^i + B i^{b+2} + C i^{b+1} + \cdots + M i^3 + N i^2, \qquad (4.3.1)$$

where A, B, C, ..., M, and N are combinatorial coefficients.

5. Determine the values of the combinatorial coefficients in $x(i)^{(P)}$ such

that the original difference equation is identically satisfied when the assumed $x(i)^{(P)}$ is substituted for $x(i)$ in the equation.

From $x(i)^{(P)}$, we would determine $x(i + 1)$ through $x(i + n)$. Since $x(i)^{(P)}$ is made up of all members of all finite families involved, no new functional forms will be created except perhaps the reinstatement of terms previously modified by multiplication by powers of i. $x(i)^{(P)}$ and the resulting $x(i + 1)$ through $x(i + n)$ are then substituted in the original equation and terms in like functional forms grouped. The coefficients in $x(i)^{(P)}$ are then calculated from the relationships formed by equating the coefficients of each functional form on the two sides of the difference equation.

Example 1   This last step is most easily illustrated in conjunction with the complete solution of a specific linear difference equation. Consider the second-order equation

$$x(i + 2) - 3x(i + 1) + 2x(i) = 4^i + 3i^2. \tag{4.3.2}$$

The homogeneous equation

$$x(i + 2) - 3x(i + 1) + 2x(i) = 0 \tag{4.3.3}$$

gives rise to the characteristic equation

$$h^2 - 3h + 2 = 0 \tag{4.3.4}$$

which has roots of 2 and 1. Thus, the homogeneous solution is

$$x(i)^{(H)} = c_1 2^i + c_2 1^i = c_1 2^i + c_2. \tag{4.3.5}$$

Since the forcing function is the sum of two different functions of i, namely 4^i and i^2, two families must be determined. 4^i is a special case of a^i which is a one-member family. Thus 4^i is the entire family. The family associated with i^2 contains i^2, i, and 1. We now check for corresponding functions in $x(i)^{(H)}$. Although 2^i appears in $x(i)^{(H)}$ it is not the same function as 4^i, so no correspondence exists for 4^i. The constant term 1, however, corresponds to the constant term in $x(i)^{(H)}$ in spite of the fact that one constant resulted from i^0 and the other from 1^i. Therefore, the entire family of i^2, i, and 1 must be multiplied by i to obtain i^3, i^2, and i, respectively. This clears the correspondence and gives us for the assumed form of the particular solution

$$x(i)^{(P)} = A4^i + Bi^3 + Ci^2 + Di. \tag{4.3.6}$$

Note that this form of $x(i)^{(P)}$ involves coefficients whose values are as yet undetermined, hence the name of the method.

Now begins the task of evaluating the coefficients. Following previous discussion we first find $x(i + 1)$ and $x(i + 2)$ as follows:

$$\begin{aligned} x(i + 1) &= A4^{i+1} + B(i + 1)^3 + C(i + 1)^2 + D(i + 1) \\ &= 4A \cdot 4^i + B(i^3 + 3i^2 + 3i + 1) + C(i^2 + 2i + 1) + D(i + 1) \\ &= 4A \cdot 4^i + Bi^3 + (3B + C)i^2 + (3B + 2C + D)i + (B + C + D). \end{aligned}$$

$$(4.3.7)$$

Note that we have grouped terms in like functions of i. In similar fashion, it is found that

$$x(i + 2) = 16A \cdot 4^i + Bi^3 + (6B + C)i^2 + (12B + 4C + D)i + (8B + 4C + 2D).$$

$$(4.3.8)$$

The expressions from (4.3.6), (4.3.7), and (4.3.8) are now substituted into the original difference equation, (4.3.2), to obtain

$$\begin{aligned} &[16A \cdot 4^i + Bi^3 + (6B + C)i^2 + (12B + 4C + D)i + (8B + 4C + 2D)] \\ &\quad - 3[4A \cdot 4^i + Bi^3 + (3B + C)i^2 + (3B + 2C + D)i + (B + C + D)] \\ &\quad + 2[A \cdot 4^i + Bi^3 + Ci^2 + Di] \\ &= 4^i + 3i^2. \end{aligned}$$

Expanding and regrouping terms in like functions of i yields

$$6A \cdot 4^i + 0 \cdot i^3 - 3Bi^2 + (3B - 2C)i + (5B + C - D) = 4i + 3i^2. \quad (4.3.9)$$

For this equation to be satisfied for all values of i, the coefficients of each function of i on each side of the equation must be equal. Note that with the vanishing of the i^3 term, we have exactly four functions of i, namely the four we started with before multiplying by i to clear the correspondence with $x(i)^{(H)}$. Terms created by such multiplication will, by the way, always vanish at this point leaving exactly the functions in the original families. This does not mean, however, that the multiplication step can be omitted when some correspondence exists. The curious reader might attempt solution of this problem without this step and see what happens.

We now have four undetermined coefficients and four relationships to use to find them. First, consider the coefficients of 4^i. On the left-hand side we have $6A$ and on the right side a one. Thus, $6A$ must equal one or $A = \frac{1}{6}$. Since i^3 has vanished, we next look at the coefficients of i^2. These are $-3B$ on the left-hand side and 3 on the right, which gives $B = -1$. The coefficients of $(3B - 2C)$ on the left and zero on the right for i gives, after substitution of -1 for B, a C value of $-\frac{3}{2}$. Finally, equating the coefficients of the constant terms of $(5B + C - D)$ on the left to zero on the right yields $D = -\frac{13}{2}$.

Note that in these calculations the computation of A was direct, but that B, C, and D were interrelated. This is because A was associated with one family and B, C, and D with another. In general, the coefficients associated with the members of a given family will be interrelated but can be calculated completely independently from the coefficients associated with all the other families.

As the result of the above calculations, the particular solution is

$$x(i)^{(P)} = \tfrac{1}{6} \cdot 4^i - i^3 - \tfrac{3}{2}i^2 - \tfrac{13}{2}i. \qquad (4.3.10)$$

The complete solution is, therefore,

$$x(i) = x(i)^{(P)} + x(i)^{(H)} = \tfrac{1}{6} \cdot 4^i - i^3 - \tfrac{3}{2}i^2 - \tfrac{13}{2}i + c_1 \cdot 2^i + c_2, \qquad (4.3.11)$$

where c_1 and c_2 must now be evaluated from boundary conditions. This will be discussed in a later section.

Example 2   Now let us look at a second example. We formulate it by simply changing the forcing function in Example 1 to $2 \sin 3i$. The equation is, therefore,

$$x(i + 2) - 3x(i + 1) + 2x(i) = 2 \sin 3i, \qquad (4.3.12)$$

whose homogeneous solution is of course the same as that for the first example given in (4.3.5).

The forcing function consists of a single function of i, namely $\sin 3i$, the members of whose family are $\sin 3i$ and $\cos 3i$. Since there are no correspondences with any of the functions in $x(i)^{(H)}$, we formulate the particular solution directly as

$$x(i)^{(P)} = A \sin 3i + B \cos 3i, \qquad (4.3.13)$$

where A and B are the undetermined coefficients whose values we seek. Proceeding as before,

$$
\begin{aligned}
x(i + 1) &= A \sin 3(i + 1) + B \cos 3(i + 1) \\
&= A(\sin 3 \cos 3i + \cos 3 \sin 3i) + B(\cos 3 \cos 3i - \sin 3 \sin 3i) \\
&= (A \cos 3 - B \sin 3) \sin 3i + (A \sin 3 + B \cos 3) \cos 3i \qquad (4.3.14)
\end{aligned}
$$

and, by a similar procedure,

$$x(i + 2) = (A \cos 6 - B \sin 6) \sin 3i + (A \sin 6 + B \cos 6) \cos 3i. \qquad (4.3.15)$$

Substituting in the original equation and grouping terms we get

$$
\begin{aligned}
&[(A \cos 6 - B \sin 6) - 3(A \cos 3 - B \sin 3) + 2A] \sin 3i \\
&\quad + [(A \sin 6 + B \cos 6) - 3(A \sin 3 + B \cos 3) + 2B] \cos 3i \\
&= 2 \sin 3i. \qquad (4.3.16)
\end{aligned}
$$

This can be further simplified by substitution of numerical values for sin 6, cos 6, sin 3, and cos 3. Recall that arguments of all trigonometric functions are in radians. Thus $\cos 6 = \cos 344° = \cos 16° = 0.964$. Similarly $\sin 6 = \sin 344° = -\sin 16° = -0.276$, and cos 3 and sin 3 are found in the same way to be -0.991 and 0.139, respectively. Substitution of these values in (4.3.16) and collecting terms yields

$$(5.937A + 0.693B) \sin 3i + (-0.693A + 5.937B) \cos 3i = 2 \sin 3i. \quad (4.3.17)$$

Equating the coefficients of $\sin 3i$ and $\cos 3i$ results in the two equations

$$5.937A + 0.693B = 2, \qquad -0.693A + 5.937B = 0$$

from which we calculate $A = 0.337$ and $B = 0.039$. Thus, the particular solution is

$$x(i)^{(P)} = 0.337 \sin 3i + 0.039 \cos 3i \qquad (4.3.18)$$

and the complete solution is

$$x(i) = 0.337 \sin 3i + 0.039 \cos 3i + c_1 2^i + c_2, \qquad (4.3.19)$$

where again c_1 and c_2 must be evaluated from boundary conditions.

Example 3 *The General Second-Order Process Controller* As a third example, consider the general second-order rapid-response controller whose system difference equation, originally stated in (2.3.32), is

$$m(i + 1) - (1 - K_0)m(i) + K_1 m(i - 1)$$
$$= (K_0 + K_1)D - K_0 \varepsilon(i) - K_1 \varepsilon(i - 1) + p(i). \quad (4.3.20)$$

Assume a pattern of perturbations such that

$$p(i) = \begin{cases} p \sin wi, & i \geq 0, \\ 0, & i < 0, \end{cases} \qquad (4.3.21)$$

and, for ease in illustration, assume all measurement errors are insignificant, i.e., let $\varepsilon(i) = 0$ for all i. Thus, putting (4.3.20) in standard form for solution, we get

$$m(i + 2) - (1 - K_0)m(i + 1) + K_1 m(i) = (K_0 + K_1)D + p \sin w(i + 1).$$
$$(4.3.22)$$

The characteristic equation is easily found to be

$$h^2 - (1 - K_0)h + K_1 = 0 \qquad (4.3.23)$$

with roots $[(1 - K_0) \pm ((1 - K_0)^2 - 4K_1)^{1/2}]/2$. Thus,

$$m(i)^{(H)} =$$

$$c_1\left[\frac{1 - K_0 + ((1 - K_0)^2 - 4K_1)^{1/2}}{2}\right]^i + c_2\left[\frac{1 - K_0 - ((1 - K_0)^2 - 4K_1)^{1/2}}{2}\right]^i$$

(4.3.24)

unless $(1 - K_0)^2 = 4K_1$, in which case the radicals vanish and we have a double real root at $(1 - K_0)/2$. Then the homogeneous solution would be

$$m(i)^{(H)} = (c_1 i + c_2)(1 - K_0)/2.$$ (4.3.25)

The forcing function consists of two finite families: a constant from $(K_0 + K_1)D$, and $\sin w(i + 1)$, $\cos w(i + 1)$ from the $p \sin w(i + 1)$ term. Note that the argument is $(i + 1)$, not i, in the sinusoids. Several possibilities exist regarding correspondences with terms in $m(i)^{(H)}$. First, if the roots of the characteristic equation are real but neither is equal to unity, there is no correspondence and

$$m(i)^{(P)} = A + B \sin w(i + 1) + C \cos w(i + 1).$$ (4.3.26)

Should one but not both roots equal unity, the constant term would have to be multiplied by i to clear the correspondence resulting in

$$m(i)^{(P)} = Ai + B \sin w(i + 1) + C \cos w(i + 1).$$ (4.3.27)

Should a double root of unity occur, $m(i)^{(H)}$ would then be given by (4.3.25) so A would have to be multiplied by i^2. On the other hand, if the two roots of the characteristic equation constituted a complex pair, the correspondence of the sine and cosine terms would have to be checked. If complex, the roots would be written as $(1 - K_0)/2 \pm j(4K_1 - (1 - K_0)^2)^{1/2}/2$. For correspondence to occur, the angle θ associated with the roots would have to equal w, the angular frequency of the forcing function sinusoid, i.e.,

$$\theta = \tan^{-1}\frac{[4K_1 - (1 - K_0)^2]^{1/2}}{1 - K_0} = w.$$ (4.3.28)

If so, the sine and cosine terms would have to be multiplied by i resulting in

$$m(i)^{(P)} = A + Bi \sin w(i + 1) + Ci \cos w(i + 1);$$ (4.3.29)

otherwise $m(i)^{(P)}$ as given by (4.3.26) would apply. Completing the solutions for each of the five cases is left as an exercise for the reader.

Later it will be shown that various system performance considerations, in particular a criterion called "system stability," will limit the useful ranges of K_0 and K_1 so that not all of the five cases just cited would actually be involved in system design.

Variation of Parameters

We first outline in general terms the steps involved in the variation of parameters. The method will then be applied to two example problems. Since this method is applicable for any forcing function, our development will be in terms of the general forcing function $f(i)$.

The variation-of-parameters method requires the prior determination of the roots of the characteristic equation and the resulting homogeneous solution $x(i)^{(H)}$. It must be emphasized, though, that in expressing $x(i)^{(H)}$, all complex and imaginary roots must be given in their original form of $a \pm jb$ rather than in trigonometric form with sine and cosine terms. Let $x(i)^{(H)}$ be expressed as

$$x(i)^{(H)} = \sum_{r=1}^{n} c_r w_r(i), \tag{4.3.30}$$

where $w_r(i)$ includes the appropriate root h raised to the ith power and any powers of i by which h^i must be multiplied in the case of multiple roots. The c_r are combinatorial coefficients. Again, all terms are given as functions of h^i whether h is real, complex, or imaginary.

To illustrate, suppose $x(i)^{(H)}$ for an eighth-order equation were:

$$x(i)^{(H)} = c_1(0.5)^i + c_2\, i^2(0.7)^i + c_3\, i(0.7)^i + c_4(0.7)^i + c_5\, i(0.2)^i \cos 1.1i$$
$$+ c_6(0.2)^i \cos 1.1i + c_7\, i(0.2)^i \sin 1.1i + c_8(0.2)^i \sin 1.1i. \tag{4.3.31}$$

This would result if the characteristic equation had a single real root at 0.5, a triple real root at 0.7, and a double pair of complex roots at $0.091 \pm j(0.178)$. In terms of the format designated in (4.3.30),

$$w_1(i) = (0.5)^i, \qquad w_2 = i^2(0.7)^i, \qquad w_3(i) = i(0.7)^i, \qquad w_4(i) = (0.7)^i,$$
$$w_5(i) = i(0.091 + j(0.178))^i, \qquad w_6(i) = (0.091 + j(0.178))^i,$$
$$w_7(i) = i(0.091 - j(0.178))^i, \qquad w_8(i) = (0.091 - j(0.178))^i.$$

Thus, as has been previously stated, for an nth-order difference equation, there will be exactly n terms in $x(i)^{(H)}$.

We now assume a complete solution (i.e., $x(i) = x(i)^{(P)} + x(i)^{(H)}$) of the form

$$x(i) = \sum_{r=1}^{n} A_r(i) w_r(i), \tag{4.3.32}$$

where the $A(i)$ are functions of i to be determined. Therefore,

$$x(i + 1) = \sum_{r=1}^{n} A_r(i + 1) w_r(i + 1). \tag{4.3.33}$$

We now subtract and add $\sum_{r=1}^{n} A_r(i)w_r(i+1)$ to obtain:

$$x(i+1) = \sum_{r=1}^{n} A_r(i+1)w_r(i+1) - \sum_{r=1}^{n} A_r(i)w_r(i+1) + \sum_{r=1}^{n} A_r(i)w_r(i+1)$$

$$= \sum_{r=1}^{n} A_r(i)w_r(i+1) + \sum_{r=1}^{n} [A_r(i+1) - A_r(i)]w_r(i+1)$$

$$= \sum_{r=1}^{n} A_r(i)w_r(i+1) + \sum_{r=1}^{n} \Delta A_r(i)w_r(i+1). \qquad (4.3.34)$$

The nature of the n unknown functions $A_r(i)$, $r = 1, \ldots, n$, is such that we must specify n independent conditions on the functions [see Hildebrand (1952) for sources of explanation and proof]. To facilitate solution, the first such condition is that the second summation in (4.3.34) vanish, i.e.,

$$\sum_{r=1}^{n} \Delta A_r(i)w_r(i+1) = 0. \qquad (4.3.35)$$

As a result,

$$x(i+1) = \sum_{r=1}^{n} A_r(i)w_r(i+1). \qquad (4.3.36)$$

Based on (4.3.36) for $x(i+1)$, we may write for $x(i+2)$

$$x(i+2) = \sum_{r=1}^{n} A_r(i+1)w_r(i+2). \qquad (4.3.37)$$

This time we subtract and add $\sum_{r=1}^{n} A_r(i)w_r(i+2)$ to (4.3.37) to obtain, in the manner of (4.3.34),

$$x(i+2) = \sum_{r=1}^{n} A_r(i)w_r(i+2) + \sum_{r=1}^{n} \Delta A_r(i)w_r(i+2). \qquad (4.3.38)$$

The second condition to be imposed on the functions $A_r(i)$ is that the second summation in (4.3.38) vanish. This gives

$$\sum_{r=1}^{n} \Delta A_r(i)w_r(i+2) = 0, \qquad (4.3.39)$$

which results in

$$x(i+2) = \sum_{r=1}^{n} A_r(i)w_r(i+2). \qquad (4.3.40)$$

This procedure is repeated a total of $n-1$ times, which gives the $n-1$ independent conditions

$$\sum_{r=1}^{n} \Delta A_r(i)w_r(i+k) = 0, \qquad k = 1, \ldots, n-1, \qquad (4.3.41)$$

and the $n - 1$ expressions

$$x(i + k) = \sum_{r=1}^{n} A_r(i)w_r(i + k), \qquad k = 1, \ldots, n - 1. \qquad (4.3.42)$$

Since we are seeking a solution to the linear difference equation (4.1.1), at this point we would like to develop an expression for $x(i + n)$ and find a way to relate the functions $A_r(i)$ to the forcing function $f(i)$. We accomplish both of these objectives as follows. We form $x(i + n)$ from $x(i + n - 1)$ as given by (4.3.42) with $k = n - 1$ to get

$$x(i + n) = \sum_{r=1}^{n} A_r(i + 1)w_r(i + n). \qquad (4.3.43)$$

As before, we subtract and add $\sum_{r=1}^{n} A_r(i)w_r(i + n)$ to obtain

$$x(i + n) = \sum_{r=1}^{n} A_r(i)w_r(i + n) + \sum_{r=1}^{n} \Delta A_r(i)w_r(i + n). \qquad (4.3.44)$$

Now as the nth condition to be imposed on the unknown functions, we require that the original difference equation be satisfied by $x(i)$ through $x(i + n - 1)$ as given by (4.3.42) and $x(i + n)$ as given by (4.3.43). Meeting this condition assures that the solution found satisfies (4.1.1). Substitution of these expressions into (4.1.1) yields

$$\sum_{k=0}^{n} b_{n-k} \left[\sum_{r=1}^{n} A_r(i)w_r(i + k) \right] + b_0 \sum_{r=1}^{n} \Delta A_r(i)w_r(i + n) = f(i). \qquad (4.3.45)$$

Rearranging the order of summation in the first part of this expression results in

$$\sum_{r=1}^{n} A_r(i) \left[\sum_{k=0}^{n} b_{n-k} w_r(i + k) \right] + b_0 \sum_{r=1}^{n} \Delta A_r(i)w_r(i + n) = f(i). \qquad (4.3.46)$$

Since each $w_r(i)$ corresponds to a root of the characteristic equation,

$$\sum_{k=0}^{n} b_{n-k} w_r(i + k) = 0 \qquad (4.3.47)$$

for all r. (Any powers of i present for a given term simply factor out.) Thus we have as our nth condition on the $A_r(i)$

$$\sum_{r=1}^{n} \Delta A_r(i)w_r(i + n) = f(i)/b_0 . \qquad (4.3.48)$$

We now have n linear equations in the n unknowns $\Delta A_r(i)$. In matrix form they can be represented as

$$W \cdot \Delta A = F, \qquad (4.3.49)$$

where

$$W = \begin{bmatrix} w_1(i+1) & w_2(i+1) & \cdots & w_n(i+1) \\ w_1(i+2) & w_2(i+2) & \cdots & w_n(i+2) \\ \vdots & \vdots & & \vdots \\ w_1(i+n) & w_2(i+n) & \cdots & w_n(i+n) \end{bmatrix},$$

$$\Delta A = \begin{bmatrix} \Delta A_1(i) \\ \Delta A_2(i) \\ \vdots \\ \Delta A_n(i) \end{bmatrix}, \quad \text{and} \quad F = \begin{bmatrix} 0 \\ 0 \\ \vdots \\ f(i)/b_0 \end{bmatrix}.$$

The solution for ΔA is thus

$$\Delta A = W^{-1} \cdot F, \tag{4.3.50}$$

but, since all elements of F are zero except the last, only the elements in the last column of W^{-1} will appear in the product, so that

$$\Delta A_r(i) = w^{(rn)} f(i)/b_0, \tag{4.3.51}$$

where $w^{(rn)}$ is the element in row r column n of W^{-1}.

It is, of course, not necessary to calculate $\Delta A_r(i)$ by direct matrix inversion. It could be done by any convenient method such as:

1. Determinants. With this approach, the denominator of all $\Delta A_r(i) = |W|$, the determinant of W. The numerator of $\Delta A_r(i)$ is the determinant of the matrix formed by replacing column r of W by the constant vector F.

2. Gauss–Jordan complete-elimination procedure.

3. Available computer programs for the solution of simultaneous linear equations.

All of these methods of course are based on mathematical methods which are in essence those used in matrix inversion; but, in (2) and (3) at least, the work is accomplished in the context of equation solution which permits some economy in calculation in that the inverse does not have to be explicitly calculated and displayed. One should also be constantly on the lookout for special properties of the $w_r(i)$ which might give rise to efficient special procedures for specific cases.

Once the $\Delta A_r(i)$ are determined, the $A_r(i)$ are calculated from

$$A_r(i) = A_r(0) + \sum_{j=0}^{i-1} \Delta A_r(j), \qquad r = 1, \ldots, n. \tag{4.3.52}$$

Substitution of (4.3.52) into (4.3.32) gives the complete solution $x(i)$. Since the boundary coefficients are still present in the form of the $A_r(0)$, they must be evaluated from boundary conditions.

Example 1 *Real Roots in Characteristic Equation* Let us now illustrate this technique with an example. Consider the equation which we already have solved by the method of undetermined coefficients

$$x(i + 2) - 3x(i + 1) + 2x(i) = 4^i + 3i^2. \qquad (4.3.53)$$

The roots of the characteristic equation were 2 and 1, both single, real roots. Therefore, we assume a solution of the form

$$x(i) = A_1(i)2^i + A_2(i)1^i = A_1(i)2^i + A_2(i). \qquad (4.3.54)$$

The resulting W matrix is

$$W = \begin{bmatrix} w_1(i+1) & w_2(i+1) \\ w_1(i+2) & w_2(i+2) \end{bmatrix} = \begin{bmatrix} 2^{i+1} & 1 \\ 2^{i+2} & 1 \end{bmatrix}, \qquad (4.3.55)$$

the ΔA vector is

$$\Delta A = \begin{bmatrix} \Delta A_1(i) \\ \Delta A_2(i) \end{bmatrix}, \qquad (4.3.56)$$

and, since $b_0 = 1$, the F vector is simply

$$F = \begin{bmatrix} 0 \\ f(i) \end{bmatrix} = \begin{bmatrix} 0 \\ 4^i + 3i^2 \end{bmatrix}. \qquad (4.3.57)$$

The resulting equations are

$$2^{i+1} \Delta A_1(i) + \Delta A_2(i) = 0, \qquad 2^{i+2} \Delta A_1(i) + \Delta A_2(i) = 4^i + 3i^2,$$

which can be written in the more convenient form for solution purposes as

$$2 \cdot 2^i \Delta A_1(i) + \Delta A_2(i) = 0, \qquad 4 \cdot 2^i \Delta A_1(i) + \Delta A_2(i) = 4^i + 3i^2.$$

Subtracting the first equation from the second yields

$$2 \cdot 2^i \Delta A_1(i) = 4^i + 3i^2$$

or

$$\Delta A_1(i) = \frac{4^i + 3i^2}{2 \cdot 2^i} = \frac{1}{2} \left[\frac{4^i}{2^i} + 3i^2 \frac{1}{2^i} \right] = \frac{1}{2} \cdot 2^i + \frac{3}{2} i^2 \left(\frac{1}{2} \right)^i. \qquad (4.3.58)$$

Substitution of (4.3.58) into the first equation results in

$$\Delta A_2(i) = -2 \cdot 2^i \Delta A_1(i) = -2 \cdot 2^i \cdot \frac{4^i + 3i^2}{2 \cdot 2^i} = -4^i - 3i^2. \qquad (4.3.59)$$

Summing (4.3.58) and (4.3.59) yields, respectively,

$$A_1(i) = A_1(0) + \sum_{k=0}^{i-1} \Delta A_1(k)$$

$$= A_1(0) + \frac{1}{2} \sum_{k=0}^{i-1} 2^k + \frac{3}{2} \sum_{k=0}^{i-1} k^2 \left(\frac{1}{2}\right)^k$$

$$= A_1(0) + \frac{1}{2} \cdot \frac{1 - 2^i}{1 - 2}$$

$$+ \frac{3}{2} \cdot \frac{(\frac{1}{2} - 1)^2 i^2 (\frac{1}{2})^i - 2(\frac{1}{2} - 1)^i (\frac{1}{2})^{i+1} + (\frac{1}{2})^{i+2} - (\frac{1}{2})^2 + (\frac{1}{2})^{i+1} - \frac{1}{2}}{(\frac{1}{2} - 1)^3}$$

$$= A_1(0) - \frac{1}{2} + \frac{1}{2} \cdot 2^i$$

$$+ \frac{\frac{3}{2}}{-(\frac{1}{8})} \left[\frac{1}{4} i^2 \left(\frac{1}{2}\right)^i + i \cdot \frac{1}{2} \cdot \left(\frac{1}{2}\right)^i + \frac{1}{4} \left(\frac{1}{2}\right)^i - \frac{1}{4} + \frac{1}{2} \left(\frac{1}{2}\right)^i - \frac{1}{2}\right]$$

$$= A_1(0) - \frac{1}{2} + \frac{1}{2} \cdot 2^i - 12[(\frac{1}{4} i^2 + \frac{1}{2} i + \frac{3}{4})(\frac{1}{2})^i - \frac{3}{4}]$$

$$= [A_1(0) - \frac{1}{2} + 9] + \frac{1}{2} \cdot 2^i - (3i^2 + 6i + 9)(\frac{1}{2})^i \tag{4.3.60}$$

and

$$A_2(i) = A_2(0) + \sum_{k=0}^{i-1} \Delta A_2(k)$$

$$= A_2(0) - \sum_{k=0}^{i-1} 4^k - 3 \sum_{k=0}^{i-1} k^2$$

$$= A_2(0) - \frac{4^i - 1}{4 - 1} - 3 \frac{(i-1)(i)(2i-1)}{6}$$

$$= A_2(0) - \frac{1}{3} \cdot 4^i + \frac{1}{3} - \frac{1}{2}(2i^3 - 3i^2 + 1)$$

$$= [A_2(0) + \frac{1}{3}] - \frac{1}{3} \cdot 4^i - i^3 + \frac{3}{2} i^2 - \frac{1}{2} i. \tag{4.3.61}$$

Note that use was made of formulas 4, 6, and 8 from Table 3.3.2 in these derivations.

We now let

$$\hat{A}_1 = A_1(0) - \frac{1}{2} + 9 \tag{4.3.62}$$

and

$$\hat{A}_2 = A_2(0) + \frac{1}{3} \tag{4.3.63}$$

in (4.3.60) and (4.3.61), respectively, and substitute for $A_1(i)$ and $A_2(i)$ in (4.3.54) to obtain

$$
\begin{aligned}
x(i) &= [\hat{A}_1 + \tfrac{1}{2} \cdot 2^i - (3i^2 + 6i + 9)(\tfrac{1}{2})^i] \cdot 2^i \\
&\quad + [\hat{A}_2 - \tfrac{1}{3} \cdot 4^i - i^3 + \tfrac{3}{2}i^2 - \tfrac{1}{2}i] \\
&= [\hat{A}_1 2^i + \tfrac{1}{2} \cdot 4^i - 3i^2 - 6i - 9] + [\hat{A}_2 - \tfrac{1}{3} \cdot 4^i - i^3 + \tfrac{3}{2}i^2 - \tfrac{1}{2}i] \\
&= \tfrac{1}{6} \cdot 4^i - i^3 - \tfrac{3}{2}i^2 - \tfrac{13}{2}i + \hat{A}_1 2^i + (\hat{A}_2 - 9) \\
&= \tfrac{1}{6} \cdot 4^i - i^3 - \tfrac{3}{2}i^2 - \tfrac{13}{2}i + c_1 2^i + c_2 ,
\end{aligned} \tag{4.3.64}
$$

where $c_1 = \hat{A}_1$ and $c_2 = \hat{A}_2 - 9$. This result is identical to (4.3.11), the result obtained using undetermined coefficients.

Let us now repeat this example with the generalized forcing function $f(i)$, that is,

$$
x(i + 2) - 3x(i + 1) + 2x(i) = f(i). \tag{4.3.65}
$$

The procedures followed above now yield

$$
\Delta A_1(i) = f(i)/2^{i+1} \tag{4.3.66}
$$

and

$$
\Delta A_2(i) = -f(i) \tag{4.3.67}
$$

which give

$$
A_1(i) = A_1(0) + \sum_{k=0}^{i-1} \frac{f(k)}{2^{k+1}} \tag{4.3.68}
$$

and

$$
A_2(i) = A_2(0) - \sum_{k=0}^{i-1} f(k). \tag{4.3.69}
$$

Substitution into (4.3.54) gives

$$
\begin{aligned}
x(i) &= \left[A_1(0) + \sum_{k=0}^{i-1} \frac{f(k)}{2^{i+1}} \right] 2^i + \left[A_2(0) - \sum_{k=0}^{i-1} f(k) \right] \\
&= \sum_{k=0}^{i-1} 2^{i-1-k} f(k) - \sum_{k=0}^{i-1} f(k) + A_1(0) 2^i + A_2(0). \tag{4.3.70}
\end{aligned}
$$

Obviously, this is as far as the solution for $x(i)$ can be carried without specific knowledge of $f(i)$. However, let us note the form that (4.3.70) has taken. Each of the summations is in the form of what is called a convolution sum. In the first summation, the convolution is between the two functions 2^i and $f(i)$. In the second, it involves 1^i and $f(i)$ which reduces to the form shown. In general, the convolution $y(i)$ of two functions $f(i)$ and $g(i)$ of discrete time is given by either

$$y(i) = \sum_{k=0}^{i} f(i - k)g(k) \qquad (4.3.71)$$

or

$$y(i) = \sum_{k=0}^{i} f(k)g(i - k) \qquad (4.3.72)$$

provided $f(i)$ and $g(i)$ both equal zero for $i < 0$. On this basis, one could write, if there were any reason to do so, the first summation of (4.3.70) as $\sum_{k=0}^{i-1} 2^k f(i - 1 - k)$.

Continuing our analysis of the form of (4.3.70), we note that the last two terms are identical to the homogeneous solution $x(i)^{(H)}$. Therefore, generalizing these observations to an nth-order difference equation with $x(i)^{(H)}$ as given by (4.3.30),

$$x(i) = \sum_{r=1}^{n} \left[\begin{array}{c} \text{convolution} \\ \text{involving } w_r(i), \\ A_r(i) \text{ and } f(i) \end{array} \right] + \sum_{r=1}^{n} c_r w_r(i), \qquad (4.3.73)$$

where the c_r are constants and again the $w_r(i)$ are in terms of powers of the roots of the characteristic equation and multipliers of powers of i and do not contain cosine and sine equivalents of complex pairs of roots. Note that the first summation in (4.3.73) corresponds generally to the particular solution $x(i)^{(P)}$, and the second summation to the homogeneous solution $x(i)^{(H)}$, although the first sum may contribute terms which add directly to terms in the second summation. This of course does not introduce any basically new terms or functions.

Example 2 *Complex Roots in Characteristic Equation* A second example will now be used to illustrate the situation in which the characteristic equation contains one or more pairs of complex or imaginary roots. To facilitate presentation, we will consider a second-order difference equation with a single pair of complex roots, although the procedures followed would still apply to higher order equations where additional roots are present. Let us assume a forcing function of the form a^i.

Given

$$x(i + 2) + 0.6x(i + 1) + 0.25x(i) = a^i, \qquad (4.3.74)$$

the characteristic equation is found to be

$$h^2 + 0.6h + 0.25 = 0 \qquad (4.3.75)$$

which has roots $-0.3 \pm j(0.4)$. In using variation of parameters, there is no need at this point to compute an R and θ to obtain terms in trigonometric form for $x(i)^{(H)}$ as would be required were undetermined coefficients to be used, since we work directly with the roots in complex form to obtain the complete solution $x(i)$. Conversion to trigonometric form will, however, be accomplished later.

Following the same procedures as in the first example, we write

$$x(i) = A_1(i)(-0.3 + j(0.4))^i + A_2(i)(-0.3 - j(0.4))^i. \qquad (4.3.76)$$

The equations to be solved for $\Delta A_1(i)$ and $\Delta A_2(i)$ are

$$(-0.3 + j(0.4))^{i+1} \Delta A_1(i) + (-0.3 - j(0.4))^{i+1} \Delta A_2(i) = 0,$$
$$(-0.3 + j(0.4))^{i+2} \Delta A_1(i) + (-0.3 - j(0.4))^{i+2} \Delta A_2(i) = a^i.$$

Multiplication of the first equation by $(-0.3 + j(0.4))$ and subtraction of the result from the second equation yields

$$[(-0.3 - j(0.4)) - (-0.3 + j(0.4))](-0.3 - j(0.4))^{i+1} \Delta A_2(i) = a^i$$

or

$$\Delta A_2(i) = a^i/[-j(0.8)][-0.3 - j(0.4)]^{i+1}. \qquad (4.3.77)$$

Similarly, multiplying the first equation by $(-0.3 - j(0.4))$ and subtracting from the second equation gives

$$\{[-0.3 - j(0.4)] - [-0.3 - j(0.4)]\}[-0.3 + j(0.4)]^{i+1} \Delta A_1(i) = a^i$$

or

$$\Delta A_1(i) = a^i/[j(0.8)][-0.3 + j(0.4)]^{i+1}. \qquad (4.3.78)$$

Therefore,

$$A_1(i) = A_1(0) + \sum_{k=0}^{i-1} \frac{a^k}{[j(0.8)][-0.3 + j(0.4)]^{k+1}}$$

$$= A_1(0) + \frac{1}{[j(0.8)][-0.3 + j(0.4)]} \sum_{k=0}^{i-1} \left[\frac{a}{-0.3 + j(0.4)}\right]^k$$

$$= A_1(0) + \frac{1}{[j(0.8)][-0.3 + j(0.4)]} \cdot \frac{\{a/[-0.3 + j(0.4)]\}^i - 1}{a/[-0.3 + j(0.4)] - 1}$$

$$= A_1(0) + \frac{\{a/[-0.3 + j(0.4)]\}^i - 1}{[j(0.8)][a + 0.3 - j(0.4)]}, \qquad (4.3.79)$$

and, similarly,

$$A_2(i) = A_2(0) + \frac{\{a/[-0.3 - j(0.4)]\}^i - 1}{[-j(0.8)][a + 0.3 + j(0.4)]}, \qquad (4.3.80)$$

which gives for $x(i)$:

$$x(i) = \left[A_1(0) + \frac{\{a/[-0.3 + j(0.4)]\}^i - 1}{(j(0.8))(a + 0.3 - j(0.4))} \right][-0.3 + j(0.4)]^i$$

$$+ \left[A_2(0) + \frac{\{a/[-0.3 - j(0.4)]\}^i - 1}{[-j(0.8)][a + 0.3 + j(0.4)]} \right][-0.3 - j(0.4)]^i$$

$$= A_1(0)[-0.3 + j(0.4)]^i + \frac{a^i - [-0.3 + j(0.4)]^i}{[j(0.8)][a + 0.3 - j(0.4)]}$$

$$+ A_2(0)[-0.3 - j(0.4)]^i - \frac{a^i - [-0.3 - j(0.4)]^i}{[j(0.8)][a + 0.3 + j(0.4)]}. \qquad (4.3.81)$$

Now, combining terms in a^i, $-0.3 + j(0.4)$, and $-0.3 - j(0.4)$, we get

$$x(i) = \left[\frac{1}{[j(0.8)][a + 0.3 - j(0.4)]} - \frac{1}{[j(0.8)][a + 0.3 + j(0.4)]} \right] a^i$$

$$+ \left[A_1(0) - \frac{1}{[j(0.8)][a + 0.3 - j(0.4)]} \right][-0.3 + j(0.4)]^i$$

$$+ \left[A_2(0) + \frac{1}{[j(0.8)][a + 0.3 + j(0.4)]} \right][-0.3 - j(0.4)]^i$$

$$= \frac{[a + 0.3 + j(0.4)] - [a + 0.3 - j(0.4)]}{[j(0.8)][a + 0.3 - j(0.4)][a + 0.3 + j(0.4)]} a^i + c_1[-0.3 + j(0.4)]^i$$

$$+ c_2[-0.3 - j(0.4)]^i, \qquad (4.3.82)$$

where

$$c_1 = A_1(0) - \frac{1}{[j(0.8)][a + 0.3 - j(0.4)]}$$

and

$$c_2 = A_2(0) + \frac{1}{[j(0.8)][a + 0.3 + j(0.4)]}.$$

It is noted that the coefficient of a^i can be further simplified to

$$\frac{1}{(a^2 + 0.6a + 0.25)}$$

and that the terms involving $[-0.3 + j(0.4)]$ and $[-0.3 - j(0.4)]$ can be combined in the manner discussed in Section 4.2 of this chapter to obtain a cosine and a sine term with magnitude $R = 0.5$ and angle $\theta = 2.217$. Therefore,

$$x(i) = \frac{a^i}{a^2 + 0.6a + 0.25} + c_3(0.5)^i \cos 2.217i + c_4(0.5)^i \sin 2.217i. \quad (4.3.83)$$

Thus we do our converting of complex roots to trigonometric form after solving for $x(i)$ when using variation of parameters but immediately after finding the homogeneous solution when we use undetermined coefficients.

4.4 Boundary Conditions

In each of the examples examined thus far, we have ended up with two constants that we said would have to be evaluated from boundary conditions. The reason we had two such constants was because all our examples involved second-order equations. Since each of these constants generally corresponds to a root of the characteristic equation, there will be exactly n of them for an nth-order equation. Therefore, to completely solve an nth-order equation, we must be able to specify exactly n boundary conditions that the equation must satisfy. We of course may not specify more than n, because then we would be overconstraining the system.

Often the boundary conditions will be initial conditions which usually would be specified in terms of the values of the dependent variable and its first $n - 1$ forward differences at $i = 0$, or the values of the dependent variable at $i = 0$ through $i = n - 1$, or some combination thereof. Obviously one can express one of these sets of conditions in terms of the other. Initial conditions are frequently used because they correspond to system start-up ($i = 0$) and can be measured or controlled.

For our example involving the equation

$$x(i + 2) - 3x(i + 1) + 2x(i) = 4^i + 3i^2 \quad (4.4.1)$$

(Example 1 in both Undetermined Coefficients and Variation of Parameters), the solution, as determined by both procedures, was found to be

$$x(i) = \tfrac{1}{6} \cdot 4^i - i^3 - \tfrac{3}{2}i^2 - \tfrac{13}{2}i + c_1 2^i + c_2. \quad (4.4.2)$$

To find c_1 and c_2 we must know or specify two conditions which (4.4.2) must satisfy. Suppose we knew that $x(0) = 0$ and $x(1) = \tfrac{2}{3}$. Then, writing $x(i)$ for $i = 0$ and $i = 1$ and equating them to the required values of 0 and $\tfrac{2}{3}$, we obtain

$$x(0) = \tfrac{1}{6} + c_1 + c_2 = 0$$

or

$$c_1 + c_2 = -\tfrac{1}{6},\qquad(4.4.3)$$

and

$$x(1) = \tfrac{1}{6}\cdot 4 - 1 - \tfrac{3}{2} - \tfrac{13}{2} + c_1\cdot 2 + c_2 = \tfrac{2}{3}$$

or

$$2c_1 + c_2 = 9.\qquad(4.4.4)$$

Subtracting (4.4.3) from (4.4.4) gives $c_1 = 55/6$ from which $c_2 = -56/6$ and

$$x(i) = \tfrac{1}{6}\cdot 4^i - i^3 - \tfrac{3}{2}i^2 - (13/2)i + (55/6)2^i - 56/6.\qquad(4.4.5)$$

Now suppose in the example which involved the same difference equation but with the forcing function $2\sin 3i$ (Example 2 in Undetermined Coefficients), we required $x(0) = 0.0390$ and $\Delta x(0) = 0.1465$. From our previous solution of this example,

$$x(i) = 0.337\sin 3i + 0.039\cos 3i + c_1 2^i + c_2.\qquad(4.4.6)$$

We immediately write

$$x(0) = 0.039 + c_1 + c_2 = 0.039,$$

from which

$$c_1 = -c_2.\qquad(4.4.7)$$

We now have a choice for our second condition. We could take the first forward difference of (4.4.6), set $i = 0$, and equate the result to 0.1465, or we could use the relationship

$$x(1) = x(0) + \Delta x(0)\qquad(4.4.8)$$

to determine that $x(1) = 0.039 + 0.1465 = 0.1855$ and use (4.4.6) with $i = 1$. In this case it appears easier to do the latter, which gives

$$x(1) = 0.337\sin 3 + 0.039\cos 3 + c_1 2 + c_2 = 0.1855.\qquad(4.4.9)$$

Now $\sin 3 = \sin 172° = \sin 8° = 0.139$ and $\cos 3 = -\cos 8° = -0.991$, resulting in

$$x(1) = (0.337)(0.139) + (0.039)(-0.991) + 2c_1 + c_2$$
$$= 0.0469 - 0.0386 + 2c_1 + c_2 = 0.0083 + 2c_1 + c_2$$

or

$$2c_1 + c_2 = 0.1772.\qquad(4.4.10)$$

Combining (4.4.7) and (4.4.10) yields $c_1 = 0.177$ and $c_2 = -0.177$.

Obviously, in situations involving only generalized forcing functions, direct evaluation of the combinatorial constants and hence the incorporation of boundary conditions in the solution is not possible because of their dependence on the convolution summations involved. Thus, in general, this step must follow the determination or specification of the form of $f(i)$.

4.5   Finding the Roots of the Characteristic Equation

It is apparent from the preceding material that determination of all n of the roots of the characteristic equation (4.2.5) is an essential step in the solution of linear difference equations with constant coefficients. We, therefore, address the task of finding roots of polynomial equations.

Available Approaches

For systems with first- or second-order transfer functions (i.e., those described by first- or second-order difference equations), finding the roots is simple. For first-order systems, the root is simply $-b_n/b_0$; and for second-order systems, the roots are readily determined from the quadratic formula. For higher order systems, however, this task may not be so easily accomplished. It is true that third- and fourth-order systems can be handled by the cubic and quartic formulas, respectively. It does not require too many confrontations with these two monsters, however, to make one seek alternative approaches. For fifth and higher order systems, no formulas are available and, furthermore, it has been proven that none exist.

Occasionally roots might be guessed at and divided out of the equation leaving a quotient of lower order which might be more easily handled. Certainly a zero root is obvious from the missing constant term in the polynomial and multiple zero roots from other missing terms in the low powers of h. In cases where the system is composed of a series of components whose individual transfer functions are known, the left-hand side of the system equation could occur originally in partially or even full factored form. Nevertheless, the system designer must have at his disposal an easily implemented method for finding roots of high order polynomial equations to assure his being able to deal with any system which may be of interest.

Numerous root-solver computer routines are available for finding roots of polynomial equations. These routines are based on any of a number of algorithms. The Newton–Raphson and Horner's methods can be used to find real roots, while other methods will identify both real and complex roots. One of the oldest of such methods is due to Lin (1941) and involves the successive extraction of quadratic factors. This method was improved upon by Friedman (1949) and Luke [see Luke and Ufford (1951)], but still runs into difficulty when the magnitudes of the roots are relatively close together. Graeffe's root-squaring method was devised to provide wider spacing among the magnitudes of the roots. Other methods now in common use include Muller's method and the Newton–Bairstow method (Conte, 1965). The Newton–Bairstow method is described below for use by those not having ready access to an existing computerized routine.

Note that all numerical approaches are applicable only for specific systems for which numerical values have been determined for all system parameters. However, even though analytic solutions and the resulting insights regarding the effects of changes in the values of various parameters are not directly obtainable, sensitivity analyses may still be performed by repeated solution of the system equation for ranges of parameter values. In many cases optimal values can be determined using appropriate numerical search methods. In other words, being limited to numerical solutions to system difference equations does not preclude synthesis and design of optimal systems nor the investigation of system sensitivities to changes in the operation of system components; although the job generally becomes much more difficult and computer dependent.

The Newton–Bairstow Approach

The Newton–Bairstow method is based upon the same basic concept as Lin's method (1941) of removing by division a quadratic factor from polynomials of degree $n > 2$ to yield a quotient polynomial of degree $n - 2$. If the quotient polynomial is of degree $n > 2$, a quadratic factor is then removed from it, leaving a resulting quotient polynomial of degree $n - 4$. This procedure of removing quadratic factors from quotient polynomials is repeated until a quotient polynomial of first or second degree is obtained. At this point, the set of roots of the original polynomial is found by applying the quadratic formula to each of the quadratic factors obtained and, if n is odd, from the single linear factor remaining at the conclusion of the process. The advantage of this approach is that if the coefficients of the original polynomial are real numbers, which they will be in system equations of real systems, then the coefficients of all quadratic factors and all quotient polynomials obtained will be real regardless of whether any or all of the roots are real or complex. Thus all calculations can be done with real arithmetic, a decided advantage for either hand or computer calculations.

The problem, of course, is to find the quadratic factor to be removed at each stage, and the authors of the various methods approach this problem using a variety of iterative schemes. As the name implies, the Newton–Bairstow method is derived from the basic method of Newton for finding roots of general equations. Newton makes use of linear approximations based on first partial derivatives of the function involved to modify current estimates of a root of the function. The method is explained in detail by Nielsen (1956) and other numerical analysis texts. Although they provide background and motivation for the Newton–Bairstow method, knowledge of these details is not required for implementation, so they are not included here.

Bairstow applied Newton's method to the parameters of the remainder term resulting from the division of a polynomial of order $n > 2$ by an assumed quadratic factor. This is based on the fact that when the trial divisor is a true factor of the polynomial, the remainder will be zero. Thus Bairstow's approach is one of driving the remainder to zero. By relating the parameters of the remainder terms to those of the trial quadratic factors, he was able to devise an iterative technique which converged on a true quadratic factor of the polynomial. Because of the nature of polynomials and their partial derivatives with respect to the parameters in the trial divisors, the set of equations used in the calculations are all of a relatively simple nature. The popularity of the method is due in large part to this simplicity and to the relatively rapid convergence often obtained. It must be admitted, however, that unfortunate guesses regarding the parameters of the quadratic factor used to start each stage can have severe adverse effects on the number of iterations required to obtain a specified degree of precision of convergence.

In the following section the steps of the Newton–Bairstow algorithm will be presented and then applied to several illustrative problems. No particular explanations for the reasons behind each step are included. The interested reader is referred to the work of Conte (1965, Chapter 2) for an account of the derivation of the method.

The Newton–Bairstow Algorithm

Let $b(h)$ represent the left-hand side of the characteristic equation (4.2.5), i.e.,

$$b(h) = b_0 h^n + b_1 h^{n-1} + \cdots + b_{n-1}h + b_n, \tag{4.5.1}$$

where $n > 2$, and let $h^2 - \alpha h - \beta$ be a quadratic factor of $b(h)$. Therefore, if we divide $b(h)$ by this factor, the quotient $c(h)$ will be a polynomial of degree $n - 2$, i.e.,

$$c(h) = c_0 h^{n-2} + c_1 h^{n-3} + \cdots + c_{n-3}h + c_{n-2}, \tag{4.5.2}$$

and there will be no remainder. If, however, an approximation to this quadratic factor is divided into $b(h)$ instead of the true factor, a linear remainder term will result. The relationship among $b(h)$, $c(h)$, and the remainder when the qth approximation is used can be expressed as

$$b(h) = (h^2 - \alpha_q h - \beta_q)c(h) + c_{n-1}(h - \alpha_q) + c_n, \qquad (4.5.3)$$

where α_q and β_q are the approximations to α and β, respectively, in the qth approximate quadratic factor. The somewhat unusual form for expressing the remainder $c_{n-1}(h - \alpha_q) + c_n$ is to facilitate description of the algorithm. The algorithm begins with the arbitrary selection of initial approximation values α_0 and β_0 and proceeds to adjust these values iteratively until two successive pairs (α_q, β_q) and $(\alpha_{q+1}, \beta_{q+1})$ are sufficiently close together to satisfy the accuracy criteria of the analyst.

Specifically, the steps of the algorithm are:

1.   Choose initial approximations α_0 and β_0 (i.e., $q = 0$).
2.   Using the current approximations α_q and β_q, compute the coefficients c_0 through c_n of (4.5.2) and (4.5.3) recursively as follows:

$$
\begin{aligned}
c_0 &= b_0, \\
c_1 &= b_1 &&+ \alpha_q c_0, \\
c_2 &= b_2 &&+ \alpha_q c_1 &&+ \beta_q c_0, \\
&\vdots \\
c_k &= b_k &&+ \alpha_q c_{k-1} &&+ \beta_q c_{k-2}, \\
&\vdots \\
c_{n-1} &= b_{n-1} &&+ \alpha_q c_{n-2} &&+ \beta_q c_{n-3}, \\
c_n &= b_n &&+ \alpha_q c_{n-1} &&+ \beta_q c_{n-2}.
\end{aligned}
\qquad (4.5.4)
$$

3.   Using the c_k computed in step 2 and the current approximations α_q and β_q, compute the coefficients d_0 through d_{n-1} recursively as follows:

$$
\begin{aligned}
d_0 &= c_0, \\
d_1 &= c_1 &&+ \alpha_q d_0, \\
d_2 &= c_2 &&+ \alpha_q d_1 &&+ \beta_q d_0, \\
&\vdots \\
d_k &= c_k &&+ \alpha_q d_{k-1} &&+ \beta_q d_{k-2}, \\
&\vdots \\
d_{n-2} &= c_{n-2} &&+ \alpha_q d_{n-3} &&+ \beta_q d_{n-4}, \\
d_{n-1} &= c_{n-1} &&+ \alpha_q d_{n-2} &&+ \beta_q d_{n-3}.
\end{aligned}
\qquad (4.5.5)
$$

Note the similarity in the form of Eqs. (4.5.4) and (4.5.5) in that the d_k are computed from the c_k just as the c_k were computed from the b_k. Note, however, that there is no d_n term.

4. Compute the corrections $\delta\alpha_q$ and $\delta\beta_q$ in the current approximations by solving the following pair of linear simultaneous equations:

$$d_{n-2}\,\delta\alpha_q + d_{n-3}\,\delta\beta_q = -c_{n-1},$$
$$d_{n-1}\,\delta\alpha_q + d_{n-2}\,\delta\beta_q = -c_n. \tag{4.5.6}$$

5. Using the values of $\delta\alpha_q$ and $\delta\beta_q$ found in step 4, compute a new set of approximations α_{q+1} and β_{q+1} by

$$\alpha_{q+1} = \alpha_q + \delta\alpha_q,$$
$$\beta_{q+1} = \beta_q + \delta\beta_q. \tag{4.5.7}$$

6. Repeat steps 2 through 5 until convergence occurs.

7. If $c(h)$ is of degree greater than 2, repeat steps 1 through 6 using $c(h)$ as the polynomial involved.

8. Repeat step 7 as long as the resulting quotient polynomial is of degree greater than 2.

9. Compute the roots of all quadratic factors by use of the quadratic formula. If the original n is odd, the final quotient polynomial will be of the form $c_0 h + c_1$, in which case the remaining root will be $h = -c_1/c_0$.

Examples Using the Newton–Bairstow Method

Example 1 The following contrived example illustrates in detail the operation of the algorithm. It must be confessed that the "original" polynomial $b(h)$ was actually derived by working backwards from a set of values α_1, β_1, α_0, and β_0, selected to result in rapid, absolute convergence using entirely integer arithmetic. Nevertheless, we state: "given" the fourth-order polynomial

$$b(h) = h^4 - 8h^3 + 34h^2 - 40h - 23, \tag{4.5.8}$$

let us use the Newton–Bairstow method to find the roots of $b(h) = 0$. With what will prove to be amazing foresight, we let $\alpha_0 = 1$ and $\beta_0 = -1$, i.e., we assume a quadratic factor for (4.5.8) of $h^2 - h + 1$. The corresponding c_k values are calculated from (4.5.4) to be:

$$\begin{aligned}
c_0 &= b_0 = 1, \\
c_1 &= b_1 + \alpha_0 c_0 = -8 + 1 \cdot 1 = -7, \\
c_2 &= b_2 + \alpha_0 c_1 + \beta_0 c_0 = 34 + 1(-7) + (-1)1 = 26, \\
c_3 &= b_3 + \alpha_0 c_2 + \beta_0 c_1 = -40 + 1 \cdot 26 + (-1)(-7) = -7, \\
c_4 &= b_4 + \alpha_0 c_3 + \beta_0 c_2 = -23 + 1(-7) + (-1)26 = -56,
\end{aligned} \tag{4.5.9}$$

from which the d_k values are determined from (4.5.5) as:

$$d_0 = c_0 = 1,$$
$$d_1 = c_1 + \alpha_0 d_0 = -7 + 1 \cdot 1 = -6,$$
$$d_2 = c_2 + \alpha_0 d_1 + \beta_0 d_0 = 26 + 1(-6) + (-1)1 = 19, \qquad (4.5.10)$$
$$d_3 = c_3 + \alpha_0 d_2 + \beta_0 d_1 = -7 + 1 \cdot 19 + (-1)(-6) = 18.$$

Equations (4.5.6) thus become

$$19\,\delta\alpha_0 - 6\,\delta\beta_0 = 7,$$
$$18\,\delta\alpha_0 + 19\,\delta\beta_0 = 56, \qquad (4.5.11)$$

which are easily solved to obtain $\delta\alpha_0 = 1$ and $\delta\beta_0 = 2$. Thus,

$$\alpha_1 = \alpha_0 + \delta\alpha_0 = 1 + 1 = 2$$
$$\beta_1 = \beta_0 + \delta\beta_0 = -1 + 2 = 1. \qquad (4.5.12)$$

The previous steps are now repeated using α_1 and β_1 in place of α_0 and β_0, respectively. Calculation of the c_k's from (4.5.4) now yields

$$c_0 = b_0 = 1,$$
$$c_1 = b_1 + \alpha_1 c_0 = -8 + 2 \cdot 1 = -6,$$
$$c_2 = b_2 + \alpha_1 c_1 + \beta_1 c_0 = 34 + 2(-6) + 1 \cdot 1 = 23, \qquad (4.5.13)$$
$$c_3 = b_3 + \alpha_1 c_2 + \beta_1 c_1 = -40 + 2 \cdot 23 + 1(-6) = 0,$$
$$c_4 = b_4 + \alpha_1 c_3 + \beta_1 c_2 = -23 + 2 \cdot 0 + 1 \cdot 23 = 0.$$

Recall that the remainder term as given in (4.5.3) is $c_{n-1}(h - \alpha_q) + c_n$, which for the case at hand is $c_3(h - \alpha_1) + c_4$. Since both the c_3 and c_4 calculated in (4.5.13) are zero, there is no remainder when $b(h)$ is divided by the quadratic divisor $h^2 - 2h - 1$. Thus $h^2 - 2h - 1$ is an exact quadratic factor of $b(h)$. The coefficients of the quotient polynomial $c(h)$ are the c_0, c_1, and c_2 calculated in (4.5.13), i.e.,

$$c(h) = h^2 - 6h + 23. \qquad (4.5.14)$$

Since $c(h)$ is of second order, the roots of $c(h) = 0$, as well as the roots of $h^2 - 2h - 1 = 0$, can be found by use of the quadratic formula and no further cycles of the Newton–Bairstow method are required. For the record, the four roots of $b(h) = 0$ are $1 \pm (2)^{1/2}$ and $3 \pm j(14)^{1/2}$.

Example 2   A more likely form for $b(h)$ is

$$b(h) = h^5 - 12h^4 + 34.25h^3 - 13.25h^2 - 40.5h + 20, \qquad (4.5.15)$$

the actual roots of which are -1.00, 0.50, 2.00, 2.50, and 8.00. As in most practical cases, the arithmetic involved here is not all integer and choices for α_0 and β_0 in the absence of clairvoyance or unusual knowledge of the system must be considered as arbitrary. Initial guesses of $\alpha_0 = -3$ and $\beta_0 = 2$ produced an $\alpha_9 = \alpha_{10} = 1.5000$ and a $\beta_9 = \beta_{10} = 2.5000$. That is, convergence to four decimal places occurred in 10 iterations. The corresponding roots are -1.0000 and 2.5000. Again using $\alpha_0 = -3$ and $\beta_0 = 2$ on the resulting cubic quotient polynomial produced $\alpha_7 = \alpha_8 = 2.5000$ and $\beta_7 = \beta_8 = -1.0000$. These values correspond to the roots 0.5000 and 2.0000, leaving a quotient of $h - 8.0000$ from which the fifth root is obtained by observation. On the other hand, an $\alpha_0 = 3$ and $\beta_0 = -2$ produced from Eq. (4.5.15), with the same precision as above, the pair of roots 0.5000 and 2.5000 in 6 iterations and the pair -1.0000 and 2.0000 from the resulting cubic quotient polynomial in 5 more iterations. Other α_0, β_0 combinations required from 3 to 14 iterations to obtain the first quadratic factor to four-decimal precision.

These results were obtained using a FORTRAN IV program and an IBM 370/165 computer, but could have been done on a desk calculator had necessity demanded.

We now turn our attention to the z transform and its use in formulating and solving linear difference equation models. It will be observed that in spite of the many beneficial features of transform methods, the task of finding the roots of the characteristic equation remains unchanged. Thus, the previous discussion is just as applicable whether classical or transform techniques are employed.

Exercises

4.1   Derive or prove each of the following:

a.   The contribution to the homogeneous solution of a triple real root h is of the form $(c_1 i^2 + c_2 i + c_3)h^i$.

b.   The contribution to the homogeneous solution of a pair of double complex conjugate roots is $R^i[(c_1 i + c_2) \cos \theta i + (c_3 i + c_4) \sin \theta i]$, where R and θ are as defined in (4.2.9) and (4.2.10).

4.2   Find the homogeneous solutions to each of the following linear difference equations:

a. $2 \Delta x(i) - 2x(i) = f(i)$ **b.** $\Delta^2 x(i) - 6 \Delta x(i) + 9x(i) = f(i)$
c. $x(i + 2) - 2x(i + 1) + 2x(i) = f(i)$
d. $x(i + 2) - 2x(i + 1) - 3x(i) = f(i)$
e. $x(i + 2) - 4x(i + 1) + 4x(i) = f(i)$
f. $x(i + 2) - 2x(i + 1) + x(i) = f(i)$
g. $x(i + 3) - 6x(i + 2) + 11x(i + 1) - 6x(i) = f(i)$ (*Note*: $h_1 = 1$)
h. $x(i + 3) - 0.5x(i + 2) - 0.25x(i + 1) + 0.125x(i) = f(i)$
      (*Note*: $h_1 = 0.5$)
i. $x(i + 3) - 3x(i + 2) + 3x(i + 1) - x(i) = f(i)$

4.3 (a) through (g).   Write the characteristic equation of the system difference equation derived in the corresponding part of Exercise 2.3.

4.4–4.10   The same as 4.3 using the difference equations derived in the
corresponding parts of Exercises 2.4 through 2.10, respectively.

4.11   Using the method of undetermined coefficients, solve the difference
equation given in 4.2a for each of the following forcing functions:

    a. 3   **b.** 3^i **c.** $3 + 3^i$ **d.** 2^i **e.** i **f.** $\sin 2i$
    g. $2^i \sin 2i$

4.12   Using the method of undetermined coefficients, solve the difference
equation given in 4.2b for each of the following forcing functions:

    a. 4   **b.** 4^i **c.** i **d.** $4^i \sin i$ **e.** $i + 4$

4.13   Using the method of undetermined coefficients, solve the difference
equation given in 4.2 c for each of the following forcing functions:

    a. $(2)^{1/2}$ **b.** $(2)^{i/2}$ **c.** $\sin(\pi/4)^i$ **d.** $(2)^{i/2} \sin(\pi/4)^i$
    e. $(2)^{i/2} + \sin(\pi/4)^i$

4.14   Using the method of undetermined coefficients, solve the difference
equation given in 4.2d for each of the following forcing functions:

    a. 3   **b.** i **c.** 3^i **d.** $i + 3$ **e.** $i + 3^i$ **f.** $3^i + 3$
    g. $\sin 3i$

4.15   Using the method of undetermined coefficients, solve the difference
equation given in 4.2e for $f(i) = i^3 + i2^i$.

4.16   Using the method of undetermined coefficients, solve the difference
equation given in 4.2f for $f(i) = i^3 + \sin i$.

4.17 (a) through (g).   Using variation of parameters, solve the difference
equation in 4.2a in terms of the generalized forcing function $f(i)$. Then
substitute each of the specific forcing functions given in 4.11 and derive a
closed-form solution.

4.18 through **4.20**   The same as 4.17 except using the difference equations
given in 4.2b, c, and d and the specific forcing functions in 4.12, 4.13, and
4.14, respectively.

4.21 (a) through (g).   Evaluate the boundary coefficient in the solutions
found in each part of 4.11 given that $x(0) = 1.9$.

4.22 through **4.26**   The same as 4.21 except using the solutions of 4.12
through 4.16, respectively, and the conditions $x(0) = 0$ and $x(1) = 1$.

4.27 Solve the following difference equations. Express all boundary conditions as initial conditions, i.e., in terms of $m(0)$, $m(1)$, etc., as appropriate:

  a. Eq. (2.2.5) $m(i + 1) - (1 - K)m(i) = KD + p(i)$
  b. Eq. (2.3.21) $m(i + 1) - (1 - K)m(i) = KD - K\varepsilon(i) + p(i)$
  c. The equation for the proportional-plus-difference process controller obtained by setting $N = 0$ in (2.3.30)
  d. The equation for the proportional-plus-summation process controller obtained by setting $L = 0$ in (2.3.30)
  e. Eq. (2.3.32) $m(i + 1) - (1 - K_0)m(i) + K_1 m(i - 1)$
$$= (K_0 + K_1)D - K_0 \varepsilon(i) - K_1 \varepsilon(i - 1) + p(i)$$

4.28 A company has an ordering policy for a stock item which requires k months to deliver which states that $q(i + k)$, the amount ordered at the beginning of month i for delivery at the beginning of month $i + k$, is equal to the forecast demand $p(i, k)$ for week $i + k$ plus $p(i)$, the known demand for month i minus the order arriving at the beginning of month i, $q(i)$. Thus

$$q(i + k) = p(i, k) + p(i) - q(i).$$

  a. Solve this difference equation for order size as a function of discrete time i in terms of the general order and forecast terms $p(i)$ and $p(i, k)$ for k equal to (1) 1, (2) 2, (3) 3.
  b. For a forecast based on the assumption that demand has an increasing trend and a seasonal sinusoidal component, specifically,

$$p(i, k) = a_1 + a_2(i + k) + a_3 \sin \frac{2\pi}{12} i + a_4 \cos \frac{2\pi}{12} i,$$

and that actual demand is given by

$$p(i) = b_1 + b_2 i + b_3 \sin \frac{2\pi}{12} i + b_4 \cos \frac{2\pi}{12} i,$$

find a closed-form solution for order size as a function of discrete time for
  (1) $k = 1$ given $q(0) = a_1 + a_4 = A_0$,
  (2) $k = 2$ given $q(0) = A_0$,
       $q(1) = a_1 + a_2 + a_3 \sin(\pi/6) + a_4 \cos(\pi/6) = A_1$,
  (3) $k = 3$ given $q(0) = A_0$, $q(1) = A_1$,
       $q(2) = a_1 + 2a_2 + a_3 \sin(\pi/3) + a_4 \cos(\pi/3) = A_2$.

4.29 Use the Newton–Bairstow method to find the roots of the following characteristic equations to at least four decimal places:

  a. $h^3 - h^2 - 2h + 1 = 0$
  b. $h^4 + 0.6h^3 + 0.5h^2 + 0.15h + 0.0625 = 0$

c. $h^5 + 0.6h^4 - 3.4h^3 - 0.8h^2 + 1.8h - 0.4 = 0$
d. $h^5 + 11.05h^4 + 13.55h^3 + 25.65h^2 - 48.75h - 2.5 = 0$
e. $h^5 - 78.01h^4 - 2157.22h^3 - 4138.42h^2 - 3958.4h + 40 = 0$
f. $h^5 + 0.6h^4 - 0.51h^3 - 1.666h^2 + 0.054h + 0.522 = 0$

4.30 (**a**) through (**f**).   Find the roots of the characteristic equations in 4.29 by any appropriate technique. Give the name of the technique used and indicate the nature of the approach.

Chapter V | The z Transform

The z transform and other transform techniques were developed by mathematicians over two hundred years ago. They were used by DeMoivre in his studies of probability theory during the 18th century and have found increasing application by engineers since that time [see Cadzow and Martens (1970, pp. 118–119)]. The applicability of the z transform to discrete-time controls stems from its property of transforming linear difference equations into algebraic equations which can then be manipulated using the familiar laws of algebra. It thus often serves as a most useful alternative to classical methods in both the formulation and solution of discrete linear models. It is the discrete counterpart of the Laplace transform used for linear differential equations. Essentially every step in the procedure and every formula used has an analogous step or formula in Laplace transform theory. Those familiar with the Laplace transform will be able to draw heavily on their knowledge when applying the z transform. In fact, such familiarity may permit some readers to simply skim or even skip Chapters V and VI.

5.1 The Basic Transform

The z transform may be applied to transform functions of discrete time i into functions of the transform variable z. The most common form of the transform for engineering applications is

$$Z[f(i)] = \sum_{i=0}^{\infty} f(i)z^{-i}, \tag{5.1.1}$$

where $Z[\cdot]$ is read *the z transform of*, $f(i)$ is a function of discrete time such that $f(i) = 0$ for $i < 0$, and z is the transform variable. This form is called the *one-sided z transform* to distinguish it from the mathematically more general *two-sided z transform* given by

$$Z[f(i)] = \sum_{i=-\infty}^{\infty} f(i)z^{-i}. \tag{5.1.2}$$

Since a specific starting point can almost always be defined for the operation of any man–machine system, the functions of interest to most systems analysts can be defined such that they are indeed zero for negative values of the time index i. Since, therefore, the conditions for use of the one-sided transform apply, it is the form used almost exclusively by engineers and the form to be developed herein. Many essentially mathematical problems involving existence, convergence, and the like are avoided or simplified by so doing.

It should also be pointed out that some authors define the z transform as $\sum_{i=0}^{\infty} f(i)z^i$, i.e., as a positive power series in z instead of as a negative power series as given in (5.1.1). Strictly speaking, however, the positive power form is called the *geometric transform* in most references. Obviously the two forms are interchangeable, the transform variable of one form being the reciprocal of the transform variable of the other. Still, it is essential that the reader of any given reference on the subject know precisely which transform is being used and how it is being referred to by the author. In this book the negative power series form given by (5.1.1) is used exclusively and referred to as *the z transform*. Any reference made to the positive power form will be as *the geometric transform*.

Convergence

The usefulness of any transform technique depends on the convergence of the transform. In the case of the z transform, convergence means that the infinite sum given by (5.1.1) must be finite. This can also be stated as

$$\lim_{k \to \infty} \sum_{i=0}^{k} f(i)z^{-i} = S, \tag{5.1.3}$$

where S is a finite quantity. Convergence or lack of convergence, called divergence, depends on the $f(i)$ involved.

Several theorems relate to the convergence of infinite series.† The applicability of each of these as a test for the convergence of the z transform depends on the specific $f(i)$ involved; however, at least one theorem should apply to almost any form of $f(i)$ likely to be encountered in engineering and systems

---

† Such theorems can be found in most texts on advanced algebra [see, for example, Taylor (1955, Chapter XVII)].

analysis work. Thus questions regarding the applicability of the z transform in any given case can essentially always be readily determined in straightforward fashion should the need arise.

We illustrate the process of testing for convergence where the test is based on the following theorem, usually referred to as d'Alembert's ratio test:

Theorem 5.1   Let $\sum_{i=0}^{\infty} u_i$ be a series with all its terms different from zero. Then the series is absolutely convergent if there is a positive number $r < 1$ such that $|u_{i+1}/u_i| < r$ for all sufficiently large values of i. In particular, this condition is satisfied if the limit

$$L = \lim_{i \to \infty} u_{i+1}/u_i \qquad (5.1.4)$$

exists and $L < 1$. If $L > 1$, the series is divergent, and if $L = 1$, no conclusion can be drawn.

By absolute convergence is meant that the series $\sum_{i=0}^{\infty} |u_i|$ converges. A separate theorem states that if a series is absolutely convergent it is also convergent. Thus, the conditions for convergence expressed by Theorem 5.1 are sufficient but not necessary.

Now let $f(i) = a^i$, the z transform for which is

$$Z[a^i] = \sum_{i=0}^{\infty} a^i z^{-i} = \sum_{i=0}^{\infty} (a/z)^i. \qquad (5.1.5)$$

Thus

$$u_i = (a/z)^i \qquad (5.1.6)$$

and

$$|u_{i+1}/u_i| = |(a/z)^{i+1}/(a/z)^i| = |a/z|. \qquad (5.1.7)$$

Thus, for convergence $|a|/|z| < 1$ or

$$|a| < |z|. \qquad (5.1.8)$$

However, there is no physical or mathematical reason for not thinking of z as being as large as we please, certainly large enough to allow (5.1.3) to exist for any given a.

The result just obtained can be shown to hold for essentially all the common functions encountered in the study of man–machine systems, i.e., the conditions required for convergence of (5.1.1) for the kinds of $f(i)$ occurring in conjunction with such systems can always be met by considering z large enough. For this reason, the existence of the z transform is almost always assumed without test in engineering work, and it will be so assumed in this book.

We, therefore, will not pursue tests of convergence any further. Instead let us look at the derivation of a few z transforms.

z Transform Derivations

Example 1 $f(i) = a$, "a" a constant.   Substitution of the constant a into (5.1.1) yields

$$Z[a] = \sum_{i=0}^{\infty} az^{-i} = a \sum_{i=0}^{\infty} z^{-i} = a\left[\frac{-1}{1/z - 1}\right], \qquad (5.1.9)$$

assuming convergence. This is usually expressed in the more convenient form

$$Z[a] = az/(z - 1) \qquad (5.1.10)$$

by multiplying numerator and denominator by $-z$.

Example 2 $f(i) = a^i$, "a" a constant.   Equation (5.1.1) yields

$$Z[a^i] = \sum_{i=0}^{\infty} a^i z^{-i} = \sum_{i=0}^{\infty} \left(\frac{a}{z}\right)^i = \frac{-1}{a/z - 1} = \frac{z}{z - a}. \qquad (5.1.11)$$

Convergence is again assumed as it will be from here on without further reference to it.

Example 3 $f(i) = \sin ai$, "a" a constant.   Two methods will be demonstrated in deriving the z transform of $\sin ai$. The first makes use of the identity

$$\sin x = \frac{e^{jx} - e^{-jx}}{2j} \qquad (5.1.12)$$

and the second the fact that $\sin x$ is the imaginary part of the complex expression

$$e^{jx} = \cos x + j \sin x. \qquad (5.1.13)$$

Obviously (5.1.12) and (5.1.13) stem from the same basic relationships, but the different forms in which they are expressed require different manipulations in the derivation of the transform.

(a)   *Method 1*   In this approach we substitute (5.1.12) for $\sin ai$ in the basic transform (5.1.1) to obtain

$$Z[\sin ai] = \sum_{i=0}^{\infty} (\sin ai)z^{-i} = \sum_{i=0}^{\infty} \frac{e^{jai} - e^{-jai}}{2j} z^{-i}$$

$$= \frac{1}{2j}\left[\sum_{i=0}^{\infty} e^{jai}z^{-i} - \sum_{i=0}^{\infty} e^{-jai}z^{-i}\right]. \qquad (5.1.14)$$

Now using the result (5.1.11) from Example 2,

$$Z[\sin ai] = \frac{1}{2j}\left[\frac{z}{z - e^{ja}} - \frac{z}{z - e^{-ja}}\right], \tag{5.1.15}$$

which when put over a common denominator yields

$$Z[\sin ai] = \frac{1}{2j} \cdot \frac{(z^2 - ze^{-ja}) - (z^2 - ze^{ja})}{z^2 - z(e^{ja} + e^{-ja}) + 1}$$

$$= \frac{1}{2j} \cdot \frac{z(e^{ja} - e^{-ja})}{z^2 - z(e^{ja} + e^{-ja}) + 1}. \tag{5.1.16}$$

The $(e^{ja} - e^{-ja})$ in the numerator and the $2j$ in the denominator can be replaced by $\sin a$. Further, from the identity

$$\cos x = (e^{jx} + e^{-jx})/2, \tag{5.1.17}$$

the $(e^{ja} + e^{-ja})$ in the denominator can be replaced by $2\cos a$. Therefore,

$$Z[\sin ai] = \frac{z \sin a}{z^2 - 2z \cos a + 1}. \tag{5.1.18}$$

   (b)   *Method 2*   Here we note that $\sin ai$ is the imaginary part of e^{jai}; and then find $Z[\sin ai]$ as the imaginary part of $Z[e^{jai}]$. From Example 2,

$$Z[e^{jai}] = z/(z - e^{ja}), \tag{5.1.19}$$

which must now be expressed in the form

$$\mathrm{Re}\{Z[e^{jai}]\} + j\,\mathrm{Im}\{Z[e^{jai}]\},$$

where $\mathrm{Re}\{\cdot\}$ is read *the real part of* and $\mathrm{Im}\{\cdot\}$ is read *the imaginary part of.* We first substitute the form (5.1.13) in (5.1.19) to obtain

$$Z[e^{jai}] = \frac{z}{z - (\cos a + j \sin a)} = \frac{z}{(z - \cos a) - j \sin a} \tag{5.1.20}$$

and rationalize the denominator by multiplying both numerator and denominator by the complex conjugate of the denominator, $(z - \cos a) + j \sin a$. This yields

$$Z[e^{jai}] = \frac{(z^2 - z \cos a) + jz \sin a}{z^2 - 2z \cos a + \cos^2 a + \sin^2 a}$$

$$= \frac{z^2 - z \cos a}{z^2 - 2z \cos a + 1} + j \frac{z \sin a}{z^2 - 2z \cos a + 1}, \tag{5.1.21}$$

the last step being based on the identity

$$\cos^2 x + \sin^2 x = 1. \tag{5.1.22}$$

Now the imaginary part of $Z[e^{jai}]$ is seen to be $(z \sin a)/(z^2 - 2z \cos a + 1)$, which is identical to the expression for $Z[\sin ai]$ found by Method 1.

Note that just as $\cos x$ is the real part of e^{jx} as given by (5.1.13), $Z[\cos ai]$ is the real part of $Z[e^{jai}]$. Therefore, Method 2 yields

$$Z[\cos ai] = \frac{z^2 - z \cos a}{z^2 - 2z \cos a + 1} \tag{5.1.23}$$

in addition to the transform for $\sin ai$.

One could of course work out the z transform for any $f(i)$ of interest using the definitional formula (5.1.1) and appropriate summation formulas. Once the z transform for a given function has been worked out, there is some potential benefit in filing the result away for future reference. Because of the diversity of functions usually encountered in studies of man–machine systems, however, tables including even the more common ones would be cumbersome and their access difficult. Fortunately, the z transform has many useful properties which can be used to greatly expand the available list of transforms once a few basic ones are developed. We will now examine several of these properties, the results of which form the basis for Table 5.3.1. Table 5.3.2 follows, containing many useful transform pairs. More will be said about tables in Section 5.3.

5.2   Properties of the z Transform

Linearity

Since the z transform is a summing operation, all of the properties developed in Chapter III for summations apply to the z transform. Of primary importance is the fact that summation is a linear operation. Thus the associative, distributive, and commutative laws of algebra apply, so that given functions of discrete time $f(i)$ and $g(i)$ and constants a and b,

$$\begin{aligned}
Z[af(i) + bg(i)] &= \sum_{i=0}^{\infty} [af(i) + bg(i)]z^{-i} \\
&= a \sum_{i=0}^{\infty} f(i)z^{-i} + b \sum_{i=0}^{\infty} g(i)z^{-i} \\
&= aZ[f(i)] + bZ[g(i)].
\end{aligned} \tag{5.2.1}$$

Real Forward Translation

Let the z transform of some function $f(i)$ of discrete time i be represented by $F(z)$, i.e.,

$$F(z) = Z[f(i)] = \sum_{i=0}^{\infty} f(i)z^{-i}. \tag{5.2.2}$$

Then for k a positive integer, $f(i + k)$ is the function translated forward k units of time and the z transform of the translated function is

$$Z[f(i + k)] = \sum_{i=0}^{\infty} f(i + k)z^{-i} = z^k \sum_{i=0}^{\infty} f(i + k)z^{-(i+k)}. \tag{5.2.3}$$

We now let $u = i + k$ and note that when $i = 0$, $u = k$, and when $i = \infty$, $u = \infty$. Substituting for i in (5.2.3) gives

$$Z[f(i + k)] = z^k \sum_{u=k}^{\infty} f(u)z^{-u}. \tag{5.2.4}$$

We now add and subtract the quantity $z^k \sum_{u=0}^{k-1} f(u)z^{-u}$ which yields:

$$Z[f(i + k)] = z^k \sum_{u=0}^{\infty} f(u)z^{-u} - z^k \sum_{u=0}^{k-1} f(u)z^{-u}, \tag{5.2.5}$$

or, since the symbol used as the summation index is immaterial as far as the final result is concerned,

$$Z[f(i + k)] = z^k \sum_{i=0}^{\infty} f(i)z^{-i} - z^k \sum_{i=0}^{k-1} f(i)z^{-i}$$

$$= z^k F(z) - \sum_{i=0}^{k-1} f(i)z^{k-i}. \tag{5.2.6}$$

Thus, the z transform of a function translated forward k units of discrete time is z^k times the transform of the original function minus a series of k terms each involving one of the values of the function at times $i = 0$ through $i = k - 1$ weighted by z^{k-i}.

Specifically,

$$Z[f(i + 1)] = zF(z) - zf(0),$$
$$Z[f(i + 2)] = z^2 F(z) - z^2 f(0) - zf(1),$$

and

$$Z[f(i + 3)] = z^3 F(z) - z^3 f(0) - z^2 f(1) - zf(2).$$

For $f(i) = a^i$, the latter expression becomes

$$Z[a^{i+3}] = z^3 \frac{z}{z-a} - z^3 a^0 - z^2 a^1 - za^2$$

$$= \frac{z^4}{z-a} - z^3 - z^2 a - za^2$$

which the interested reader can easily show to be equal to

$$Z[a^{i+3}] = a^3 Z[a^i] = a^3 \frac{z}{z-a}.$$

Real Backward Translation

Again let $F(z)$ be the z transform of $f(i)$ and k be a positive integer. Then $f(i - k)$ is $f(i)$ translated backwards k units of time. The z transform of $f(i - k)$ is

$$Z[f(i-k)] = \sum_{i=0}^{\infty} f(i-k)z^{-i} = z^{-k} \sum_{i=0}^{\infty} f(i-k)z^{-(i-k)}. \qquad (5.2.7)$$

Now let $v = i - k$ and note that for $i = 0$, $v = -k$, and for $i = \infty$, $v = \infty$. Substituting into (5.2.7) yields

$$Z[f(i-k)] = z^{-k} \sum_{v=-k}^{\infty} f(v)z^{-v} \qquad (5.2.8)$$

but, since we have assumed $f(i) = 0$ for $i < 0$, the terms in the summation for $v = -k$ through $v = -1$ vanish, leaving, after returning to the index i,

$$Z[f(i-k)] = z^{-k} \sum_{i=0}^{\infty} f(i)z^{-i} = z^{-k} F(z). \qquad (5.2.9)$$

Note that in the more general case where nonzero values of $f(i)$ could exist for negative i,

$$Z[f(i-k)] = z^{-k} F(z) + \sum_{i=-k}^{-1} f(i)z^{-(i+k)}. \qquad (5.2.10)$$

For the case in which

$$f(i) = \begin{cases} a^i & \text{for} \quad i \geq 0, \\ 0 & \text{for} \quad i < 0, \end{cases}$$

$$Z[f(i-3)] = Z[a^{i-3}] = z^{-3} \frac{z}{z-a} = \frac{1}{z^2(z-a)}.$$

If, however, $f(i) = a^i$ for all i positive or negative,

$$Z[f(i - 3)] = z^{-3}\frac{z}{z - a} + z^0 a^{-3} + z^1 a^{-2} + z^2 a^{-1}$$

$$= \frac{1}{z^2(z - a)} + a^{-3} + za^{-2} + z^2 a^{-1}.$$

Forward Differencing

For $F(z)$ the z transform of $f(i)$ and k a positive integer, the z transform of the kth forward difference of $f(i)$ can be expressed in terms of $F(z)$ as follows:

$$Z[\Delta^k f(i)] = Z\left[\sum_{p=0}^{k}(-1)^p C_p^k f(i + k - p)\right]$$

$$= \sum_{p=0}^{k}(-1)^p C_p^k Z[f(i + k - p)]. \qquad (5.2.11)$$

Substituting (5.2.6) for $Z[f(i + k - p)]$ yields

$$Z[\Delta^k f(i)] = \sum_{p=0}^{k}(-1)^p C_p^k\left[z^{k-p}F(z) - \sum_{i=0}^{k-p-1}f(i)z^{k-p-i}\right]$$

$$= F(z)\sum_{p=0}^{k}(-1)^p C_p^k z^{k-p} + \sum_{p=0}^{k}(-1)^{p+1}C_p^k\sum_{i=0}^{k-p-1}f(i)z^{k-p-i}.$$

$$(5.2.12)$$

Note that the coefficient of $F(z)$ is the expression for the binomial series $(z - 1)^k$ so the first term of (5.2.12) reduces to $(z - 1)^k F(z)$. The second part of (5.2.12) involves the "initial condition" terms, i.e., terms in $f(0)$ through $f(k - 1)$. It is convenient to combine terms involving the same values of $f(i)$. It can be shown that the coefficient of the term $f(i)$ for $i = 0$ through $i = k - 1$ is

$$\sum_{p=0}^{k-1-i}(-1)^{p+1}C_p^k z^{k-p-i}, \qquad (5.2.13)$$

so the over-all transform is given by

$$Z[\Delta^k f(i)] = (z - 1)^k F(z) + \sum_{i=0}^{k-1}f(i)\left[\sum_{p=0}^{k-1-i}(-1)^{p+1}C_p^k z^{k-p-i}\right]. \qquad (5.2.14)$$

It might be helpful to examine specifically the coefficients of a few of the initial condition terms. Starting with $i = 0$, the coefficient of $f(0)$ is

$$\sum_{p=0}^{k-1}(-1)^{p+1}C_p^k z^{k-p} = -z^k + kz^{k-1} + \cdots + (-1)^{k-1}\frac{k!}{(k-2)!2!}z^2 + (-1)^k kz.$$

The coefficient of $f(1)$ is

$$\sum_{p=0}^{k-2}(-1)^{p+1}C_p{}^k z^{k-p-1} = -z^{k-1} + kz^{k-2} + \cdots + (-1)^{k-1}\frac{k!}{(k-2)!2!}z.$$

Note that each term in this coefficient is of the same form as the corresponding term in the coefficient of $f(0)$ except that the power of z is one less. There is also one fewer term, but the lowest power of z in both cases is the first power. This format continues for increasing i until we get to $i = k - 3$. This corresponds to $k - 1 - i = 2$ and $k - p - i = 3 - p$, so the coefficient of $f(k - 3)$ is

$$\sum_{p=0}^{2}(-1)^{p+1}C_p{}^3 z^{3-p} = -z^3 + kz^2 - \frac{k(k-1)}{2}z.$$

Similarly, the coefficient of $f(k - 2)$ is

$$\sum_{p=0}^{1}(-1)^{p+1}C_p{}^2 z^{2-p} = -z^2 + kz,$$

and the coefficient of $f(k - 1)$ is

$$\sum_{p=0}^{0}(-1)^{p+1}C_p{}^1 z^{1-p} = -z.$$

Note that all of these coefficients are incomplete binomial expansions so it is not possible to write a general closed-form expression for them in terms of elementary functions.

It might also prove valuable to write out the complete transforms of a few forward differences. Specifically,

$$Z[\Delta f(i)] = (z - 1)F(z) - zf(0),$$
$$Z[\Delta^2 f(i)] = (z - 1)^2 F(z) + (-z^2 + 2z)f(0) - zf(1),$$

and

$$Z[\Delta^3 f(i)] = (z - 1)^3 F(z) + (-z^3 + 3z^2 - 3z)f(0) + (-z^2 + 3z)f(1) - zf(2).$$

In the latter case if $f(i) = a^i$,

$$Z[\Delta^3 a^i] = (z - 1)^3\,\frac{z}{z - a} + (-z^3 + 3z^2 - 3z) + (-z^2 + 3z)a - za^2.$$

Backward differencing can be handled directly as above or as a translated forward difference as has previously been mentioned. By any appropriate method, it is found that for $f(i) = 0$, $i < 0$,

$$Z[\nabla^k f(i)] = \left(\frac{z-1}{z}\right)^k F(z). \qquad (5.2.15)$$

This is easily verified for the first backward difference

$$\nabla f(i) = f(i) - f(i-1) \tag{5.2.16}$$

for which

$$Z[\nabla f(i)] = Z[f(i)] - Z[f(i-1)] = F(z) - \frac{1}{z}F(z) = \left(1 - \frac{1}{z}\right)F(z) = \frac{z-1}{z}F(z).$$

$$\tag{5.2.17}$$

Summation

Let $F(z)$ be the z transform of $f(i)$ and define $g(n)$ as

$$g(n) = \sum_{i=0}^{n} f(i). \tag{5.2.18}$$

Let $G(z)$ be the z transform of $g(n)$, i.e.,

$$G(z) = Z\left[\sum_{i=0}^{n} f(i)\right]. \tag{5.2.19}$$

From (5.2.18) we note that

$$g(n-1) = \sum_{i=0}^{n-1} f(i)$$

so that

$$f(n) = g(n) - g(n-1) = \nabla g(n), \tag{5.2.20}$$

the first backward difference of $g(n)$. Therefore, using (5.2.17),

$$F(z) = Z[f(n)] = Z[\nabla g(n)] = \frac{z-1}{z}G(z). \tag{5.2.21}$$

Note that no term in $g(-1)$ is present in the transform of $g(n)$ since by definition $\sum_{i=0}^{-1} f(i)$ is zero. Combining (5.2.21) and (5.2.19), we get

$$Z\left[\sum_{i=0}^{n} f(i)\right] = \frac{z}{z-1}F(z). \tag{5.2.22}$$

We verify (5.2.22) for the case of $f(i) = a^i$. Since $F(z)$ in this case is $z/(z-a)$,

$$Z\left[\sum_{i=0}^{n} f(i)\right] = \frac{z}{z-1} \cdot \frac{z}{z-a} = \frac{z^2}{(z-1)(z-a)}.$$

This result could of course also be obtained by summing $f(i)$ to obtain

$$\sum_{i=0}^{n} f(i) = \frac{a^i}{a-1}\bigg|_0^{n+1} = \frac{a^{n+1}-1}{a-1}$$

and taking the z transform of this result as follows:

$$Z\left[\frac{a^{n+1}-1}{a-1}\right] = \frac{1}{a-1}\{Z[a^{n+1}] - Z[1]\}$$

$$= \frac{1}{a-1}\{zZ[a^n] - za^0 - Z[1]\}$$

$$= \frac{1}{a-1}\left\{z\frac{z}{z-a} - z - \frac{z}{z-1}\right\}$$

$$= \frac{z}{a-1} \cdot \frac{z(z-1) - (z-a)(a-1) - (z-a)}{(z-a)(z-1)}$$

$$= \frac{z}{a-1} \cdot \frac{z^2 - z - z^2 + az + z - a + z + a}{(z-a)(z-1)}$$

$$= \frac{z^2}{(z-a)(z-1)}. \tag{5.2.23}$$

The benefit of the summation property of the z transform should be apparent.

Multiplication by i

Given that $F(z)$ is the z transform of $f(i)$, we seek the transform of $if(i)$.

$$Z[if(i)] = \sum_{i=0}^{\infty} if(i)z^{-i} = -z\sum_{i=0}^{\infty} f(i)(-iz^{-i-1})$$

$$= -z\sum_{i=0}^{\infty} f(i)\frac{d}{dz}(z^{-i}). \tag{5.2.24}$$

Since both differentiation and summation are linear operations, their order can be interchanged to obtain

$$Z[if(i)] = -z\frac{d}{dz}\sum_{i=0}^{\infty} f(i)z^{-i} = -z\frac{d}{dz}[F(z)]. \tag{5.2.25}$$

Multiplication by i^2 and higher integer powers of i can of course be obtained by successive application of (5.2.25). Thus,

$$Z[ia^i] = -z\frac{d}{dz}Z[a^i] = -z\frac{d}{dz}\left[\frac{z}{z-a}\right] = -z\frac{(z-a)-z}{(z-a)^2} = \frac{az}{(z-a)^2}$$

and

$$Z[i^2a^i] = -z\frac{d}{dz}Z[ia^i] = -z\frac{d}{dz}\left[\frac{az}{(z-a)^2}\right]$$

$$= -z\frac{(z-a)^2a - 2az(z-a)}{(z-a)^4} = \frac{az(z+a)}{(z-a)^3}.$$

Power Weighting (Complex Translation)

For $F(z)$ the z transform of $f(i)$ and a a constant, we wish to find the z transform of the power weighted function $a^if(i)$. We proceed as follows:

$$Z[a^if(i)] = \sum_{i=0}^{\infty} a^if(i)z^{-i} = \sum_{i=0}^{\infty} f(i)\left(\frac{z}{a}\right)^{-i} = F\left(\frac{z}{a}\right). \qquad (5.2.26)$$

Thus, the z transform of $a^if(i)$ is found by substituting z/a for z in $F(z)$. For example, the z transform of b^ia^i, where a and b are constants, is simply

$$Z[b^ia^i] = \frac{z/a}{z/a - b} = \frac{z}{z - ab}. \qquad (5.2.27)$$

This result can be easily verified by direct application of the transformation formula.

Partial Differentiation

Let $f(i, p)$ be a function of discrete time i and some continuous parameter p, and let $F(z, p)$ be the z transform of $f(i, p)$. Then, the z transform of the partial derivative of $f(i, p)$ with respect to p is found as follows:

$$Z\left[\frac{\partial f(i, p)}{\partial p}\right] = \sum_{i=0}^{\infty} \frac{\partial f(i, p)}{\partial p}z^{-i} = \frac{\partial}{\partial p}\left[\sum_{i=0}^{\infty} f(i, p)z^{-i}\right] = \frac{\partial}{\partial p}[F(z, p)], \quad (5.2.28)$$

the derivation being dependent on the interchangeability of the order of summation and partial differentiation. As an example, consider the function a^i, where i is the index of discrete time and a can be thought of as a parameter. Now suppose we want to find the z transform of $\partial a^i/\partial a$. From (5.2.28), this is

$$Z\left[\frac{\partial a^i}{\partial a}\right] = \frac{\partial}{\partial a}[F(z, a)] = \frac{\partial}{\partial a}\left[\frac{z}{z - a}\right] = \frac{z}{(z - a)^2}.$$

An alternative approach would of course be to take the partial first and then transform it. The doubting reader may try this alternative to convince himself of the value of (5.2.28).

Convolution

In Chapter IV we defined the convolution of two functions of discrete time $f(i)$ and $g(i)$, whose values are zero for $i < 0$, as

$$y(i) = \sum_{k=0}^{i} f(i - k)g(k) = \sum_{k=0}^{i} f(k)g(i - k). \tag{5.2.29}$$

Because both $f(i)$ and $g(i)$ vanish for $i < 0$, these expressions may be put in the equivalent form

$$y(i) = \sum_{k=0}^{\infty} f(i - k)g(k) = \sum_{k=0}^{\infty} f(k)g(i - k) \tag{5.2.30}$$

which is more convenient for our current purposes. Note that all terms in the summation for $k > i$ will be zero. Let $F(z)$ and $G(z)$ be the z transforms of $f(i)$ and $g(i)$, respectively. The z transform of $y(i)$ can now be determined as follows:

$$Z[y(i)] = \sum_{i=0}^{\infty} \left[\sum_{k=0}^{\infty} f(i - k)g(k) \right] z^{-i} = \sum_{k=0}^{\infty} g(k) \sum_{i=0}^{\infty} f(i - k)z^{-i}$$

$$= \sum_{k=0}^{\infty} g(k) \cdot Z[f(i - k)]. \tag{5.2.31}$$

Using the formula for a real backward translation of a function which is zero for $i < 0$, $Z[f(i - k)]$ is $z^{-k}F(z)$, which when substituted in (5.2.31) gives

$$Z[y(i)] = \sum_{k=0}^{\infty} g(k)z^{-k}F(z) = F(z) \sum_{k=0}^{\infty} g(k)z^{-k} = F(z)G(z). \tag{5.2.32}$$

Thus, the z transform of the convolution of two functions whose values are zero for $i < 0$ is simply the product of the transforms of the functions. To illustrate, suppose

$$f(i) = \begin{cases} a^i & \text{for } i \geq 0, \\ 0 & \text{for } i < 0, \end{cases} \quad \text{and} \quad g(i) = \begin{cases} b^i & \text{for } i \geq 0, \\ 0 & \text{for } i < 0. \end{cases}$$

Then, the z transform of the convolution $y(i)$ is

$$Z[y(i)] = \frac{z}{z - a} \cdot \frac{z}{z - b} = \frac{z^2}{(z - a)(z - b)}.$$

Note that for general $f(i)$ and $g(i)$ defined as above with $b = 1$,

$$y(i) = \sum_{k=0}^{i} f(k)g(i - k) = \sum_{k=0}^{i} f(k)1^{i-k} = \sum_{k=0}^{i} f(k)$$

and

$$Z[y(i)] = F(z)G(z) = F(z)\frac{z}{z - 1}$$

which is identical to the previously developed formula for the transform of a summation (5.2.22). Thus, summation is a special case of convolution in which one of the functions involved is unity for $i \geq 0$ and zero for $i < 0$.

It will be found that the convolution formula is one of the most useful relationships for systems analysis and forms the basis of the "transfer function" which will be introduced in Chapter VI.

Initial Value Theorem

Given $F(z)$ as the z transform of $f(i)$, then if the limit of $F(z)$ as $z \to \infty$ exists,

$$\lim_{i \to 0} f(i) = \lim_{z \to \infty} F(z). \tag{5.2.33}$$

The proof of this theorem is readily apparent from the following:

$$F(z) = \sum_{i=0}^{\infty} f(i)z^{-i} = f(0) + \frac{f(1)}{z} + \frac{f(2)}{z^2} + \cdots .$$

Obviously, as $z \to \infty$ all terms vanish except $f(0)$. For $F(z) = z/(z - a)$, $\lim_{z \to \infty} F(z) = 1$, which we know to be correct since $z/(z - a)$ is the transform of a^i and $a^0 = 1$.

Although a systems designer would often know or be able to specify the initial value of a system output, there could still be many occasions where it might not be known directly. The initial value theorem allows determination of this initial value without having to transform back to the time domain to determine the actual function of discrete time.

Final Value Theorem

If $(z - 1)F(z)$, where $F(z)$ is the z transform of $f(i)$, is analytic for $z \geq 1$ (i.e., all derivatives of $(z - 1)F(z)$ exist for all $z \geq 1$), then

$$\lim_{i \to \infty} f(i) = \lim_{z \to 1} (z - 1)F(z). \tag{5.2.34}$$

The proof follows. We first express the transform as

$$Z[f(i)] = \lim_{k \to \infty} \sum_{i=0}^{k} f(i)z^{-i}.$$

Therefore,

$$Z[f(i + 1) - f(i)] = \lim_{k \to \infty} \sum_{i=0}^{k} [f(i + 1) - f(i)]z^{-i}.$$

This is of course the transform of the first forward difference of $f(i)$, which has been shown to be $(z - 1)F(z) - zf(0)$. Thus

$$(z - 1)F(z) - zf(0) = \lim_{k \to \infty} \sum_{i=0}^{k} [f(i + 1) - f(i)]z^{-i}.$$

Now let $z \to 1$ on both sides of the equation. This yields

$$\lim_{z \to 1} (z - 1)F(z) - f(0) = \lim_{k \to \infty} \sum_{i=0}^{k} [f(i + 1) - f(i)]$$

$$= \lim_{k \to \infty} \{[f(1) - f(0)] + [f(2) - f(1)] + \cdots$$

$$+ [f(k + 1) - f(k)]\}.$$

Note that all $f(i)$ for $i = 1$ through k appear with both a plus and minus sign, so cancel, leaving

$$\lim_{z \to 1} (z - 1)F(z) - f(0) = \lim_{k \to \infty} [f(k + 1) - f(0)] = \lim_{k \to \infty} f(k + 1) - f(0).$$

Canceling the $f(0)$ terms on each side, noting that in expressing the limit there is no difference between the arguments $k + 1$ and k, and changing to index i, we obtain

$$\lim_{i \to \infty} f(i) = \lim_{z \to 1} (z - 1)F(z),$$

the required form. It must be emphasized, however, that this theorem depends on the existence of the limit $(z - 1)F(z)$ as $z \to 1$.

For example, consider the function $f(i) = \sum_{k=0}^{i} a^k$, whose transform has been shown to be

$$F(z) = \frac{z^2}{(z - 1)(z - a)}.$$

By the final value theorem,

$$\lim_{i \to \infty} f(i) = \lim_{z \to 1} (z - 1) \frac{z^2}{(z - a)(z - 1)} = \lim_{z \to 1} \frac{z^2}{(z - a)} = \frac{1}{1 - a},$$

if $F(z)$ is analytic for $z \to 1$. Note that $F(z)$ is unbounded at $z = 1$ and $z = a$. In terms of complex variable theory, $F(z)$ is said to have " poles " at 1 and a. Obviously $F(z)$ is not analytic at these two points. The multiplication by $(z - 1)$, however, permits the theorem to hold even though $F(z)$ has a single pole at $z = 1$.

Now let us examine the pole at a. From standard summation,

$$f(i) = \left. \frac{a^k}{a - 1} \right|_0^{i+1} = \frac{a^{i+1} - 1}{a - 1}.$$

For $|a| < 1$,

$$\lim_{i \to \infty} f(i) = 1/(1 - a),$$

which is the value obtained from the final value theorem. If, however, $a > 1$, then as i gets very large $f(i)$ approaches $a^{i+1}/(a - 1)$ which continues to increase without bound as $i \to \infty$. Under these conditions, the theorem does not hold. For $a = 1$, $f(i) = \sum_{k=0}^{i} 1 = i + 1$, which also increases without bound as $i \to \infty$ and again the theorem does not hold.

In summary, the final value theorem may be used to find $\lim_{i \to \infty} f(i)$ if the transform $F(z)$ is analytic (has no poles) on or outside of the unit circle in the z plane with the exception that a single pole may exist at $z = 1$. Pole locations should always be checked before use of the theorem since erroneous results can be obtained if the conditions for application do not apply. The designer must also be wary of periodic functions which may approach a sinusoid with constant frequency, phase, and amplitude. Obviously this condition cannot be described by the final value theorem. In spite of these limitations, however, the final value theorem can be extremely useful in providing the designer with a knowledge of the limiting properties of his system output (we later refer to this as " steady-state " output) directly from the z transform of the output without the necessity of first converting back to the time domain.

5.3 Tables

A number of authors have collected transform pairs and properties of transforms into tables. Several of these have been published, among which are those by Azeltine (1958), Ragazzini and Zadah (1955), Ragazzini and Franklin (1958), and Truxal (1958). One of the most useful tables from the point of view of the systems engineer is the one compiled by Beightler et al. (1961). All of these tables, however, mix the general properties of the z transform with transforms of specific functions. A more logical and, it is felt, useful approach is taken by Cadzow and Martens (1970) who summarize the general properties in one table and list in a separate table some of the more common specific transforms. We adopt this latter procedure here. Table 5.3.1 summarizes the eleven properties of the z transform discussed in the previous section. Table 5.3.2 presents specific transform pairs including many of the more commonly occurring functions and example implementations of some of the previously listed properties.

Tables such as these are often more useful in performing the inverse transformation back to the i domain. Inverse transformation will be discussed in the next chapter.

Table 5.3.1

Properties of the z Transform

Property	Time function	z Transform
Linearity	$af(i) + bg(i)$	$aF(z) + bG(z)$
Real forward translation	$f(i + k)$	$z^k F(z) - \sum_{i=0}^{k-1} f(i) z^{k-i}$
Real backward translation	$f(i - k)$	$z^{-k} F(z)$, for $f(i) = 0$ for $i < 0$
Forward differencing	$\Delta^k f(i)$	$(z-1)^k F(z)$ $+ \sum_{i=0}^{k-1} f(i) \left[\sum_{p=0}^{k-1-i} (-1)^{p+1} C_p{}^k z^{k-p-i} \right]$
Backward differencing	$\nabla^k f(i)$	$\left(\dfrac{z-1}{z} \right)^k F(z)$, for $f(i) = 0$ for $i < 0$
Summation	$\sum_{i=0}^{n} f(i)$	$\dfrac{z}{z-1} F(z)$
Multiplication by i	$if(i)$	$-z \dfrac{d}{dz} [F(z)]$
Power weighting (complex translation)	$a^i f(i)$	$F(z/a)$
Partial differentiation	$\dfrac{\partial}{\partial p} [f(i, p)]$	$\dfrac{\partial}{\partial p} [F(z, p)]$
Convolution	$\sum_{k=0}^{i} f(k) g(i - k)$	$F(z) \cdot G(z)$ for $f(i) = 0$ for $i < 0$ and $g(i) = 0$ for $i < 0$
Initial value theorem	$\lim_{i \to 0} f(i) = \lim_{z \to \infty} F(z)$	
Final value theorem	$\lim_{i \to \infty} f(i) = \lim_{z \to 1} (z - 1)F(z)$ if $(z - 1)F(z)$ is analytic for $z \geq 1$	

Table 5.3.2^a

Selected One-Sided z-Transform Pairs

Item number	$f(i)$	$F(z)$
1	$f(i) = \begin{cases} a & \text{for} \quad i = k \\ 0 & \text{for} \quad i \neq k \end{cases}$	az^{-k}
2	1	$\dfrac{z}{z-1}$
3	$u(i-k)$, i.e., unit step at $i = k$, $f(i) = \begin{cases} 0 & \text{for} \quad i < k \\ 1 & \text{for} \quad i \geq k \end{cases}$	$z^{-k} \cdot \dfrac{z}{z-1} = \dfrac{z^{1-k}}{z-1}$
4	$f(i+1)$	$zF(z) - zf(0)$
5	$f(i+2)$	$z^2 F(z) - z^2 f(0) - zf(1)$
6	$f(i-1)$	$z^{-1} F(z)$
7	$f(i-2)$	$z^{-2} F(z)$
8	$\Delta f(i)$	$(z-1)F(z) - zf(0)$
9	$\Delta^2 f(i)$	$(z-1)^2 F(z) + (-z^2 + 2z)f(0) - zf(1)$
10	$\nabla f(i)$	$\dfrac{z-1}{z} F(z)$
11	$\nabla^2 f(i)$	$\left(\dfrac{z-1}{z}\right)^2 F(z)$
12	i	$\dfrac{z}{(z-1)^2}$
13	i^2	$\dfrac{z(z+1)}{(z-1)^3}$
14	i^3	$\dfrac{z(z^2 + 4z + 1)}{(z-1)^4}$
15	a^i	$\dfrac{z}{z-a}$
16	$a^i f(i)$	$F(z/a)$

^{a} Many of the items in this table were adapted with permission of the editor from Beightler *et al.* (1961).

Table 5.3.2   (*Continued*)

Item number	$f(i)$	$F(z)$
17	ia^i	$\dfrac{az}{(z-a)^2}$
18	$i^2 a^i$	$\dfrac{az(z+a)}{(z-a)^3}$
19	$i^3 a^i$	$\dfrac{az(z^2 + az + a^2)}{(z-a)^4}$
20	$\sin ai$	$\dfrac{z \sin a}{z^2 - 2z \cos a + 1}$
21	$\cos ai$	$\dfrac{z^2 - z \cos a}{z^2 - 2z \cos a + 1}$
22	$\sinh ai$	$\dfrac{z \sinh a}{z^2 - 2z \cosh a + 1}$
23	$\cosh ai$	$\dfrac{z^2 - z \cosh a}{z^2 - 2z \cosh a + 1}$
24	$b^i \sin ai$	$\dfrac{bz \sin a}{z^2 - 2bz \cos a + b^2}$
25	$b^i \cos ai$	$\dfrac{z^2 - bz \cos a}{z^2 - 2bz \cos a + b^2}$
26	$\dbinom{i+k}{k} a^i$	$\left(\dfrac{z}{z-a}\right)^{k+1}$
27	$\left(\dfrac{1}{a^{k-1}}\right)\dbinom{i}{k-1} a^i$	$\dfrac{z}{(z-a)^k}$
28	$\dbinom{b}{i} c^i a^{b-i}$	$\left(\dfrac{az+c}{z}\right)^b$
29	$\dfrac{a^i}{i}$	$\ln z - \ln(z-a)$
30	$\dfrac{a^i}{i!}$	$e^{a/z}$
31	$\dfrac{(\ln a)^i}{i!}$	$a^{1/z}$

5.4 Transformation of Linear Difference Equations

The z transform offers a useful alternative to classical methods for the solution of linear difference equations. Furthermore, after familiarity is gained with the transforms which correspond to common system elements and relationships, much design and analysis can be accomplished in the "z domain," i.e., with the transforms themselves without the necessity of converting back to the time domain.

The objective of transforming a linear difference equation involving the dependent variable $x(i)$ to the z domain is to permit solution of the transformed equation for the transform of the dependent variable $X(z)$. Once this is accomplished, inverse transformation back to the time domain yields the desired solution of the original difference equation, i.e., an expression for $x(i)$ as a function of discrete time i and the applicable system parameters. As will be seen, transformation of a linear difference equation results in an algebraic equation, which facilitates solution for $X(z)$.

Transformation of a linear difference equation is accomplished by straightforward transformation of each term in the equation. The equation may be in any desired form, expanded, difference-operator, or a conglomerate. In general, expanded form is the most convenient to use. Manipulation of the resulting transformed equation to solve for $X(z)$ follows the rules of algebra. Inverse transformation to obtain $x(i)$ will be discussed in Chapter VI after methods of inverse transformation have been presented.

We now illustrate this procedure by means of three examples.

Example 1 Given the linear difference equation already solved by classical methods in Chapter IV, viz.,

$$x(i + 2) - 3x(i + 1) + 2x(i) = 4^i + 3i^2, \tag{5.4.1}$$

we now utilize the properties of the z transform to transform the equation to the z domain.

Let $Z[x(i)]$ be represented by $X(z)$. The left-hand side is transformed using the properties of real forward translation as follows:

$$
\begin{aligned}
Z[x(i &+ 2) - 3x(i + 1) + 2x(i)] \\
&= [z^2 X(z) - zx(1) - z^2 x(0)] - 3[z X(z) - zx(0)] + 2X(z) \\
&= (z^2 - 3z + 2)X(z) - zx(1) - (z^2 - 3z)x(0). \tag{5.4.2}
\end{aligned}
$$

Transformation of the forcing function is easily accomplished using items 15 and 13 from Table 5.3.2 and the linear property of the z transform. Thus

$$Z[4^i + 3i^2] = \frac{z}{z - 4} + \frac{3z(z + 1)}{(z - 1)^3}. \tag{5.4.3}$$

The z transform of the original difference equation is, therefore,

$$(z^2 - 3z + 2)X(z) - zx(1) - (z^2 - 3z)x(0) = \frac{z}{z - 4} + \frac{3z(z + 1)}{(z - 1)^3}. \quad (5.4.4)$$

Note that the transformed equation is an algebraic equation in z.

The transformed equation is easily solved for $X(z)$ to obtain:

$$X(z) = \frac{\dfrac{z}{z - 4} + \dfrac{3z(z + 1)}{(z - 1)^3}}{z^2 - 3z + 2} + \frac{zx(1) + (z^2 - 3z)x(0)}{z^2 - 3z + 2} \quad (5.4.5)$$

which could be further simplified if desired.

Note that $X(z)$ is a function of two categories of terms. The first, which are grouped into the first fraction on the right-hand side, are those representing the forcing functions in the original equation. The other, grouped into the second fraction on the right-hand side, are those involving the initial condition terms resulting from the transformation. Accordingly, the first fraction is called the *forcing function transform* or the *system transform* and the second fraction is called the *initial condition transform*.

It should be further noted that the coefficient of $X(z)$ in (5.4.4), which is of course the denominator of both transforms, is the same polynomial that formed the characteristic equation when this same problem was solved by classical methods in Chapter IV.

Example 2 *Proportional-Plus-Difference Process Controller* The expanded form of the system difference equation for the proportional-plus-difference controller has been given in (4.2.49) as

$$m(i + 1) - (1 - K - L)m(i) - Lm(i - 1)$$
$$= KD - (K + L)\varepsilon(i) + L\varepsilon(i - 1) + p(i). \quad (5.4.6)$$

In discussing this controller in Chapter IV, we stated that although the presence of terms involving arguments less than i did not affect the formulation of the characteristic equation, it was still desirable from the standpoint of finding the particular solution to increment all discrete-time arguments so that the lowest remaining value was i itself. A similar situation exists when z transform methods are used, as may be seen by the following.

Let $M(z)$, $E(z)$, and $P(z)$ represent the z transforms of $m(i)$, $\varepsilon(i)$, and $p(i)$, respectively. Using the appropriate formulas for real forward and backward translation, term-by-term transformation of (5.4.6) yields

$$[zM(z) - zm(0)] - (1 - K - L)M(z) - Lz^{-1}M(z)$$

$$= KD\frac{z}{z - 1} - (K + L)E(z) + Lz^{-1}E(z) + P(z). \quad (5.4.7)$$

Had we first, however, incremented the arguments in (5.4.6) by one to obtain

$$m(i + 2) - (1 - K - L)m(i + 1) - Lm(i)$$
$$= KD - (K + L)\varepsilon(i + 1) + L\varepsilon(i) + p(i + 1), \quad (5.4.8)$$

transformation would have yielded

$$[z^2 M(z) - z^2 m(0) - zm(1)] - (1 - K - L)[zM(z) - zm(0)] - LM(z)$$

$$= KD \frac{z}{z - 1} - (K + L)[zE(z) - z\varepsilon(0)] + LE(z) + [zP(z) - zp(0)]. \quad (5.4.9)$$

Obviously, (5.4.7) and (5.4.9) are not the same. Their differences can be analyzed more easily if we multiply (5.4.7) by z to obtain

$$[z^2 M(z) - z^2 m(0)] - (1 - K - L)zM(z) - LM(z)$$

$$= KD \frac{z^2}{z - 1} - (K + L)zE(z) + LE(z) + zP(z). \quad (5.4.10)$$

The discrepancies involve the absence in (5.4.10) of the initial condition terms $zm(1)$, $zm(0)$, $z\varepsilon(0)$, and $zp(0)$ which are present in (5.4.9) and the z^2 multiplier in the KD term in (5.4.10) as opposed to z in (5.4.9).

It so happens that (5.4.9) is correct, but (5.4.10) lacks proper consideration of boundary terms. This situation arose when we chose to ignore the initial condition terms for negative i which are associated with real backward translation. Had we used the complete formula, (5.2.10), we would have obtained

$$[zM(z) - zm(0)] - (1 - K - L)M(z) - L[z^{-1}M(z) + m(-1)]$$

$$= KD \frac{z}{z - 1} - (K + L)E(z) + L[z^{-1}E(z) + \varepsilon(-1)] + P(z) \quad (5.4.11)$$

which, when multiplied by z, gives

$$[z^2 M(z) - z^2 m(0)] - (1 - K - L)zM(z) - L[M(z) + zm(-1)]$$

$$= KD \frac{z^2}{z - 1} - (K + L)zE(z) + L[E(z) + z\varepsilon(-1)] + zP(z). \quad (5.4.12)$$

Although it is far from obvious at this point, it can be shown that the solution of (5.4.12) is identical to that of (5.4.9) except that the effective time of system start-up is modeled as $i = -1$ instead of $i = 0$. Thus, the effect of working directly with equations containing functions with arguments less than i is to shift the effective time origin to the point on the i scale corresponding to the lowest value of discrete-time argument present, which could be a source of confusion in interpreting results. In addition, if one omits any of the appropriate boundary conditions involving negative values of i, erroneous answers will be obtained. Likewise, incrementing such that the lowest argument present is greater than i has the effect of shifting the effective start-up time in the other direction as well as increasing the solution effort required through adding unneeded boundary terms. Thus, it is highly recommended that regardless of the form of the equation and no matter what solution procedure is to be applied, all time arguments be incremented by whatever amount is necessary to make the lowest resulting value equal to i.

Example 3 *Proportional-Plus-Difference-Plus-Summation Process Controller*   One of the advantages of using the z transform in the solution of linear difference equations is in its flexibility of application. It can be applied directly to any linear difference equation regardless of the form in which the equation is expressed. To illustrate, consider the system equation for the proportional-plus-difference-plus-summation process controller in its operator form. This may be derived from combining (2.3.2), (2.3.17), and (2.3.29) to obtain

$$\Delta m(i) = K[D - m(i) - \varepsilon(i)] + L\,\Delta[D - m(i-1) - \varepsilon(i-1)]$$
$$+ N\sum_{n=0}^{i}[D - m(n) - \varepsilon(n)] + p(i). \quad (5.4.13)$$

As previously recommended, we increment the discrete-time argument by one so that i is the lowest value present and expand the bracketed terms. Thus

$$\Delta m(i+1) = KD - Km(i+1) - K\varepsilon(i+1) - L\,\Delta m(i) - L\,\Delta\varepsilon(i)$$
$$+ N\sum_{n=0}^{i+1}D - N\sum_{n=0}^{i+1}m(n) - N\sum_{n=0}^{i+1}\varepsilon(n) + p(i+1), \quad (5.4.14)$$

which we transform directly using the properties of the z transform involving translation, differencing, and summation to obtain

$$z(z-1)M(z) - z(z-1)m(0) - zm(1)$$

$$= KD\frac{z}{z-1} - K[zM(z) - zm(0)] - K[zE(z) - z\varepsilon(0)]$$

$$- L[(z-1)M(z) - zm(0)] - L[(z-1)E(z) - z\varepsilon(0)]$$

$$+ ND\left[z\frac{z}{z-1}\cdot\frac{z}{z-1} - z\right] - N\left[z\frac{z}{z-1}M(z) - zm(0)\right]$$

$$- N\left[z\frac{z}{z-1}E(z) - z\varepsilon(0)\right] + zP(z) - zp(0). \quad (5.4.15)$$

All these steps are straightforward, but nevertheless some explanation of the transformation of $\Delta m(i+1)$ and of the summation terms is in order.

$\Delta m(i + 1)$ involves both a first forward difference and a real forward translation. Similarly, the summation runs from 0 to $i + 1$. This represents a one-period forward translation of the usual summation, the formula for which is based on limits of 0 to i. Therefore, it is necessary both to multiply the z transform of the function involved by $z/(z - 1)$ to represent the summation through i and then to invoke the formula for real forward translation with an additional multiplication by z and subtraction of initial condition terms. Specifically,

$$Z[\Delta m(i + 1)] = zZ[\Delta m(i)] - z\,\Delta m(0)$$
$$= z\{(z - 1)Z[m(i)] - zm(0)\} - z[m(1) - m(0)]$$
$$= z(z - 1)M(z) - z^2 m(0) - zm(1) + zm(0)$$
$$= z(z - 1)M(z) - z(z - 1)m(0) - zm(1), \qquad (5.4.16)$$

$$Z\left[N\sum_{n=0}^{i+1} D\right] = zZ\left[N\sum_{n=0}^{i} D\right] - zN\sum_{n=0}^{0} D = z\,\frac{z}{z - 1}\,Z[ND] - zND$$

$$= z\,\frac{z}{z - 1}\,ND\,\frac{z}{z - 1} - zND = ND\left[z\cdot\frac{z}{z - 1}\cdot\frac{z}{z - 1} - z\right],$$

$$(5.4.17)$$

and

$$Z\left[\sum_{n=0}^{i+1} m(n)\right] = zZ\left[N\sum_{n=0}^{i} m(n)\right] - zN\sum_{n=0}^{0} m(n)$$

$$= z\,\frac{z}{z - 1}\,Z[Nm(i)] - zNm(0)$$

$$= z\,\frac{z}{z - 1}\,NM(z) - zNm(0)$$

$$= N\left[z\,\frac{z}{z - 1}\,M(z) - zm(0)\right]. \qquad (5.4.18)$$

The summation involving $\varepsilon(i)$ is handled exactly as the one involving $m(i)$.

To solve (5.4.15) for $M(z)$, we first group terms to obtain

$$\left[z(z - 1) + Kz + L(z - 1) + N\,\frac{z^2}{z - 1}\right]M(z)$$

$$= \left[K\,\frac{z}{z - 1} + N\,\frac{z^3}{(z - 1)^2} - Nz\right]D + \left[-Kz - L(z - 1) - N\,\frac{z^2}{z - 1}\right]E(z)$$
$$+ zP(z) + [z(z - 1) + Kz + Lz + Nz]m(0) + zm(1)$$
$$+ [Kz + Lz + Nz]\varepsilon(0) - zp(0). \qquad (5.4.19)$$

Because of the presence of the term corresponding to the transform of the summation, the coefficient of the independent variable $M(z)$ is no longer a simple polynomial. However, by putting all terms in the coefficient over the common denominator $z - 1$ and expressing the resulting numerator as a polynomial in z, we obtain

$$\frac{z^3 - (2 - K - L - N)z^2 + (1 - K - 2L)z + L}{z - 1}, \tag{5.4.20}$$

which in general is a reasonably convenient form for such complicated cases. Furthermore, comparison of (5.4.20) with (4.2.59), the expanded form of the system difference equation for the proportional-plus-difference-plus-summation controller derived in Section 4.2, shows that the numerator of (5.4.20) is of exactly the same form as the characteristic equation of (4.2.29). Thus, no matter what form the difference equation is in when it is transformed, if all the terms in the coefficient of the independent variable are put over a common denominator, the resulting numerator will be identical to the characteristic equation found by classical procedures. Therefore, the coefficient of the independent variable contains all the information about the nature of the system that the characteristic equation contains. In Example 1, where the coefficient of $X(z)$ was the simple polynomial $z^2 - 3z + 2$, it and the characteristic equation contained the same system information. In our current example, however, because of the $z - 1$ in the denominator, the coefficient of $M(z)$ contains more information than the characteristic equation; in fact, it completely describes the inherent nature of the system. More will be said about this in Section 7 of Chapter VI under the heading of Transfer Functions.

   To complete the solution of (5.4.15) for $M(z)$, we of course divide (5.4.19) by (5.4.20), which actually means multiplying by $z - 1$ and dividing by the numerator polynomial. Letting $H(z)$ represent the numerator of (5.4.20) to facilitate presentation, this results in

$$M(z) = \frac{(K + 2N)z^2 - (K + N)z}{(z - 1)H(z)} D + \frac{-(K + L + N)z^2 + (K + 2L)z - L}{H(z)} E(z)$$

$$+ \frac{z^2 - z}{H(z)} P(z) + \frac{z^3 - (2 - K - L - N)z^2 + (1 - K - L - N)z}{H(z)} m(0)$$

$$+ \frac{z^2 - z}{H(z)} m(1) + \frac{(K + L + N)z^2 - (K + L + N)z}{H(z)} \varepsilon(0)$$

$$+ \frac{-z^2 + z}{H(z)} p(0). \tag{5.4.21}$$

As noted in Example 1, the terms on the right can be divided into two categories. The terms involving the forcing functions D, $E(z)$, and $P(z)$ comprise the forcing function transform or system transform. The terms involving the initial conditions $m(0)$, $m(1)$, $\varepsilon(0)$, and $p(0)$ comprise the initial condition transform.

5.5 The *z* Transform as a Probability Generating Function

Probability generating functions have long been used in the calculation of probabilities of values of random variables. When the random variable involved is restricted to the nonnegative integers, the z transform of the probability distribution function can serve as a probability generating function. In this section we show briefly how the z transform is used to generate probabilities and moments and then illustrate its use for this purpose in systems analysis with an example in the context of production and inventory control.

Properties

Let $f(i)$ represent the probability distribution function of the nonnegative, integer, random variable i. The z transform $F(z)$ may be written as

$$F(z) = f(0) + f(1)z^{-1} + f(2)z^{-2} + f(3)z^{-3} + \cdots. \tag{5.5.1}$$

Letting z increase without bound, which is equivalent to setting z^{-1} to zero, gives

$$F(z^{-1} = 0) = f(0), \tag{5.5.2}$$

which is the probability of i equaling zero. Now differentiate $F(z)$ with respect to z^{-1} to obtain

$$\frac{d}{dz^{-1}} F(z) = 1f(1) + 2f(2)z^{-1} + 3f(3)z^{-2} + 4f(4)z^{-3} + \cdots. \tag{5.5.3}$$

Setting z^{-1} equal to zero now yields

$$\frac{d}{dz^{-1}} F(z^{-1} = 0) = f(1). \tag{5.5.4}$$

The second derivative of $F(z)$ with respect to z^{-1},

$$\frac{d^2}{d(z^{-1})^2} F(z) = 2 \cdot 1f(2) + 3 \cdot 2f(3)z^{-1} + 4 \cdot 3f(4)z^{-2} + \cdots, \tag{5.5.5}$$

when evaluated at zero, gives

$$\frac{d^2}{d(z^{-1})^2} F(z^{-1} = 0) = 2! f(2),$$ (5.5.6)

so that

$$f(2) = \frac{1}{2!} \cdot \frac{d^2}{d(z^{-1})^2} F(z^{-1} = 0).$$ (5.5.7)

Continuing in this manner we find that in general $f(i)$, the probability of the random variable taking on the value i, is

$$f(i) = \frac{1}{i!} \cdot \frac{d^i}{d(z^{-1})^i} F(z^{-1} = 0).$$ (5.5.8)

Thus, if the $F(z)$ for an unknown $f(i)$ can be obtained, the probability of as many values of i as desired may be determined by successive differentiation of $F(z)$ without the necessity for transformation back to the i domain.

Generation of Moments

The z transform can also be used to generate the moments of distributions. Unfortunately, this process is somewhat cumbersome since it involves converting from factorial moments to power moments.

From (5.5.1), (5.5.3), and (5.5.5) it may be noted that $\hat{F}^{(k)}(z)$, the kth derivative of $F(z)$ with respect to z^{-1}, is in the form

$$\hat{F}^{(k)}(z) = \frac{d^k}{d(z^{-1})^k} F(z)$$

$$= k! f(k) + \frac{(k+1)!}{1!} f(k+1)z^{-1} + \frac{(k+2)!}{2!} f(k+2)z^{-2} + \cdots$$

$$= \sum_{i=k}^{\infty} i^{(k)} f(i) z^{-(i-k)},$$ (5.5.9)

where $i^{(k)}$ is the factorial function $i(i-1)\cdots(i-k+1)$. Since $i^{(k)} = 0$ for $k > i$, (5.5.9) can be written

$$\hat{F}^{(k)}(z) = \sum_{i=0}^{\infty} i^{(k)} f(i) z^{-(i-k)}.$$ (5.5.10)

Now if we let $z = 1$, we have

$$\hat{F}^{(k)}(1) = \sum_{i=0}^{\infty} i^{(k)} f(i),$$ (5.5.11)

which is recognized as the expected value of $i^{(k)}$, or, by definition, the kth factorial moment of $f(i)$.

Conversion to power moments follows directly the procedures set forth in Section 3.2 for conversion from factorial functions to power functions. Thus, for $k = 1$,

$$E(i) = E(i^{(1)}) = \hat{F}^{(1)}(1). \tag{5.5.12}$$

For $k = 2$,

$$E(i^2) = E(i^{(2)}) + E(i^{(1)}) = \hat{F}^{(2)}(1) + \hat{F}^{(1)}(1), \tag{5.5.13}$$

from which the variance of i, σ_i^2, is

$$\sigma_i^2 = E(i^2) - [E(i)]^2 = \hat{F}^{(2)}(1) + \hat{F}^{(1)}(1) - [\hat{F}^{(1)}(1)]^2. \tag{5.5.14}$$

Thus, these and other moments can also be determined without the necessity for transformation back to the i domain.

Compound Distributions

The transform approach can also be useful in determining compound distributions involving more than one random variable. Specifically, let r and s be two random variables with probability functions $g(r)$ and $h(s)$, respectively, and suppose we seek the distribution function $f(i)$, where $i = r + s$. From standard probability relationships, we know that

$$f(i) = \sum_{r=0}^{i} g(r)h(i - r) \tag{5.5.15}$$

which is the convolution of $g(r)$ and $h(s)$. Thus, $F(z)$, the z transform of $f(i)$, is readily found from

$$F(z) = G(z) \cdot H(z). \tag{5.5.16}$$

Once $F(z)$ is obtained, one can then proceed to find $f(i)$ by inverse transformation (to be discussed in the next chapter), or to calculate numerical values of $f(0), f(1)$, etc., as far as time and interest dictate, or to determine any moments of interest.

Use in Systems Analysis

The main benefits in systems work to be obtained from the z transform in its role as a probability generating function accrue not so much from its use in numerically calculating probabilities or even moments, but in providing a mechanism for describing moments or even distributions of important performance characteristics subject to random phenomena.

To illustrate, let us refer back to the production–inventory control system of Section 2.4 in which the production facility is characterized by a delay of τ periods. (See Eqs. (2.4.16) and (2.4.17).) A production order $c(i - \tau)$ received at the beginning of period $i - \tau$ will be filled by production which emerges at the end of period i. Thus if $\tau = 0$, the production is complete at the end of the period in which it was ordered. Further suppose that τ is a random variable with probability function $f(\tau)$. Then the expected production emerging from the factory at the end of period i is the sum of the products of each past production order multiplied by the probability that the delay experienced by that order would be such as to bring about its completion at the end of period i. For the order received at the beginning of period i, $c(i)$, to be finished at the end of the same period, τ is taken as zero. Thus we have the product $c(i)f(0)$. For the order received at the beginning of period $i - 1$ to be finished at the end of period i, the delay must be one period. Since $\tau = 1$ with probability $f(1)$, we have the product $c(i - 1)f(1)$. In general, the order received at the beginning of period $i - \tau$ must experience a delay of τ to be ready at the end of period i. Summing over $\tau = 0, 1, \ldots$ produces the expected production $\bar{a}(i)$ to be completed at the end of period i. Specifically,

$$\bar{a}(i) = \sum_{\tau=0}^{\infty} c(i - \tau)f(\tau). \tag{5.5.17}$$

Since $c(i - \tau) = 0$ for $\tau > i$, (5.5.17) can be written

$$\bar{a}(i) = \sum_{\tau=0}^{i} c(i - \tau)f(\tau), \tag{5.5.18}$$

which we recognize as the standard convolution formula. The z transform of $\bar{a}(i)$, $\bar{A}(z)$, is thus

$$\bar{A}(z) = C(z)F(z). \tag{5.5.19}$$

To illustrate, suppose τ has a Poisson distribution with parameter μ, i.e.,

$$f(\tau) = \frac{\mu^{\tau}}{\tau!} e^{-\mu}. \tag{5.5.20}$$

From item 30 in Table 5.3.2,

$$Z[f(\tau)] = F(z) = e^{-\mu}Z[\mu^{\tau}/\tau!] = e^{-\mu}e^{\mu/z} = e^{-\mu(1 - 1/z)}, \tag{5.5.21}$$

so that

$$\bar{A}(z) = e^{-\mu(1 - 1/z)}C(z). \tag{5.5.22}$$

This example involved the expected value of production. However, by similar procedures one could formulate expressions for variance or any other moment. Any of these expressions could then be used as components in the development of any desired system equation or performance measure. An extensive discussion of transform techniques in systems analysis and probability modeling is given by Giffin (1971).

This concludes our discussion of the z transform. We now turn our attention to the methodology associated with the inverse transformation and its application to the solution of linear difference equations with constant coefficients.

Exercises

5.1 Find the z transforms of the following functions of discrete time i:

a. i	$f(i)$	b. i	$f(i)$	c. i	$f(i)$
0	1	0	1	0	5
1	2	1	2	1	4
2	3	2	1	2	3
3	2	3	0	3	2
4	1	4	-1	≥ 4	0
≥ 5	0	≥ 5	0		

5.2 (a) through (c). For the $f(i)$ in 5.1, find the z transform of $f(i+3)$.

5.3 (a) through (c). For the $f(i)$ in 5.1, find the z transform of $\Delta^2 f(i)$.

5.4 Find the z transforms of the following functions of discrete time i given that all functions of i are zero for $i < 0$.

a. $ab^i + ce^{di}$ b. $\cos wi + 3i^3 a^i$ c. $\cos(wi + \theta)$
d. $\cos w(i+4)$ e. $\cos w(i-4)$ f. $\Delta^3 \cos wi$

g. $\sum_{k=0}^{i} \cos wk$ h. $i \cos wi$ i. $e^{2i} \cos wi$

j. $\sum_{k=0}^{i} \sin w(i-k) \cos k$ k. $1/(i-1)!$ l. $(i+2) \sin w(i+2)$

m. $\Delta^2[i \sin wi]$ n. $\sum_{k=0}^{i+2} (k+2) \sin w(i+2-k)$

o. $\sum_{n=0}^{i}\left[n \sum_{k=0}^{n} a^k \sin w(n-k) \right]$ p. $\dfrac{\partial}{\partial w}[i \cos wi]$

q. $\sum_{k=0}^{i}\left[(i-k) \sum_{n=0}^{k} na^n \right]$ r. $\sum_{n=0}^{i+1}\sum_{k=0}^{n} ka^k \cosh k$

5.5   Given that $Z[f(i)] = F(z)$, that $f(i) = 0$ for $i < 0$, and that all necessary initial conditions are known, find the z transform of the following functions of $f(i)$:

a. $f(i + 4)$ b. $\Delta^4 f(i)$ c. $\displaystyle\sum_{n=0}^{i}\left[\sum_{k=0}^{n} f(k)\right]$

d. $i^2 f(i)$ e. $i^3 f(i)$ f. $\dfrac{\partial^2 [f(i, x, y)]}{\partial x\, \partial y}$

g. $a^{2i} f(i)$ h. $if(i - 2)$ i. $\displaystyle\sum_{k=0}^{i-1} ka^k \frac{\partial}{\partial b} [\Delta^2 f(i, b)]$

5.6   (a) through (i).   Transform the difference equation in the corresponding part of Exercise 4.2 and solve for $X(z)$, the z transform of $x(i)$ in terms of $F(z)$, the z transform of $f(i)$, the various parameters in the equation, and appropriate initial conditions.

5.7   (a) through (g).   Derive an expression for $M(z)$, the z transform of $m(i)$, in terms of z transforms of the three forcing functions D, $\varepsilon(i)$, and $p(i)$, the various system parameters, and the appropriate initial conditions for the system difference equations derived in the corresponding part of Exercise 2.3.

5.8 and **5.9**   The same as 5.7 except using the difference equations derived in the corresponding parts of Exercises 2.4 and 2.5, respectively. $M(z)$ is still the variable to be solved for.

5.10   a.   Transform the difference equation derived in Exercise 2.6c and solve for the z transform of Net Inventory in terms of the z transforms of Demand and Safety Stock, the various system parameters, and the appropriate initial conditions.

 b.   The same as part (a) except use the difference equation derived in 2.6d and solve for the z transform of Replenishment Order Placed.

 c.   The same as part (a) except use the difference equation derived in 2.6e and solve for the z transform of Replenishment Amount on Order.

5.11 through **5.14**   Transform the difference equations derived in the various parts of Exercises 2.7 through 2.10, respectively, and solve for the z transform of the dependent variable in terms of the z transforms of the forcing functions, the various system parameters, and the appropriate initial conditions.

5.15   (a) through (e).   Transform the difference equation derived in the corresponding part of Exercise 4.27 and solve for $M(z)$, the z transform of $m(i)$ in terms of the z transform of D, $\varepsilon(i)$, and $p(i)$, the various system parameters, and the appropriate initial conditions.

5.16 a. (1), (2), (3) and **b.** (1), (2), (3). Transform the difference equation involved in the corresponding part of Exercise 4.28 and solve for $Q(z)$, the z transform of the order quantity $q(i)$ in terms of the z transforms of the applicable forcing functions, system parameters, and initial conditions.

5.17 Find the probability generating functions of the following probability distribution functions of the nonnegative integer i:

a. $f(i) = \dfrac{n!}{i!\,(n-i)!}\, c^i (1-c)^{n-i}, \quad \begin{array}{l} i = 0, 1, \ldots, n \text{ (binomial} \\ 0 < c < 1 \quad\quad \text{distribution),} \end{array}$

b. $f(i) = \dfrac{(i+a-1)!}{i!\,(a-1)!}\, c^a (1-c)^i, \quad \begin{array}{l} i = 0, 1, \ldots \text{ (negative binomial} \\ 0 < c < 1 \quad\quad \text{distribution),} \end{array}$

c. $f(i) = \dfrac{1}{n+1}, \; i = 0, 1, \ldots, n$ (uniform distribution)

d. $f(i) = \dfrac{2i}{n(n-1)}, \; i = 0, 1, \ldots, n$ (triangular distribution)

e. $f(i) = \begin{cases} 0.2 & \text{for } i = 0 \\ 0.5 & \text{for } i = 1 \\ 0.3 & \text{for } i = 3 \end{cases}$

5.18 **(a)** through **(e)**. Assuming $n > 3$, $0 < c < 1$, and a a positive integer, use the generating functions found in 5.17 to derive expressions for $f(0)$ through $f(3)$. Check these against the original $f(i)$.

5.19 **(a)** through **(e)**. Use the generating functions found in 5.17 to derive the first, second, and third factorial moments of i.

5.20 **(a)** through **(e)**. Use the generating functions found in 5.17 to find the mean and variance of the variable i.

5.21 Each unit produced in a particular production facility must undergo a final inspection before being released. If it passes, it is sent to satisfy customer orders or to warehouse inventory. If it is rejected, it must go through rework which requires one week to complete, the same amount of time required for original production. Reworked items must again undergo inspection and will be released only if they pass. Items continue to recycle through rework until they pass inspection, taking one production period each time. If the probability of an item passing inspection is r, no matter how many times it has previously failed, find:

a. the probability distribution function for τ, the number of periods required before an item is released;

b. the probability generating function for this distribution;
c. the expected value of τ;
d. the variance of τ.

5.22 The number x of customer orders received for a given production period by the facility described in 5.21 is Poisson distributed with parameter μ. Considering both variations in the number of customer orders arriving per period and the number being reworked, find:

a. the probability generating function for the number of items released by the inspector per period;
b. the expected value of the number of items released per period;
c. the variance of the number of items released per period.

5.23 For the production facility described in 5.21, it is found that the number of customers sending in orders for each production period is Poisson distributed with parameter μ, but that the number of items y ordered per customer has a distribution given by

$$h(y) = \begin{cases} 0.2 & \text{for} \quad y = 1, \\ 0.5 & \text{for} \quad y = 2, \\ 0.3 & \text{for} \quad y = 3. \end{cases}$$

Find:

a. the probability generating function for the number of items released by the inspector per period;
b. the expected value of the number of items released per period;
c. the variance of the number of items released per period.

5.24 **(a)** through **(c)**. The same as 5.23 except that y, the number of items per customer order, follows a binomial distribution with parameters n and c (as in 5.17a).

Chapter VI | Inverse Transformation

In the last chapter we introduced the z transform and its properties and showed that it could be used to transform a linear difference equation with constant coefficients into an algebraic equation. Once the resulting algebraic equation is solved for the z transform of the dependent variable, however, it becomes necessary to transform back to the i domain if functions of discrete time are sought. Thus, even though insights regarding system performance can be gained by an experienced analyst from examination of functions in the z domain, in general one must perform the inverse transformation to obtain complete performance information.

The inverse transformation can be accomplished in any of a number of ways. These will now be discussed in turn.

6.1 Contour Integration

Given any function $F(z)$ in the z domain the corresponding function of i, $f(i)$, can be determined from

$$f(i) = \frac{1}{2\pi j} \oint F(z) z^{i-1} \, dz, \tag{6.1.1}$$

which is the contour integral over a sufficient expanse of the z plane to include all the poles of $F(z)z^{i-1}$.

Use of the formula requires reasonable mathematical sophistication, in particular a familiarity with the Cauchy residue theorem. For this reason engineers normally try to avoid direct use of this approach if at all possible. The reader is referred to a text on the theory of complex variables for a discussion of this integral and its use (Gardner and Barnes, 1942).

6.2   Table of Transform Pairs

The easiest procedure for the inverse transformation is use of a table of transform pairs, possibly in conjunction with a table of z transform properties. The tables given in Chapter V and the others referred to should be of definite assistance when a given $F(z)$ can be recognized as being of the same form as one of the table entries.

Since tables are necessarily limited in scope, it is necessary to have more general methods available to cover a wide variety of functions, especially when systems are represented by high order difference equations. In general, the analyst will use tables whenever possible, and resort to other methods when the tables fail to yield the form desired. Three such methods are presented in the following three sections.

6.3   Power Series Expansion

Since the z transform is a negative power series, expanding a function of z as a negative power series will provide a term-by-term development of the corresponding function of i since the coefficient of z^{-i} in the expansion is $f(i)$ by definition. Expansions can be found either from tables or by long division. Tables of power series are available in Jolley (1961), Mangulus (1965), Burington (1957), and elsewhere. Long division requires the determination of a sufficient number of terms in the power series by actual division to allow recognition of the form of the general expression for the coefficient of z^{-i} for any i.

As an example of power series expansion by long division, consider the function in the z domain

$$F(z) = Kz/(z - c),$$

where K and c are constants. We take advantage of the linear properties of the z transform to work with $z/(z - c)$ directly and multiply the resulting function of i by K later.

By long division:

$$
\begin{array}{r}
1 + cz^{-1} + c^2z^{-2} + c^3z^{-3} + \cdots \\
z - c \,\overline{\smash{\big)}\, z} \\
\underline{z - c} \\
c \\
\underline{c \quad\; - c^2z^{-1}} \\
c^2z^{-1} \\
\underline{c^2z^{-1} - c^3z^{-2}} \\
c^3z^{-2} \cdots
\end{array}
$$

etc.

It is apparent that the quotient is an infinite power series, the ith term of which is $c^i z^{-i}$. Even though this result is obvious, for illustrative purposes we will confirm it by inductive proof.

It is shown above that the form $c^i z^{-i}$ holds for the lowest value of i of interest to us, namely, $i = 0$. (It also shows it holds for $i = 1$, 2, and 3.) It remains to be proven that, given it holds for the general value i, it must then hold for $i + 1$. This proof proceeds as follows. Given that the ith term is $c^i z^{-i}$, the previous remainder must be $c^i z^{-i+1}$, i.e.,

$$
\begin{array}{r}
1 + \cdots + c^i z^{-i} \\
z - c \overline{)\; z } \\
\ddots \\
\overline{c^i z^{-i+1}}
\end{array}
$$

Multiplying the divisor $z - c$ by $c^i z^{-i}$ yields $c^i z^{-i+1} - c^{i+1} z^{-i}$, which, when subtracted from the previous remainder, leaves $c^{i+1} z^{-i}$ as the new remainder, i.e.,

$$
\begin{array}{r}
1 + \cdots + c^i z^{-i} \\
z - c \overline{)\; z } \\
\ddots \\
\overline{c^i z^{-i+1}} \\
c^i z^{-i+1} - c^{i+1} z^{-i} \\
\overline{c^{i+1} z^{-i}}
\end{array}
$$

Thus, the next term in the quotient must be $c^{i+1} z^{-(i+1)}$, which is the form desired in that $i + 1$ has replaced i in the form being tested. Therefore, if the assumed form holds for any i, it will hold for $i + 1$, and, since it holds for $i = 0$, it must hold for all $i \geq 0$.   Q.E.D.

Since the ith term is $c^i z^{-i}$,

$$
F(z) = K \sum_{i=0}^{\infty} c^i z^{-i}.
$$

Therefore,

$$
f(i) = K c^i,
$$

which we recognize as being correct from the fact that

$$
Z[c^i] = z/(z - c).
$$

Experience has shown that in cases in which the numerator of $F(z)$ is a polynomial, for example,

$$
F(z) = \frac{K_1 z + K_2}{z - c},
$$

it is usually easier to apply long division to each term separately and then combine the results after the various general forms for the coefficients of z^{-i} are recognized. Recognition is normally facilitated this way.

6.4   Maclaurin Series Expansion

The inverse transform can also be determined from a Maclaurin series expansion of $F(z)$ as follows:

(a)   Rewrite $F(z)$ as a function of $1/z = z^{-1}$. (For convenience, refer to the function in this form as $\hat{F}(z^{-1})$.)

(b)   Expand $\hat{F}(z^{-1})$ in a Taylor series expansion around $z^{-1} = 0$ (which, by definition, is the Maclaurin series) to obtain:

$$\hat{F}(z^{-1}) = \hat{F}(z^{-1})\Big|_{z^{-1}=0} + z^{-1}\hat{F}'(z^{-1})\Big|_{z^{-1}=0} + \frac{z^{-2}}{2!}\hat{F}''(z^{-1})\Big|_{z^{-1}=0}$$

$$+ \cdots + \frac{z^{-i}}{i!}\hat{F}^{(i)}(z^{-1})\Big|_{z^{-1}=0} + \cdots$$

$$= \hat{F}(0) + z^{-1}\hat{F}'(0) + \frac{z^{-2}}{2!}\hat{F}''(0) + \cdots + \frac{z^{-i}}{i!}\hat{F}^{(i)}(0) + \cdots, \qquad (6.4.1)$$

where $\hat{F}^{(i)}(z^{-1})$ represents the ith derivative of $\hat{F}(z^{-1})$ with respect to z^{-1}.

(c)   Recognize or derive an expression for the term involving the ith derivative of $\hat{F}(z^{-1})|_{z^{-1}=0}$ as a function of i and the appropriate system parameters. Then,

$$f(i) = \frac{1}{i!}\hat{F}^{(i)}(0). \qquad (6.4.2)$$

We illustrate with the same example as for the power series, i.e.,

$$F(z) = Kz/(z - c).$$

This is written as a function of $1/z$ by dividing both numerator and denominator by the highest power of z present in $F(z)$. In this case, we divide by z, which gives

$$\hat{F}(z^{-1}) = K\frac{1}{1 - cz^{-1}}.$$

For convenience, let $y = z^{-1}$ and write

$$\hat{F}(y) = K\frac{1}{1 - cy} = K(1 - cy)^{-1}$$

Since $f(i) = (1/i!)\hat{F}^{(i)}(0)$, we must find the successive derivatives of $\hat{F}(y)$ and evaluate them at $y = 0$. This results in the following:

$$\hat{F}(y) = K(1 - cy)^{-1}, \qquad\qquad\qquad\qquad \hat{F}(0) = K,$$
$$\hat{F}'(y) = K(-1)(1 - cy)^{-2}(-c) = Kc(1 - cy)^{-2}, \qquad \hat{F}'(0) = Kc,$$
$$\hat{F}''(y) = Kc(-2)(1 - cy)^{-3}(-c) = 2Kc^2(1 - cy)^{-3}, \qquad \hat{F}''(0) = 2Kc^2,$$
$$\hat{F}'''(y) = 2Kc(-3)(1 - cy)^{-4}(-c) = 3 \cdot 2Kc^3(1 - cy)^{-4}, \qquad \hat{F}'''(0) = 3!\,Kc^3.$$

It is apparent at this point that

$$\hat{F}^{(i)}(y) = i!\, Kc^i(1 - cy)^{-(i+1)}$$

and

$$\hat{F}^{(i)}(0) = i!\, Kc^i.$$

Again for illustrative purposes we confirm this form by induction. The above derivations show that the form $i!\,Kc^i$ holds for $i = 0$, the lowest value of interest. We must now show that if it holds for a given i it will also hold for $i + 1$. Given that

$$\hat{F}^{(i)}(y) = i!\, Kc^i(1 - cy)^{-(i+1)},$$

which means, of course, that $\hat{F}^{(i)}(0) = i!\, Kc^i$,

$$\frac{d}{dy}[\hat{F}^{(i)}(y)] = \hat{F}^{(i+1)}(y) = i!\, Kc^i[-(i + 1)](1 - cy)^{-(i+2)}(-c)$$

$$= (i + 1)!\, Kc^{i+1}(1 - cy)^{-(i+2)}.$$

This is the form desired since $i + 1$ replaces i in the form being tested. Thus the form assumed for $\hat{F}^{(i)}(y)$ and hence the form for $\hat{F}^{(i)}(0)$ hold for all $i \geq 0$

Q.E.D.

As a result,

$$f(i) = \frac{1}{i!}\, \hat{F}^{(i)}(0) = \frac{1}{i!}\, i!\, Kc^i = Kc^i.$$

A second example may be helpful. Consider

$$F(z) = \frac{z^2}{z^2 - 2z + 1},$$

from which

$$\hat{F}(z^{-1}) = \frac{1}{1 - 2z^{-1} + z^{-2}}.$$

Letting $y = 1/z$, we obtain

$$\hat{F}(y) = \frac{1}{1 - 2y + y^2} = (1 - y)^{-2},$$

which gives

$$\hat{F}(0) = 1.$$

The first derivative of $\hat{F}(y)$ with respect to y is

$$\hat{F}'(y) = -2(1 - y)^{-3}(-1) = 2(1 - y)^{-3},$$

from which it is seen that

$$\hat{F}'(0) = 2.$$

Further differentiation yields

$$\hat{F}''(y) = 3 \cdot 2(1 - y)^{-4}$$

which gives

$$\hat{F}''(0) = 3!$$

and, in general,

$$\hat{F}^{(i)}(y) = (i + 1)(1 - y)^{-(i+2)}$$

and

$$\hat{F}^{(i)}(0) = (i + 1)!,$$

which can be verified by induction. Thus,

$$f(i) = \frac{1}{i!} \hat{F}^{(i)}(0) = \frac{1}{i!}(i + 1)! = i + 1.$$

As in the case of a power series expansion, should the numerator of $F(z)$ be a polynomial, it is often more convenient to attempt to develop a Maclaurin series for each term in the numerator and combine the results afterward.

Although power series and Maclaurin series expansions worked well for the simple forms of $F(z)$ just considered, they both become quite cumbersome for even relatively complicated functions of z. An even more severe limitation in these methods is the difficulty experienced in recognizing the form of the general term (the ith term) in the quotient of the long division for the power series method and the form of the general derivative (the ith derivative) in the Maclaurin series expansion. Although luck and experience help, a more reliable method is required for general use. A method finding wide application is partial fraction expansion, which is discussed in the next section.

6.5 Partial Fraction Expansion

Partial fraction expansion can be used when $F(z)$ is in the form

$$F(z) = \frac{A(z)}{B(z)} = \frac{a_0 z^m + a_1 z^{m-1} + \cdots + a_{m-1} z}{z^n + b_1 z^{n-1} + \cdots + b_{n-1} z + b_n}, \qquad (6.5.1)$$

i.e., the ratio of two polynomials in z. Note that in this text $F(z)$ is expressed in such a way that the coefficient of the highest power of z in the denominator, z^n, is always unity. It will also be observed subsequently that the order of $B(z)$ will be equal to or greater than the order of $A(z)$, i.e., $n \geq m$. This is a necessity for a realizable system. This is due to the fact that where $m > n$, long division would yield terms in positive powers of z, which correspond mathematically to negative values of i and physically to performance of the system prior to its being turned on at $i = 0$. Since we are dealing exclusively with the one-sided transform, negative i are not defined in terms of the transform. Furthermore, no system constructed from physical components can exhibit predictable behavior in anticipation of being turned on, i.e., its performance is not describable prior to start-up. In addition, note that no constant term a_m appears in the numerator of (6.5.1). This is a very common occurrence, but it is by no means essential for physical realizability. We will use the form shown to facilitate presentation of the method and will discuss later the effects of a constant term or terms with negative powers of z in the numerator.

Although the form of the $F(z)$ for which this method can be used is restricted, it is found that many systems yield z transforms of this form, hence partial fraction expansion is one of the most widely used methods available for finding functions of discrete time for z transforms. A major contribution of the partial fraction expansion is to break $F(z)$ into smaller, easily recognized basic parts which can be found in simple tables of transform pairs for final inverse transformation.

Procedure

The partial fraction expansion proceeds as follows:

a. Put $F(z) = A(z)/B(z)$ in standard form, i.e., the coefficient of the highest power of z in $B(z)$ is made equal to unity.

For example,

$$F(z) = \frac{z^2 + 2z}{3z^2 - 4z - 7} = \frac{\frac{1}{3}z^2 + \frac{2}{3}z}{z^2 - \frac{4}{3}z - \frac{7}{3}},$$

or, alternatively,

$$F(z) = \frac{1}{3} \cdot \frac{z^2 + 2z}{z^2 - \frac{4}{3}z - \frac{7}{3}}.$$

In the second form one can work with the z terms only and then multiply the final result by $\frac{1}{3}$ afterwards if it is convenient to do so.

b.   Define: $\bar{F}(z) = z^{-1} F(z)$.

For the $F(z)$ above, this becomes

$$\bar{F}(z) = \frac{\frac{1}{3}z + \frac{2}{3}}{z^2 - \frac{4}{3}z - \frac{7}{3}}.$$

c.   Express the standardized $B(z)$ in factored form, i.e.,

$$B(z) = (z - z_1)(z - z_2) \cdots (z - z_n). \tag{6.5.2}$$

This is equivalent to finding the roots of the equation

$$B(z) = 0. \tag{6.5.3}$$

There must, of course, be exactly n roots (factors) if $B(z)$ is an nth-order polynomial. Note that finding the roots of $B(z) = 0$ is exactly the same as finding the roots of the characteristic equation in the classical solution of a linear difference equation. Methods for accomplishing this were discussed in Chapter IV.

d.   Expand $\bar{F}(z)$ into n terms as follows:

(1)   If all the roots of $B(z) = 0$ are single roots,

$$\bar{F}(z) = \frac{K_1}{z - z_1} + \frac{K_2}{z - z_2} + \cdots + \frac{K_n}{z - z_n}, \tag{6.5.4}$$

where the K's are constants whose values can be determined from

$$K_r = \lim_{z \to z_r} (z - z_r)\bar{F}(z), \tag{6.5.5}$$

where the subscript r refers to root r.

(2)   Each root of multiplicity p contributes p terms to the expansion with the form:

$$\frac{K_{r1}}{(z - z_r)^p} + \frac{K_{r2}}{(z - z_r)^{p-1}} + \cdots + \frac{K_{rp}}{z - z_r}, \tag{6.5.6}$$

where

$$K_{r1} = \lim_{z \to z_r} (z - z_r)^p \bar{F}(z), \tag{6.5.7}$$

$$K_{r2} = \lim_{z \to z_r} \left\{ \frac{1}{1!} \cdot \frac{d}{dz} [(z - z_r)^p \bar{F}(z)] \right\}, \tag{6.5.8}$$

$$K_{r3} = \lim_{z \to z_r} \left\{ \frac{1}{2!} \cdot \frac{d^2}{dz^2} [(z - z_r)^p \bar{F}(z)] \right\}, \tag{6.5.9}$$

and, in general,

$$K_{rp} = \lim_{z \to z_r} \left\{ \frac{1}{(p-1)!} \cdot \frac{d^{p-1}}{dz^{p-1}} [(z - z_r)^p \bar{F}(z)] \right\}. \tag{6.5.10}$$

(3) Conjugate complex pairs may be further manipulated from the form obtained in (6.5.4) for single roots or in (6.5.6) for multiple roots into a more useful form as follows:

Given the pair of roots $a \pm jb$, two terms appear in the expansion. These are

$$\frac{K_{r+}}{z - a - jb} + \frac{K_{r-}}{z - a + jb}. \tag{6.5.11}$$

Now,

$$K_{r+} = \lim_{z \to a + jb} \left[(z - a - jb) \frac{A'(z)}{(z - a - jb)(z - a + jb)B'(z)} \right]$$

$$= \lim_{z \to a + jb} \frac{A'(z)}{(z - a + jb)B'(z)}, \tag{6.5.12}$$

where

$$A'(z) = A(z)/z, \tag{6.5.13}$$

and

$$B'(z) = \frac{B(z)}{(z - a - jb)(z - a + bj)}. \tag{6.5.14}$$

Substitution of $a + jb$ for z in (6.5.12) yields

$$K_{r+} = \frac{A'(a + jb)}{(a + jb - a + jb)B'(a + jb)}$$

$$= \frac{\operatorname{Re} A'(a + jb) + j \operatorname{Im} A'(a + jb)}{(2jb)[\operatorname{Re} B'(a + jb) + j \operatorname{Im} B'(a + jb)]}, \tag{6.5.15}$$

where the symbols Re and Im are as previously defined. We now rationalize the denominator of K_{r+} by multiplying both numerator and denominator by the conjugate of the denominator to obtain:

$$K_{r+} = \frac{1}{2b} \cdot \frac{\text{Re } A'(a+jb) + j \text{ Im } A'(a+jb)}{j[\text{Re } B'(a+jb) + j \text{ Im } B'(a+jb)]}$$

$$\cdot \frac{-j[\text{Re } B'(a+jb) - j \text{ Im } B'(a+jb)]}{-j[\text{Re } B'(a+jb) - j \text{ Im } B'(a+jb)]}$$

$$= \frac{-\text{Re } A'(a+jb) \cdot \text{Im } B'(a+jb) + \text{Im } A'(a+jb) \cdot \text{Re } B'(a+jb)}{2b\{[\text{Re } B'(a+jb)]^2 + [\text{Im } B'(a+jb)]^2\}}$$

$$+ j \frac{-\text{Re } A'(a+jb) \cdot \text{Re } B'(a+jb) - \text{Im } A'(a+jb) \cdot \text{Im } B'(a+jb)}{2b\{[\text{Re } B'(a+jb)]^2 + [\text{Im } B'(a+jb)]^2\}}.$$

$$(6.5.16)$$

By the same procedure it may be shown that

$$K_{r-} = \frac{\text{Re } A'(a-jb) \cdot \text{Im } B'(a-jb) - \text{Im } A'(a-jb) \cdot \text{Re } B'(a-jb)}{2b\{[\text{Re } B'(a-jb)]^2 + [\text{Im } B'(a-jb)]^2\}}$$

$$- j \frac{-\text{Re } A'(a-jb) \cdot \text{Re } B'(a-jb) - \text{Im } A'(a-jb) \cdot \text{Im } B'(a-jb)}{2b\{[\text{Re } B'(a-jb)]^2 + [\text{Im } B'(a-jb)]^2\}}.$$

$$(6.5.17)$$

However, since for any linear function F of a complex variable $a+jb$, $F(a+jb)$ and $F(a-jb)$ are complex conjugates, Re $A'(a+jb)$ = Re $A'(a-jb)$, Re $B'(a+jb)$ = Re $B'(a-jb)$, Im $A'(a+jb)$ = $-$Im $A'(a-jb)$, and Im $B'(a+jb)$ = $-$Im $B'(a-jb)$. Substitution of the appropriate functions of $(a+jb)$ for those of $(a-jb)$ in (6.5.17) results in an expression for K_{r-} which is the complex conjugate of K_{r+} (verification is left to the reader). Therefore, the easiest way to find the pair of coefficients K_{r+} and K_{r-} is to evaluate one or the other from (6.5.16) or (6.5.17). The remaining coefficient is then the complex conjugate of the one calculated.

For future reference, let

$$s = \text{Re } K_{r+} = \text{Re } K_{r-} \qquad (6.5.18)$$

and

$$d = \text{Im } K_{r+} = \text{Im } K_{r-}, \qquad (6.5.19)$$

so that we can conveniently express K_{r+} and K_{r-}, respectively, as

$$K_{r+} = s + jd \qquad (6.5.20)$$

and

$$K_{r-} = s - jd. \qquad (6.5.21)$$

These formulas can be used directly for single pairs of complex or imaginary roots if desired. For multiple complex pairs one would have to begin with the appropriate formula for K_{rp}, etc., and repeat the above analysis. Pairs of K's for terms in like powers of $z - z_r$ will still be complex conjugates.

e. Multiply the expanded $\bar{F}(z)$ by z to recover $F(z)$. All terms will now have a Kz in the numerator assuming we started with an $F(z)$ as given in (6.5.1).

f. Inverse transformation.

(1) Terms corresponding to all single roots and to the pth term in the sequences for roots of multiplicity p are now of the form (see (6.5.4) and (6.5.6)):

$$Kz/(z - z_r) \tag{6.5.22}$$

or, for

$$z_r = c, \tag{6.5.23}$$

$$Kz/(z - c). \tag{6.5.24}$$

The inverse transform of such terms is known to be

$$K \cdot c^i. \tag{6.5.25}$$

(2) The next-to-last term in the sequence for roots of multiplicity p are now of the form (again see (6.5.6)):

$$\frac{Kz}{(z - z_r)^2} = \frac{Kc}{(z - c)^2}. \tag{6.5.26}$$

Long division produces

$$K(z^{-1} + 2cz^{-2} + 3c^2z^{-3} + \cdots + ic^{i-1}z^{-i} + \cdots).$$

Note that the z^{-0} term is missing, meaning that the complete transform is not present. This is quickly ascertained to make no difference since the general term is

$$Kic^{i-1}z^{-i}. \tag{6.5.27}$$

Thus the coefficient of z^{-0} is zero anyhow, and we may write

$$Kz/(z - c)^2 = K\sum_{i=0}^{\infty} ic^{i-1}z^{-i}. \tag{6.5.28}$$

Thus the inverse of these terms can be identified as being of the form

$$Kic^{i-1} \quad \text{or} \quad (K/c)ic^i. \tag{6.5.29}$$

(3)   The second-from-last terms in the sequences for roots of multiplicity p are now all of the form

$$Kz/(z - c)^3. \tag{6.5.30}$$

Long division produces

$$K\left[z^{-2} + 3cz^{-3} + 6c^2z^{-4} + \cdots + \left(\sum_{j=1}^{i-1} j\right)c^{i-2}z^{-i} + \cdots\right].$$

Since

$$\sum_{j=1}^{i-1} j = i(i-1)/2, \tag{6.5.31}$$

the general term may be written

$$K[i(i-1)/2]c^{i-2}z^{-i}. \tag{6.5.32}$$

Note that the coefficients of the missing z^{-1} and z^{-0} terms will both be zero because of the i and $i - 1$ factors present. Therefore, we may write

$$\frac{Kc}{(z-c)^3} = K\sum_{i=0}^{\infty} \frac{i(i-1)}{2}c^{i-2}z^{-i} \tag{6.5.33}$$

and the inverse transform of these terms is of the form

$$K\frac{i(i-1)}{2}c^{i-2} = \frac{K}{c^2} \cdot \frac{i(i-1)}{2}c^i. \tag{6.5.34}$$

(4)   In general, any term of the form

$$Kz/(z-c)^k \tag{6.5.35}$$

can be shown by long division to equal

$$\sum_{i=0}^{\infty} \frac{i!}{(i-k+1)!(k-1)!}c^{i-k+1}z^{-i},$$

so that the inverse is of the form

$$\frac{K}{c^{k-1}} \cdot \frac{i!}{(i-k+1)!(k-1)!}c^i. \tag{6.5.36}$$

Note that the transformations just accomplished correspond to item 27 in Table 5.3.2.

(5)   Conjugate complex pairs of roots can be further treated as follows: Each single pair is now of the form

$$\frac{(s+jd)z}{z-a-jb} + \frac{(s-jd)z}{z-a+jb}, \tag{6.5.37}$$

where expressions for s and d have been previously defined in (6.5.18) and (6.5.19), respectively. The inverse transform of these terms is, therefore,

$$(s + jd)(a + jb)^i + (s - jd)(a - jb)^i. \tag{6.5.38}$$

Now putting the roots in the form $a + jb = \mathrm{Re}^{j\theta}$ and $a - jb = \mathrm{Re}^{-j\theta}$, where $R = (a^2 + b^2)^{1/2}$ and $\theta = \tan^{-1} b/a$, allows us to write

$$(a \pm jb)^i = R^i e^{\pm j\theta i} = R^i \cos \theta i \pm jR^i \sin \theta i \tag{6.5.39}$$

and the pair becomes

$$
\begin{aligned}
R^i[(s + jd)&(\cos \theta i + j \sin \theta i)] + R^i[(s - jd)(\cos \theta i - j \sin \theta i)] \\
&= R^i[s \cos \theta i - d \sin \theta i + jd \cos \theta i + js \sin \theta i] \\
&\quad + R^i[s \cos \theta i - d \sin \theta i - jd \cos \theta i - js \sin \theta i] \\
&= 2R^i(s \cos \theta i - d \sin \theta i). \tag{6.5.40}
\end{aligned}
$$

Note that s and d each have a 2 in their denominators which will cancel the 2 above. It should be further noted that the two complex terms have been combined to produce two *real* oscillatory (sinusoidal) terms.

To illustrate, consider the pair of roots treated earlier in Section 4.2, viz., $-0.3 \pm j(0.4)$. We have shown that for this pair $R = 0.5$ and $\theta = 127°$. Thus, the inverse transform that would result were this a pair of roots of a given $B(z) = 0$, is

$$2(0.5)^i[s \cos 2.217i - d \sin 2.217i],$$

which is of the same form previously obtained by classical methods.

For multiple roots one would start with the appropriate inverse transform and follow the same procedure as above. In all cases, both a sine and cosine term would result weighted by R^i, where $R = (a^2 + b^2)^{1/2}$. Additional factors in powers and factorials of i would also be present as for any multiple root.

Inverse Transformation

Example 1 As our first example application of the partial fraction expansion, let us continue the inversion of

$$F(z) = \frac{z^2 + 2z}{3z^2 - 4z - 7}$$

begun earlier in this section. This had been put into the standard form

$$F(z) = \frac{\frac{1}{3}z^2 + \frac{2}{3}z}{z^2 - \frac{4}{3}z - \frac{7}{3}}$$

from which

$$\bar{F}(z) = \frac{F(z)}{z} = \frac{\frac{1}{3}z + \frac{2}{3}}{z^2 - \frac{4}{3}z - \frac{7}{3}}.$$

The denominator in factored form is found from the quadratic formula to be $(z - \frac{7}{3})(z + 1)$, so $\bar{F}(z)$ may be expressed as

$$\bar{F}(z) = \frac{\frac{1}{3}z + \frac{2}{3}}{(z - \frac{7}{3})(z + 1)} = \frac{K_1}{z - \frac{7}{3}} + \frac{K_2}{z + 1}.$$

The coefficients K_1 and K_2 are evaluated using (6.5.5) as follows:

$$K_1 = \lim_{z \to 7/3} \left(z - \frac{7}{3}\right) \frac{\frac{1}{3}z + \frac{2}{3}}{(z - \frac{7}{3})(z + 1)}$$

$$= \lim_{z \to 7/3} \frac{\frac{1}{3}z + \frac{2}{3}}{z + 1} = \frac{\frac{1}{3} \cdot \frac{7}{3} + \frac{2}{3}}{\frac{7}{3} + 1} = \frac{13}{30},$$

and, similarly,

$$K_2 = \lim_{z \to -1} \frac{\frac{1}{3}z + \frac{2}{3}}{z - \frac{7}{3}} = \frac{\frac{1}{3}(-1) + \frac{2}{3}}{-1 - \frac{7}{3}} = -\frac{1}{10}.$$

Thus,

$$\bar{F}(z) = \frac{13}{30} \cdot \frac{1}{z - \frac{7}{3}} - \frac{1}{10} \cdot \frac{1}{z + 1},$$

so that

$$F(z) = \frac{13}{30} \cdot \frac{z}{z - \frac{7}{3}} - \frac{1}{10} \cdot \frac{z}{z + 1}.$$

The inverse transforms of $z/(z - \frac{7}{3})$ and $z/(z + 1)$ are easily recognized to be $(\frac{7}{3})^i$ and $(-1)^i$, respectively. Should we per chance not immediately recognize these forms, they could be quickly found as item 15 in Table 5.3.2 or determined from Eqs. (6.5.22) and (6.5.25). In any event,

$$f(i) = \tfrac{13}{30}(\tfrac{7}{3})^i - \tfrac{1}{10}(-1)^i.$$

Example 2   *Real Roots in* $B(z) = 0$ For a more complicated example of inverse transformation based on partial fraction expansion, consider the difference equation

$$x(i + 2) - 3x(i + 1) + 2x(i) = 4^i + 3i^2 \qquad (6.5.41)$$

which was solved by classical methods in Chapter IV (see Example 1 for both undetermined coefficients and variation of parameters). For boundary conditions of $x(0) = 0$ and $x(1) = \frac{2}{3}$, the solution was found to be (see Eq. (4.4.5))

$$x(i) = \tfrac{1}{6} \cdot 4^i - i^3 - \tfrac{3}{2}i^2 - \tfrac{13}{2}i + \tfrac{55}{6} \cdot 2^i - \tfrac{56}{6}. \qquad (6.5.42)$$

Equation (6.5.41) was transformed in Section 5.4 to obtain the z transform of $X(z)$. This result, given (5.4.5), was

$$X(z) = \frac{z/(z-4) + 3z(z+1)/(z-1)^3}{z^2 - 3z + 2} + \frac{zx(1) + (z^2 - 3z)x(0)}{z^2 - 3z + 2}. \quad (6.5.43)$$

We will now take the inverse transformation of (6.5.43) and compare the result with $x(i)$ given by (6.5.42).

Although substitution of boundary conditions can be performed at any time, for this problem calculations are simplified by substituting at this point. This is due primarily to the fact that $x(0) = 0$ which considerably reduces the complexity of the initial condition transform. In many physical problems, the entire initial condition transform will vanish. Substitution for $x(0)$ and $x(1)$ yields

$$X(z) = \frac{z/(z-4) + 3z(z+1)/(z-1)^3}{z^2 - 3z + 2} + \frac{\frac{2}{3}z}{z^2 - 3z + 2}. \quad (6.5.44)$$

which in factored form is

$$X(z) = \frac{z/(z-4) + 3z(z+1)/(z-1)^3}{(z-2)(z-1)} + \frac{\frac{2}{3}z}{(z-2)(z-1)}. \quad (6.5.45)$$

There are several possible approaches to grouping the terms in $X(z)$ for purposes of partial fraction expansion. The grouping which would lead to the minimum number of terms in the expansion and hence the fewest possible coefficients to evaluate involves putting the entire transform over the common denominator $(z-4)(z-2)(z-1)^4$ which gives

$$X(z) = \frac{z(z-1)^3 + 3z(z+1)(z-4) + \frac{2}{3}z(z-4)(z-1)^3}{(z-4)(z-2)(z-1)^4}. \quad (6.5.46)$$

Dividing by z gives

$$\overline{X}(z) = \frac{X(z)}{z} = \frac{(z-1)^3 + 3(z+1)(z-4) + \frac{2}{3}(z-4)(z-1)^3}{(z-4)(z-2)(z-1)^4} \quad (6.5.47)$$

which is expanded to yield

$$\overline{X}(z) = \frac{K_1}{z-4} + \frac{K_2}{z-2} + \frac{K_3}{(z-1)^4} + \frac{K_4}{(z-1)^3} + \frac{K_5}{(z-1)^2} + \frac{K_6}{z-1}. \quad (6.5.48)$$

Even though only six terms are present in the expansion, it is easy to see from (6.5.47) that evaluation of the coefficients will involve relatively complicated expressions, especially for K_4, K_5, and K_6 which require successive calculation of the first three derivatives of $(z-1)^4 \overline{X}(z)$.

The following alternative, even though 10 terms result instead of 6, is felt to be somewhat easier for this problem. The alternate procedure involves separating terms instead of grouping over a single common denominator. For the problem at hand, this gives

$$\bar{X}(z) = \frac{1}{(z-4)(z-2)(z-1)} + \frac{3(z+1)}{(z-2)(z-1)^4} + \frac{\frac{2}{3}}{(z-2)(z-1)}. \quad (6.5.49)$$

Because of the linear nature of the transform, each of the three resulting terms can now be transformed individually and the results summed to obtain $x(i)$. Note that, in general, selection of groupings depends on the specific problem involved, particularly on the number and nature of the roots of $B(z) = 0$, and on the personal preferences of the analyst.

The first term on the right-hand side of (6.5.49), which we refer to for convenience as $\bar{X}_1(z)$, is now expanded to obtain

$$\bar{X}_1(z) = \frac{1}{(z-4)(z-2)(z-1)} = \frac{K_1}{z-4} + \frac{K_2}{z-2} + \frac{K_3}{z-1}. \quad (6.5.50)$$

The coefficients are evaluated using (6.5.5) as follows:

$$K_1 = \lim_{z \to 4} \left[(z-4) \frac{1}{(z-4)(z-2)(z-1)} \right] = \lim_{z \to 4} \left[\frac{1}{(z-2)(z-1)} \right] = \frac{1}{2 \cdot 3} = \frac{1}{6},$$

$$K_2 = \lim_{z \to 2} \left[\frac{1}{(z-4)(z-1)} \right] = \frac{1}{(-2)(1)} = -\frac{1}{2},$$

and

$$K_3 = \lim_{z \to 1} \left[\frac{1}{(z-4)(z-2)} \right] = \frac{1}{(-3)(-1)} = \frac{1}{3}.$$

Note that in deriving K_2 and K_3 we omitted the obvious first step of writing in $(z-z_r)\bar{X}_1(z)$ and then canceling the $(z-z_r)$ factors. This shortcut will be observed throughout the rest of this example and in general throughout this book.

The second term, which will be referred to as $\bar{X}_2(z)$, is now

$$\bar{X}_2(z) = \frac{3(z+1)}{(z-2)(z-1)^4} = \frac{K_4}{z-2} + \frac{K_5}{(z-1)^4} + \frac{K_6}{(z-1)^3} + \frac{K_7}{(z-1)^2} + \frac{K_8}{z-1}. \quad (6.5.51)$$

K_4 is found using (6.5.5) to obtain

$$K_4 = \lim_{z \to 2} \left[\frac{3(z+1)}{(z-1)^4} \right] = \frac{3 \cdot 3}{1^4} = 9$$

and K_5 is found using (6.5.7) with $p = 4$, which gives

$$K_5 = \lim_{z \to 1} \left[(z - 1)^4 \frac{3(z + 1)}{(z - 2)(z - 1)^4} \right] = \lim_{z \to 1} \left[\frac{3(z + 1)}{z - 2} \right] = \frac{3 \cdot 2}{-1} = -6.$$

K_6 through K_8 require differentiating $(z - 1)^4 \overline{X}_2(z)$ in the manner of (6.5.8) through (6.5.10). For K_6 we use (6.5.8), which yields

$$K_6 = \lim_{z \to 1} \left\{ \frac{1}{1!} \frac{d}{dz} \left[\frac{3(z + 1)}{z - 2} \right] \right\} = \lim_{z \to 1} \left\{ 1 \left[\frac{-9}{(z - 2)^2} \right] \right\} = \frac{-9}{(-1)^2} = -9.$$

Similarly,

$$K_7 = \lim_{z \to 1} \left\{ \frac{1}{2!} \frac{d}{dz} \left[\frac{-9}{(z - 2)^2} \right] \right\} = \lim_{z \to 1} \left\{ \frac{1}{2} \left[\frac{18}{(z - 2)^3} \right] \right\} = \frac{18}{2(-1)^3} = -9$$

and

$$K_8 = \lim_{z \to 1} \left\{ \frac{1}{3!} \frac{d}{dz} \left[\frac{18}{(z - 2)^3} \right] \right\} = \lim_{z \to 1} \left\{ \frac{1}{6} \left[\frac{-54}{(z - 2)^4} \right] \right\} = \frac{-54}{6(-1)^4} = -9.$$

It will be left to the reader to verify that K_9 and K_{10}, obtained from $\overline{X}_3(z)$, the third term on the right-hand side of (6.5.49), are $K_9 = \frac{2}{3}$ and $K_{10} = -\frac{2}{3}$.
Substituting these results in (6.5.49) gives

$$\overline{X}(z) = \frac{1}{6} \cdot \frac{1}{z - 4} - \frac{1}{2} \cdot \frac{1}{z - 2} + \frac{1}{3} \cdot \frac{1}{z - 1} + 9 \cdot \frac{1}{z - 2} - 6 \cdot \frac{1}{(z - 1)^4}$$

$$- 9 \cdot \frac{1}{(z - 1)^3} - 9 \cdot \frac{1}{(z - 1)^2} - 9 \cdot \frac{1}{z - 1} + \frac{2}{3} \cdot \frac{1}{z - 2} - \frac{2}{3} \cdot \frac{1}{z - 1}.$$

$$(6.5.52)$$

Combining terms involving like functions of z provides the form desired, viz.,

$$\overline{X}(z) = \frac{1}{6} \cdot \frac{1}{z - 4} + \left(-\frac{1}{2} + 9 + \frac{2}{3} \right) \frac{1}{z - 2} - 6 \cdot \frac{1}{(z - 1)^4} - 9 \cdot \frac{1}{(z - 1)^3}$$

$$- 9 \cdot \frac{1}{(z - 1)^2} + \left(\frac{1}{3} - 9 - \frac{2}{3} \right) \frac{1}{z - 1}$$

$$= \frac{1}{6} \cdot \frac{1}{z - 4} + \frac{55}{6} \cdot \frac{1}{z - 2} - 6 \cdot \frac{1}{(z - 1)^4} - 9 \cdot \frac{1}{(z - 1)^3}$$

$$- 9 \cdot \frac{1}{(z - 1)^2} - \frac{56}{6} \cdot \frac{1}{z - 1},$$

$$(6.5.53)$$

which we multiply by z to obtain $X(z)$ in expanded form:

$$X(z) = \frac{1}{6} \cdot \frac{z}{z-4} + \frac{55}{6} \cdot \frac{z}{z-2} - 6 \cdot \frac{z}{(z-1)^4}$$

$$- 9 \cdot \frac{z}{(z-1)^3} - 9 \cdot \frac{z}{(z-1)^2} - \frac{56}{6} \cdot \frac{z}{z-1}. \quad (6.5.54)$$

The inverse of $X(z)$ is now obtained by finding the inverse of each term of $X(z)$ using the relationships developed in step **f** above. This yields

$$x(i) = \frac{1}{6} \cdot 4^i + \frac{55}{6} \cdot 2^i - 6\left(\frac{1}{1^3}\right)\frac{i!}{(i-3)!3!} \cdot 1^i - 9\left(\frac{1}{1^2}\right)\frac{i!}{(i-2)!2!} 1^i$$

$$- 9\left(\frac{1}{1^1}\right)\frac{i!}{(i-1)!1!} 1^i - \frac{56}{6} \cdot 1^i$$

$$= \frac{1}{6} \cdot 4^i + \frac{55}{6} \cdot 2^i - \frac{6}{6} i(i-1)(i-2) - \frac{9}{2} i(i-1) - 9i - \frac{56}{6}$$

$$= \frac{1}{6} \cdot 4^i + \frac{55}{6} \cdot 2^i - i^3 - \frac{3}{2} \cdot i^2 - \frac{13}{2} \cdot i - \frac{56}{6}, \quad (6.5.55)$$

which agrees with the solution found by classical methods.

Even though the procedure used in this example involved 10 terms in the expansion and hence 10 coefficients to evaluate instead of the minimum number of 6, the relative ease with which the evaluations were able to be obtained working with $\bar{X}_1(z)$, $\bar{X}_2(z)$, and $\bar{X}_3(z)$ separately instead of the whole of $\bar{X}(z)$ makes this approach well worthwhile. One might, however, question the use of the z transform at all for the solution of linear difference equations with constant coefficients when one compares the effort just expended in solving this example with the effort required by the method of undetermined coefficients in Chapter IV. It is only fair to say that in many cases, especially when the forcing functions involved permit use of undetermined coefficients, classical methods can be much more easily applied than transform methods and should indeed be used. However, the z transform presents a very flexible and widely applicable technique not only for direct solution of difference equations but also in modeling and manipulating discrete linear systems. This will become more apparent later.

Example 3   *Complex Roots in $B(z) = 0$*   As our final example consider the difference equation

$$x(i+2) + 0.6x(i+1) + 0.25x(i) = a^i$$

previously solved in Section 4.3 by variation of parameters. Transformation to the z domain yields

$$(z^2 + 0.6z + 0.25)X(z) = \frac{z}{z - a} + (z^2 + 0.6z)x(0) + zx(1). \quad (6.5.56)$$

The coefficient of $X(z)$ is observed to be identical to (4.3.75), the characteristic equation for (6.5.56), the roots of which are $-0.3 \pm j(0.4)$. Therefore

$$X(z) = \frac{1}{(z - a)(z + 0.3 - j(0.4))(z + 0.3 + j(0.4))}$$

$$+ \frac{(z + 0.6)x(0) + x(1)}{(z + 0.3 - j(0.4))(z + 0.3 + j(0.4))}. \quad (6.5.57)$$

We will expand each of the two terms separately. First

$$\overline{X}_1(z) = \frac{1}{(z - a)(z + 0.3 - j(0.4))(z + 0.3 + j(0.4))}$$

$$= \frac{K_1}{z - a} + \frac{K_2}{z + 0.3 - j(0.4)} + \frac{K_3}{z + 0.3 + j(0.4)}. \quad (6.5.58)$$

K_1 is found from (6.5.5) as follows:

$$K_1 = \lim_{z \to a} \frac{1}{z^2 + 0.6z + 0.25} = \frac{1}{a^2 + 0.6a + 0.25}. \quad (6.5.59)$$

Note that the quadratic form $z^2 + 0.6z + 0.25$ was used instead of the individual roots thus simplifying calculations by avoiding complex numbers. This procedure is recommended whenever both members of a complex pair of roots are present. K_2 and K_3, being the coefficients associated with a pair of complex roots, are most easily evaluated from (6.5.16) and its conjugate, respectively. From (6.5.58), the numerator $A'(z)$ is simply unity. Therefore, Re $A'(z) = 1$ and Im $A'(z) = 0$ for all z. $B'(z)$, the quotient obtained by dividing $B(z)$ by the product of the pair of complex roots, is simply $z - a$. Thus $B'(-0.3 + j(0.4))$ is $-0.3 + j(0.4) - a$, the real part of which is Re $B'(-0.3 + j(0.4)) = -0.3 - a$ and the imaginary part Im $B'(-0.3 + j(0.4)) = 0.4$. Substitution into (6.5.16) yields

$$K_2 = \frac{[-(1)(0.4) + (0)(-0.3 - a)] + j[-(1)(-0.3 - a) - (0)(0.4)]}{2(0.4)[(-0.3 - a)^2 + (0.4)^2]}$$

$$= \frac{-0.5 + j(0.375 + 1.25a)}{a^2 + 0.6a + 0.25}. \quad (6.5.60)$$

Since K_3 is the complex conjugate of K_2, it is simply

$$K_3 = \frac{-0.5 - j(0.375 + 1.25a)}{a^2 + 0.6a + 0.25}.$$ (6.5.61)

Next,

$$\overline{X}_2(z) = \frac{(z + 0.6)x(0) + x(1)}{(z + 0.3 - j(0.4))(z + 0.3 + j(0.4))}$$

$$= \frac{K_4}{z + 0.3 - j(0.4)} + \frac{K_5}{z + 0.3 + j(0.4)}.$$ (6.5.62)

Now, $A'(-0.3 + j(0.4)) = (0.3 + j(0.4))x(0) + x(1)$, so Re $A'(-0.3 + j(0.4)) = 0.3x(0) + x(1)$ and Im $A'(-0.3 + j(0.4)) = 0.4x(0)$. Furthermore, $B'(z) = 1$, since the pair of complex roots constitute the entire denominator of $\overline{X}_2(z)$. Thus, Re $B'(-0.3 + j(0.4)) = 1$ and Im $B'(-0.3 + j(0.4)) = 0$. It can be shown that substitution into (6.5.16) yields $K_4 = 0.5x(0) - j(0.375x(0) + 1.25x(1))$, which of course means that $K_5 = 0.5x(0) + j(0.375x(0) + 1.25x(1))$. Combining the above results, we obtain

$$X(z) = \frac{1}{a^2 + 0.6a + 0.25} \cdot \frac{z}{z - a} + \frac{-0.5 + j(0.375 + 1.25a)}{a^2 + 0.6a + 0.25} \cdot \frac{z}{z + 0.3 - j(0.4)}$$

$$+ \frac{-0.5 - j(0.375 + 1.25a)}{a^2 + 0.6a + 0.25} \cdot \frac{z}{z + 0.3 + j(0.4)}$$

$$+ [0.5x(0) - j(0.375x(0) + 1.25x(1))] \frac{z}{z + 0.3 - j(0.4)}$$

$$+ [0.5x(0) + j(0.375x(0) + 1.25x(1))] \frac{z}{z + 0.3 + j(0.4)}.$$ (6.5.63)

Note that we could combine the coefficients of the two terms in $z/(z + 0.3 - j(0.4))$ and $z/(z + 0.3 + j(0.4))$ thus reducing (6.5.63) to just three terms. However, since the inverse transformation of terms involving complex roots is done in pairs using (6.5.40), it seems preferable in this case not to do so.

The inverse transformation of $z/(z - a)$ is of course a^i. The next two terms are transformed using (6.5.40) in which $R = 0.5$, $\theta = 2.217$ rad, $s = -0.5/(a^2 + 0.6a + 0.25)$ and $d = (0.375 + 1.25a)/(a^2 + 0.6a + 0.25)$. This results in

$$2 \cdot (0.5)^i \left[\frac{-0.5}{a^2 + 0.6a + 0.25} \cos 2.217i - \frac{0.375 + 1.25a}{a^2 + 0.6a + 0.25} \sin 2.217i \right]$$

or

$$\frac{-(0.5)^i}{a^2 + 0.6a + 0.25} [\cos 2.217i + (0.75 + 2.5a) \sin 2.217i].$$

The last two terms have the same R and θ since the same roots are involved, but now $s = 0.5x(0)$ and $d = -(0.375x(0) + 1.25x(1))$. The inverse transformation of these two terms therefore yields

$$(0.5)^i[x(0) \cos 2.217i + (0.75x(0) + 2.5x(1)) \sin 2.217i].$$

The entire inverse is thus

$$x(i) = \frac{a^i}{a^2 + 0.6a + 0.25} + (0.5)^i \left\{ \left[x(0) - \frac{1}{a^2 + 0.6a + 0.25} \right] \cos 2.217i \right.$$

$$\left. + \left[0.75x(0) + 2.5x(1) - \frac{0.75 + 2.5a}{a^2 + 0.6a + 0.25} \right] \sin 2.217i \right\}. \quad (6.5.64)$$

Note that the solution of this problem by variation of parameters was (see (4.3.83))

$$x(i) = \frac{a^i}{a^2 + 0.6a + 0.25} + (0.5)^i \{ c_3 \cos 2.217i + c_4 \sin 2.217i \}. \quad (6.5.65)$$

Thus both methods produced the same form of solution although the z transform solution is expressed directly in terms of the initial conditions $x(0)$ and $x(1)$ whereas the solution by classical methods contains parameters c_3 and c_4 which must be evaluated from initial or other boundary conditions. The two solutions can be shown to be identical as follows. In (6.5.65) let $i = 0$ to obtain

$$x(0) = \frac{1}{a^2 + 0.6a + 0.25} + c_3.$$

Therefore,

$$c_3 = x(0) - \frac{1}{a^2 + 0.6a + 0.25}$$

which is the coefficient of $\cos 2.217i$ in (6.5.64). Now let $i = 1$ in (6.5.65). Then, since $\cos 2.217 = -0.6$ and $\sin 2.217 = 0.8$,

$$x(1) = \frac{a}{a^2 + 0.6a + 0.25} + 0.5 \left\{ \left[x(0) - \frac{1}{a^2 + 0.6a + 0.25} \right] (-0.6) + c_4(0.8) \right\}.$$

Solving for c_4 yields

$$c_4 = 0.75x(0) + 2.5x(1) - \frac{0.75 + 2.5a}{a^2 + 0.6a + 0.25},$$

which is the coefficient of $\sin 2.217i$ in (6.5.64). Thus both methods yield the same result, as they should.

Although this example would have most easily been solved by undetermined coefficients since its forcing function is a member of a finite family, in situations where the method of undetermined coefficients is not applicable a choice must be made between variation of parameters and the z transform. Although many factors influence this choice, one item to consider is the set of known or desired boundary conditions which are to be used. Classical methods will always yield solutions with n coefficients in the homogeneous solution, which must, as an additional step, be evaluated from n boundary conditions. A great deal of choice, however, may be exercised in selecting the set to be used. The z transform solution, however, will always involve n *initial* conditions, the specific set being determined by the form of the difference equation. If it is in expanded form, the set will consist of $x(0)$ through $x(n-1)$. If it is in difference-operator form, the set will be $\Delta x(0)$ through $\Delta x(n-1)$. If the difference equation is in some mixed form, some combination of initial $x(i)$'s and $\Delta x(i)$'s will be involved. The point is that with the z transform a specific set of boundary terms specifically dictated by the form of the difference equation will be present in the solution. If this set happens to be the set of interest to the system designer, the solution is essentially complete. If not, additional conversions must take place.

6.6   The Special Case of $z_n = 0$

In our discussion of the partial fraction expansion we limited our attention to those $F(z) = A(z)/B(z)$ for which the lowest power of z in the numerator, $A(z)$, was the first power. This is equivalent to limiting $B(z)$ to functions for which the characteristic equation $B(z) = 0$ has no zero roots; for, if it did, these could be divided into $A(z)$ to reduce the order of $A(z)$ by the multiplicity of the zero root of $B(z) = 0$. This would of course create a constant term in the numerator if it were a single zero root and create terms in negative powers of z were it a multiple root.

To illustrate, suppose

$$F(z) = \frac{z^3 + 2z^2 + 3z}{z(z-1)(z+2)}.$$

Although we might normally tend to express this in the somewhat more simplified form of

$$F(z) = \frac{z^2 + 2z + 3}{(z-1)(z+2)}$$

by dividing out the common factor z, if the corresponding $f(i)$ were to be determined by partial fraction expansion, $\bar{F}(z)$ would have to be expressed as

$$\bar{F}(z) = \frac{F(z)}{z} = \frac{z + 2 + 3z^{-1}}{(z-1)(z+2)} = \frac{z^2 + 2z + 3}{z(z-1)(z+2)}.$$

Thus, the expansion would have to include a term in z as well as in $z - 1$ and $z + 2$. Similarly, for

$$F(z) = \frac{z^3 + 2z^2 + 3z}{z^2(z-1)(z+2)},$$

one could write either

$$F(z) = \frac{z^2 + 2z + 3}{z(z-1)(z+2)} \quad \text{or} \quad F(z) = \frac{z + 2 + 3z^{-1}}{(z-1)(z+2)},$$

although the latter would be an unusual choice. No matter which of the three forms is used, however, one must express $\bar{F}(z) = F(z)/z$ as

$$\bar{F}(z) = \frac{z^2 + 2z + 3}{z^2(z-1)(z+2)}$$

for purposes of partial fraction expansion.

Actually, expansions involving zero roots proceed in exactly the same manner as any other. The results, however, require a little interpretation and some supplementary techniques for expressing $f(i)$. Consider, for example,

$$F(z) = \frac{z^3 + 2z^2 + 3z}{z(z-1)(z+2)}$$

for which

$$\bar{F}(z) = \frac{z^2 + 2z + 3}{z(z-1)(z+2)} = \frac{K_1}{z} + \frac{K_2}{z-1} + \frac{K_3}{z+2}.$$

Now

$$K_1 = \lim_{z \to 0} [z\bar{F}(z)] = \lim_{z \to 0} \left[\frac{z^2 + 2z + 3}{(z-1)(z+2)}\right] = \frac{3}{(-1)(2)} = -\frac{3}{2}.$$

$$K_2 = \lim_{z \to 1} \left[\frac{z^2 + 2z + 3}{z(z+2)}\right] = \frac{6}{1 \cdot 3} = 2,$$

and

$$K_3 = \lim_{z \to -2} \left[\frac{z^2 + 2z + 3}{z(z-1)}\right] = \frac{3}{(-2)(-3)} = \frac{1}{2},$$

so

$$\bar{F}(z) = -\frac{3}{2} \cdot \frac{1}{z} + 2 \cdot \frac{1}{z-1} + \frac{1}{2} \cdot \frac{1}{z+2}$$

and

$$F(z) = -\frac{3}{2} + 2 \cdot \frac{z}{z-1} + \frac{1}{2} \cdot \frac{z}{z+2}.$$

The first term of $F(z)$ is seen to be the constant $-\frac{3}{2}$, which is more precisely written as $-\frac{3}{2}z^{-0}$. The inverse is, therefore, $-\frac{3}{2}$ for $i = 0$ and 0 for $i \neq 0$. Note this is decidedly different from the second term of $F(z)$ which is a constant 2 for *all* $i \geq 0$. The problem of finding a convenient means of expressing a function which is nonzero for a single value of i and is zero elsewhere is alleviated by means of the unit-impulse function, or delta function.

We define $\delta(i - k)$ as

$$\delta(i - k) = \begin{cases} 1 & \text{for } i - k = 0 \quad (\text{i.e., } i = k), \\ 0 & \text{elsewhere.} \end{cases} \tag{6.6.1}$$

Therefore, $\delta(i - 0)$, or more simply $\delta(i)$, is unity for $i = 0$ and zero for all other i. Thus, we can write $f(i)$, the inverse transform of $F(z)$, as

$$f(i) = -\frac{3}{2} \cdot \delta(i) + 2 + \frac{1}{2}(-2)^i.$$

Note that at $i = 0$, $f(0)$ is evaluated by $f(0) = -\frac{3}{2} + 2 + \frac{1}{2} = 1$, but at $i = 1, f(1) = 0 + 2 + \frac{1}{2}(-2) = 1$.

For the example above wherein $F(z)$ had the double root at zero, $\bar{F}(z)$ is

$$\bar{F}(z) = \frac{z^2 + 2z + 3}{z^2(z-1)(z+2)} = \frac{K_1}{z^2} + \frac{K_2}{z} + \frac{K_3}{z-1} + \frac{K_4}{z+2}.$$

Evaluation of the coefficients by appropriate formulas yields

$$\bar{F}(z) = -\frac{3}{2} \cdot \frac{1}{z^2} - \frac{7}{4} \cdot \frac{1}{z} + 2 \cdot \frac{1}{z-1} - \frac{1}{4} \cdot \frac{1}{z+2}$$

and, therefore,

$$F(z) = -\frac{3}{2} \cdot z^{-1} - \frac{7}{4} \cdot z^{-0} + 2 \cdot \frac{z}{z-1} - \frac{1}{4} \cdot \frac{z}{z+2}.$$

In this case, the inverse of the first term is $-\frac{3}{2}$ for $i = 1$ and 0 elsewhere. Hence it can be expressed as $-\frac{3}{2}\delta(i - 1)$. The second term is $-\frac{7}{4}$ for $i = 0$ and 0 elsewhere, so its inverse is $-\frac{7}{4}\delta(i - 0)$ or simply $-\frac{7}{4}\delta(i)$. Thus, $f(i)$, the inverse of $F(z)$, is

$$f(i) = -\frac{3}{2}\delta(i - 1) - \frac{7}{4}\delta(i) + 2 - \frac{1}{4}(-2)^i.$$

In general, for an $F(z) = A(z)/B(z)$ expressed such that the lowest power of z in $A(z)$ is the first power, if the corresponding $B(z) = 0$ has a zero root of multiplicity p, then the inverse function $f(i)$ will contain p terms of the form

$$K_1 \delta(i - p + 1) + K_2 \delta(i - p + 2) + \cdots + K_{p-1} \delta(i - 1) + K_p \delta(i),$$

where K_1 through K_p are evaluated by the same procedures that would apply to the evaluation of the coefficients associated with any root of multiplicity p.

6.7 Transfer Functions

Basic Definition

In discussing the transformation of a linear difference equation in Section 5.4, particular attention was directed to the coefficient of the transform of the dependent variable. It was shown in Transformation Example 1 that when this coefficient was a simple polynomial in z it was equivalent to the characteristic equation. It was further shown that when this coefficient was more complicated, as in Transformation Example 3, when all terms in the coefficient were put over a common denominator the numerator was equivalent to the characteristic equation. Examination of these coefficients in the two examples cited and others with which we have dealt also shows them to be made up entirely of basic system parameters from the left-hand sides of the system difference equations, e.g., the proportional, difference, and summation coefficients (K, L, and N) in Example 3. They are completely independent of all forcing functions and all initial conditions. Thus they describe the inherent nature, the internal structure of the system represented without regard for exogenous influences either at start-up or any other time. It is the characterization of the system in the z domain.

Since this coefficient contains all the available information concerning the system itself, it is convenient to provide it a name to facilitate description of the system and its performance. Actually the name is given to the reciprocal of the coefficient. It is called the *system transfer function* or usually just the *transfer function*, and in general is represented by the symbol $G(z)$. To illustrate, consider the z transform of the simple second-order difference equation used in Transformation Example 1 given in (5.4.4). By the definition just given,

$$G(z) = \frac{1}{z^2 - 3z + 2}. \tag{6.7.1}$$

Let us also define by $F(z)$ the z transform of the forcing function, i.e.,

$$F(z) = \frac{z}{z - 4} + \frac{3z(z + 1)}{(z - 1)^3}, \tag{6.7.2}$$

and by $I(z)$ the z transform of the initial condition terms, i.e.,

$$I(z) = -[-zx(1) - (z^2 - 3z)x(0)]. \tag{6.7.3}$$

Then (5.4.4) can be written in general terms as

$$\frac{1}{G(z)} X(z) - I(z) = F(z), \tag{6.7.4}$$

which is solved for $X(z)$ by adding $I(z)$ to both sides and multiplying by $G(z)$ to get

$$X(z) = F(z)G(z) + I(z)G(z). \tag{6.7.5}$$

Exactly the same form would have resulted if in (5.4.19) we had let $M(z) = X(z)$, $G(z) = 1/[z(z - 1) + Kz + L(z - 1) + Nz^2/(z - 1)]$, $F(z)$ represent the terms involving D, $E(z)$, and $P(z)$, and $I(z)$ represent the terms in $m(0)$, $m(1)$, $\varepsilon(0)$, and $p(0)$. Thus (6.7.5) is a standard form for representing the z domain solution of a linear difference equation. Obviously $F(z)G(z)$ is what we have referred to as the system transform and $I(z)G(z)$ what we have called the initial condition transform.

Use of the Transfer Function in Systems Analysis and Design

To obtain the complete solution for the independent variable $x(i)$ in terms of discrete time and the appropriate forcing function $f(i)$ and initial conditions, one would simply take the inverse transformation of (6.7.5). Since the inverse transformation of the system and initial condition transforms are clearly distinguishable, one can think of the former as a " forcing function solution " and the latter as an " initial condition solution." However, for systems in which the start-up transients due to the initial conditions are not likely to be of significant magnitude or duration the initial condition solution may be of little importance in the analysis or design of a system. In such cases, which experience indicates is the vast majority, analysis and design is done using the forcing function solution almost exclusively. Thus the system transform is the basis for most system decisions.

Let $S(z)$ represent the system transform. Then

$$S(z) = F(z)G(z) \tag{6.7.6}$$

and represents the response of a system, subsystem, or component with transfer function $G(z)$ to a forcing function $f(i)$ whose z transform is $F(z)$ without regard to the effects of starting conditions. For a system initially at rest, i.e., for which all initial conditions are zero, the initial condition transform is identically zero and $X(z) = S(z)$. In such cases,

$$X(z) = F(z)G(z) \qquad (I(z) = 0). \tag{6.7.7}$$

Many authors use the notation of (6.7.6) under the heading of "Transfer Function Analysis" without making explicit the assumption of the system being initially at rest. Readers should always try to remain aware of the fact that when transfer function analysis is used to produce a complete output, this assumption is implicit. When the assumption does not hold, however, the effects of initial condition terms can easily be determined by constructing the initial condition transform, transforming to the i domain, and adding the results to the forcing function solution found from the inverse of $S(z)$.

Systems Analysis

Wilde and Beightler (1967)† state that "... analysis [is the process of] understanding how the world behaves." The transfer function obviously plays an important role in analyzing the behavior of a system in a particular environment, whether or not initial condition effects are included, since $G(z)$ is an explicit part of both transforms. The major benefit in defining the transfer function as we have is that, since it depends only on the inherent structure of the system and not on any environmental effects, $G(z)$ can be determined for any given system strictly from the properties of the system itself and then used as needed with any $F(z)$ or $I(z)$ of interest to the analyst.

For example, from (5.4.9), the transform of the system difference equation for the proportional-plus-difference rapid-response process controller, the complete coefficient of the transform of the dependent variable $M(z)$ is $z^2 - (1 - K - L)z - L$. Therefore, for this system,

$$G(z) = \frac{1}{z^2 - (1 - K - L)z - L}, \tag{6.7.8}$$

which can now be used to analyze the performance of any rapid-response proportional-plus-difference controller with any assumed patterns of perturbation or measurement error and, if appropriate, any set of applicable initial conditions. Likewise, from (5.4.20), the transfer function for the rapid-response proportional-plus-difference-plus-summation controller is

$$G(z) = \frac{z - 1}{z^3 - (2 - K - L - N)z^2 + (1 - K - 2L)z + L}. \tag{6.7.9}$$

Systems Design

In design or synthesis, one attempts to find a system which, when confronted with a particular forcing function $f(i)$, will produce a specified output $x(i)$. It is thus a decision-making task. Unless previous experience provides the information needed to structure the system, one cannot even write the

---

† See Chapter I of this text for an excellent discussion of the processes of analysis and synthesis and the roll of optimization in each.

system equation. One can, however, take the z transforms of $f(i)$ and $x(i)$ and, neglecting initial conditions, i.e., letting $X(z) = S(z)$, substitute the results in (6.7.6) for $F(z)$ and $S(z)$, respectively. The system transfer function is then found by solving (6.7.6) for $G(z)$, giving

$$G(z) = S(z)/F(z), \qquad (6.7.10)$$

which actually serves as a definition for the transfer function from the design viewpoint.

An experienced designer will often be able to structure his system directly from the $G(z)$ calculated in (6.7.10). Others may gain insight from transforming $G(z)$ back to the i domain and examining the resulting function of discrete time, $g(i)$. In many cases, however, $g(i)$ will simply represent a desired result which must be approached empirically by trial and error using various experimental system configurations. Usually, all of these approaches will be used.

In illustrating the design task, we are constrained to an exceedingly simple, perhaps obvious problem, since system design, even for seemingly elementary situations, is a time-consuming effort involving much art as well as science. The design task is made somewhat simpler if we limit attention to just one general category of system configuration, a restriction often dictated by system objectives and the physical properties of available system components. For our purposes here, suppose we wish to design a rapid-response process controller for which the output $m(i)$ is of the form $D + BA^i$ when the perturbation is a step function of magnitude p occurring at $i = 0$, i.e.,

$$p(i) = \begin{cases} 0 & \text{for} \quad i < 0, \\ p & \text{for} \quad i \geq 0. \end{cases}$$

Note that for $|A| < 1$, $m(i)$ approaches D as i increases, usually a very desirable property for a process controller to possess. For simplicity we will assume all measurement errors are negligible, i.e., $\varepsilon(i) = 0$ for all i, and that the desired level of system operation, D, equals zero. Equating D to zero, a practice we will use almost exclusively from here on in all discussions of process controllers, is equivalent to shifting the origin of the scale of measuring process output to the actual desired level. Thus all values of $m(i)$ represent *deviations* from the desired level rather than values on a physical scale.

For reference purposes, the general system equation for the rapid-response process controller was given in (2.3.17) as

$$\Delta m(i) = g_D[D - m(0) - \varepsilon(0), \ldots, D - m(i) - \varepsilon(i)] + p(i). \quad (6.7.11)$$

For $D = \varepsilon(0) = \cdots = \varepsilon(i) = 0$ and $p(i) = p$, this becomes

$$m(i + 1) - m(i) = g_D[m(0), \ldots, m(i)] + p. \qquad (6.7.12)$$

The design task is to find the function g_D which makes the solution of (6.7.12) $m(i) = BA^i$, of course neglecting the effects of initial conditions. Here A is to be in the range $0 < A < 1$ while B is unspecified and is to be evaluated as part of the design process. Let $G(z)$ be the z transform of g_D. $G(z)$ is thus the sought-after system transfer function of the rapid-response controller. Thus, from (6.7.10),

$$G(z) = M(z)/P(z) \tag{6.7.13}$$

because of the assumption that $S(z) = M(z)$ for design purposes. To meet the desired conditions, $M(z) = Bz/(z - A)$ when $P(z) = pz/(z - 1)$. Substitution in (6.7.13) yields

$$G(z) = \frac{Bz/(z - A)}{pz/(z - 1)} = \frac{B}{p} \cdot \frac{z - 1}{z - A}. \tag{6.7.14}$$

Although the experienced designer could possibly interpret the result in the form given, it might be helpful for the rest of us to work in a context with which we are more familiar, namely, as the coefficient of the dependent variable in the transform of a linear difference equation. Recall that this coefficient is the reciprocal of $G(z)$. Therefore the transformed equation would be in the form

$$\frac{p}{B} \cdot \frac{z - A}{z - 1} M(z) = P(z). \tag{6.7.15}$$

In order to obtain the basic format of the difference equation (6.7.12), the coefficient of $M(z)$ must contain a z term to represent the $m(i + 1)$ on the left-hand side of the equation. Since no such term appears in (6.7.15) as given it must be introduced. We note from (6.7.14) that the denominator of $G(z)$ would contain a factor z if $M(z)$ did not contain a z in its numerator. This z is easily eliminated by writing the desired $m(i)$ in the form

$$m(i) = B \cdot A^i = BA \cdot A^{i-1} \tag{6.7.16}$$

from which

$$M(z) = BA \cdot \frac{1}{z} \cdot \frac{z}{z - A} = \frac{BA}{z - A} \tag{6.7.17}$$

and

$$G(z) = \frac{BA/(z - A)}{pz/(z - 1)} = \frac{BA}{p} \cdot \frac{z - 1}{z(z - A)} = \frac{BA}{p} \cdot \frac{z - 1}{z^2 - zA}. \tag{6.7.18}$$

Thus, (6.7.15) can be written in the form

$$\frac{p}{BA} \cdot \frac{z^2 - zA}{z - 1} M(z) = P(z). \tag{6.7.19}$$

Subtracting and adding a z in the numerator of $1/G(z)$ gives

$$\frac{1}{G(z)} = \frac{p}{BA} \cdot \frac{z^2 - z + z - zA}{z - 1} = \frac{p}{BA} \left[z + (1 - A) \frac{z}{z - 1} \right], \tag{6.7.20}$$

which, when substituted into (6.7.19), yields

$$\frac{p}{BA} \left[zM(z) + (1 - A) \frac{z}{z - 1} M(z) \right] = P(z). \tag{6.7.21}$$

The inverse transform of $zM(z)$ is of course $m(i + 1)$ and the inverse of $z/(z - 1)$ is $\sum_{n=0}^{i} m(n)$. Thus in the discrete-time domain we have

$$\frac{p}{BA} \left[m(i + 1) + (1 - A) \sum_{n=0}^{i} m(n) \right] = p(i). \tag{6.7.22}$$

We must now subtract and add $m(i)$ inside the brackets on the left to obtain

$$\frac{p}{BA} \left[m(i + 1) - m(i) + m(i) + (1 - A) \sum_{n=0}^{i} m(n) \right] = p(i) \tag{6.7.23}$$

or

$$\frac{p}{BA} [m(i + 1) - m(i)] = \frac{p}{BA} \left[-m(i) - (1 - A) \sum_{n=0}^{i} m(n) \right] + p(i), \tag{6.7.24}$$

which would be in the desired form of (6.7.12) were the factor p/BA equal to unity. Since the value of B was not specified, we are free to set its value as most expedient in meeting our design objectives. Thus we let $B = p/A$ which makes (6.7.24)

$$m(i + 1) - m(i) = -m(i) - (1 - A) \sum_{n=0}^{i} m(n) + p(i), \tag{6.7.25}$$

the system difference equation for a proportional-plus-summation controller with proportionality constant $K = 1$ and summation constant $N = 1 - A$.

The immediate reaction of some readers to the succession of mysterious steps which led to this result may be one of bewilderment. How does one know how to manipulate $G(z)$ into forms which can be interpreted physically? In practice, the answer is experience coupled with a basic knowledge of the applicable theory to assist in recognizing alternative forms and procedures. For the example just presented, this experience was gained by trying several approaches on scrap paper before writing it up.

6.8   Solution of Difference Equations with Generalized Forcing Functions

The transfer function has been shown to be a central factor in systems analysis and design and in finding complete solutions to linear difference equations when specific forcing functions are known or assumed. Both analysts and designers, however, must often deal with systems which must operate in a variety of environments or in environments in which some or all of the exogenous factors are unknown. In such situations direct application of (6.7.5) is not possible even for known $G(z)$, yet it is still desirable to have an expression for the dependent variable in the i domain. This expression must of course be in terms of a generalized forcing function $f(i)$.

From (6.7.5), we know that only the system transform is affected. The initial condition solution can be found (or ignored) as in any problem. Therefore, we limit our attention here to the system transform $S(z)$ and its inverse transform which we define as $s(i)$. Since $S(z)$ is the product of $F(z)$ and $G(z)$, then $s(i)$ is the convolution of $f(i)$ and $g(i)$, the inverse transform of $G(z)$. This convolution can be expressed in general terms in either of the following forms:

$$s(i) = \sum_{k=0}^{i} f(k)g(i-k) = \sum_{k=0}^{i} f(i-k)g(k). \qquad (6.8.1)$$

It is interesting to note from (6.8.1) that for a unit-impulse forcing function $f(i) = \delta(i)$, $s(i)$ becomes

$$s(i) = 1 \cdot g(i) + \sum_{k=1}^{i} 0 \cdot g(i-k) = g(i); \qquad (6.8.2)$$

i.e., the forcing function solution for a unit-impulse forcing function is identically equal to the inverse transform of the system transfer function $G(z)$. Therefore, $g(i)$ is commonly called the *unit-impulse response* of the system.

We now illustrate the derivation of the generalized forcing function solution $s(i)$ for both a first-order and a second-order simple polynomial system transfer function.

First-Order Transfer Function

Suppose the difference equation to be solved is of first order and is in expanded form. Then

$$G(z) = \frac{1}{b_0 z + b_1}, \qquad (6.8.3)$$

which in standard form for partial fraction expansion is

$$G(z) = \frac{1}{b_0} \cdot \frac{z}{z(z + b_1/b_0)}. \qquad (6.8.4)$$

Therefore

$$\bar{G}(z) = \frac{1}{b_0} \cdot \frac{1}{z(z + b_1/b_0)} = \frac{K_1}{z} + \frac{K_2}{z + b_1/b_0}. \qquad (6.8.5)$$

The constants K_1 and K_2 are found to be $1/b_1$ and $-1/b_1$, respectively, which gives

$$\bar{G}(z) = \frac{1}{b_1} \cdot \frac{1}{z} - \frac{1}{b_1} \cdot \frac{1}{z + b_1/b_0} \qquad (6.8.6)$$

and

$$G(z) = \frac{1}{b_1} - \frac{1}{b_1} \cdot \frac{z}{z + b_1/b_0}. \qquad (6.8.7)$$

Therefore, the inverse transformation of the transfer function is

$$g(i) = \frac{1}{b_1} \delta(i) - \frac{1}{b_1} \left(-\frac{b_1}{b_0} \right)^i. \qquad (6.8.8)$$

Note that for $i = 0$,

$$g(0) = \frac{1}{b_1} - \frac{1}{b_1} = 0, \qquad (6.8.9)$$

while for $i > 0$,

$$g(i) = 0 - \frac{1}{b_1} \left(-\frac{b_1}{b_0} \right)^i = \frac{1}{b_0} \left(-\frac{b_1}{b_0} \right)^{i-1}, \qquad (6.8.10)$$

which is the inverse of $(1/b_0)z/(z + b_1/b_0)$ delayed by one time period. Equation (6.8.10) could, of course, have been obtained directly by taking this inverse, which is $(1/b_0)(-b_1/b_0)^i$, and applying a real backward translation of one period. In so doing, however, it would have to be carefully remembered that since the form $(1/b_0)(-b_1/b_0)^i$ holds only for $i \geq 0$, when $i - 1$ is substituted for i in effecting the backward translation, the resulting form holds only for $i - 1 \geq 0$ or $i \geq 1$. This is taken care of automatically by the partial fraction expansion approach and the impulse function.

Either way we obtain

$$g(i) = \begin{cases} \frac{1}{b_0} \left(-\frac{b_1}{b_0} \right)^{i-1} & \text{for} \quad i \geq 1, \\ 0 & \text{elsewhere,} \end{cases} \qquad (6.8.11)$$

which gives

$$s(i) = \sum_{k=0}^{i-1} f(k) \frac{1}{b_0} \left(-\frac{b_1}{b_0} \right)^{i-1-k} = \sum_{k=0}^{i-1} f(i-1-k) \frac{1}{b_0} \left(-\frac{b_1}{b_0} \right)^k. \quad (6.8.12)$$

The summation limits of $i - 1$ and the argument $i - 1 - k$ can be verified by the interested reader.

Second-Order Transfer Function

Suppose now the difference equation were of second order and in expanded form. Then

$$G(z) = \frac{1}{b_0 z^2 + b_1 z + b_2}. \quad (6.8.13)$$

In standard form for partial fraction expansion this becomes

$$G(z) = \frac{1}{b_0} \cdot \frac{1}{z^2 + (b_1/b_0)z + b_2/b_0} = \frac{1}{b_0} \cdot \frac{1}{(z - z_1)(z - z_2)}, \quad (6.8.14)$$

where z_1 and z_2 are the roots of $z^2 + (b_1/b_0)z + (b_2/b_0) = 0$ and are found from the quadratic formula to be

$$z_1 = \frac{-b_1 + (b_1{}^2 - 4b_0 b_2)^{1/2}}{2b_0} \quad (6.8.15)$$

and

$$z_2 = \frac{-b_1 - (b_1{}^2 - 4b_0 b_2)^{1/2}}{2b_0}. \quad (6.8.16)$$

Consider first the case where z_1 and z_2 are unequal. This will occur when $b_1{}^2 \neq 4b_0 b_2$ and includes the two subcases of two real unequal roots (when $b_1{}^2 > 4b_0 b_2$) and a pair of complex roots (for $b_1{}^2 < 4b_0 b_2$). No attempt is made here to differentiate between the two subcases. For $z_1 \neq z_2$,

$$\bar{G}(z) = \frac{1}{b_0} \cdot \frac{1}{z(z - z_1)(z - z_2)} = \frac{K_1}{z} + \frac{K_2}{z - z_1} + \frac{K_3}{z - z_2}. \quad (6.8.17)$$

The coefficients are found to be $K_1 = 1/b_0 z_1 z_2$, $K_2 = 1/b_0 z_1 (z_1 - z_2)$, and $K_3 = 1/b_0 z_2 (z_2 - z_1)$. Thus, the transfer function in expanded form is

$$G(z) = \frac{1}{b_0 z_1 z_2} + \frac{1}{b_0 z_1 (z_1 - z_2)} \cdot \frac{z}{z - z_1} + \frac{1}{b_0 z_2 (z_2 - z_1)} \cdot \frac{z}{z - z_2} \quad (6.8.18)$$

and the inverse transform is

$$g(i) = \frac{1}{b_0 z_1 z_2} \delta(i) + \frac{1}{b_0 z_1(z_1 - z_2)} z_1{}^i + \frac{1}{b_0 z_2(z_2 - z_1)} z_2{}^i$$

$$= \frac{1}{b_0 z_1 z_2} \delta(i) + \frac{1}{b_0(z_1 - z_2)} (z_1^{i-1} - z_2^{i-1}). \tag{6.8.19}$$

For $i = 0$,

$$g(0) = \frac{1}{b_0 z_1 z_2} + \frac{1}{b_0(z_1 - z_2)} \left(\frac{1}{z_1} - \frac{1}{z_2} \right)$$

$$= \frac{1}{b_0 z_1 z_2} + \frac{1}{b_0(z_1 - z_2)} \cdot \left(\frac{z_2 - z_1}{z_1 z_2} \right) = 0, \tag{6.8.20}$$

and for $i > 0$,

$$g(i) = \frac{1}{b_0} \left[\frac{z_1^{i-1} - z_2^{i-1}}{z_1 - z_2} \right]. \tag{6.8.21}$$

Therefore,

$$s(i) = \sum_{k=0}^{i-1} f(k) \frac{1}{b_0} \left[\frac{z_1^{i-1-k} - z_2^{i-1-k}}{z_1 - z_2} \right]$$

$$= \sum_{k=0}^{i-1} f(i - 1 - k) \frac{1}{b_0} \left[\frac{z_1^{k} - z_2^{k}}{z_1 - z_2} \right], \tag{6.8.22}$$

which can also be expressed fully in terms of the coefficients of the difference equation (i.e., b_0, b_1, and b_2) by substitution of (6.8.15) and (6.8.16) for z_1 and z_2, respectively. Again, the summation limits of $i - 1$ and the argument $i - 1 - k$ can be verified by the interested reader.

For the case in which $b_1{}^2 = 4b_0 b_2$ a double root exists at $-b_1/2b_0$ and the partial fraction expansion of $\bar{G}(z)$ takes the form

$$\bar{G}(z) = \frac{1}{b_0} \cdot \frac{1}{z(z + b_1/2b_0)^2} = \frac{K_1}{z} + \frac{K_2}{(z + b_1/2b_0)^2} + \frac{K_3}{z + b_1/2b_0}. \tag{6.8.23}$$

The K's are found to be $K_1 = 4b_0/b_1{}^2$, $K_2 = -2/b_1$, and $K_3 = -4b_0/b_1{}^2$, so the transfer function is

$$G(z) = \frac{4b_0}{b_1{}^2} - \frac{2}{b_1} \cdot \frac{z}{(z + b_1/2b_0)^2} - \frac{4b_0}{b_1{}^2} \cdot \frac{z}{z + b_1/2b_0}. \tag{6.8.24}$$

Inverse transformation yields

$$g(i) = \frac{4b_0}{b_1{}^2} \delta(i) - \frac{2}{b_1} \cdot \frac{1}{-b_1/2b_0} i \left(-\frac{b_1}{2b_0} \right)^i - \frac{4b_0}{b_1{}^2} \left(-\frac{b_1}{2b_0} \right)^i$$

$$= \frac{4b_0}{b_1{}^2} \left[\delta(i) + i \left(-\frac{b_1}{2b_0} \right)^i - \left(-\frac{b_1}{2b_0} \right)^i \right]. \tag{6.8.25}$$

For $i = 0$,

$$g(0) = \frac{4b_0}{b_1^2} [1 + 0 - 1] = 0, \tag{6.8.26}$$

and for $i = 1$,

$$g(1) = \frac{4b_0}{b_1^2} \left[0 + \left(-\frac{b_1}{2b_0} \right) - \left(-\frac{b_1}{2b_0} \right) \right] = 0. \tag{6.8.27}$$

For $i \geq 2$,

$$g(i) = \frac{4b_0}{b_1^2} (i - 1) \left(-\frac{b_1}{2b_0} \right)^i = \frac{1}{b_0} (i - 1) \left(-\frac{b_1}{2b_0} \right)^{i-2}, \tag{6.8.28}$$

which results in

$$
\begin{aligned}
s(i) &= \sum_{k=0}^{i=2} f(k) \frac{1}{b_0} (i - 1 - k) \left(-\frac{b_1}{2b_0} \right)^{i-2-k} \\
&= \sum_{k=0}^{i-2} f(i - 2 - k) \frac{1}{b_0} (k + 1) \left(-\frac{b_1}{2b_0} \right)^{k}.
\end{aligned} \tag{6.8.29}
$$

The methods just illustrated can of course be extended with appropriate increases in complexity to handle any order and any form of difference equation.

6.9 Conclusion

As has been pointed out, the z transform provides a completely general and extremely flexible procedure for solution of linear difference equations with constant coefficients. As such it has proven most useful to designers of discrete, linear, time-invariant systems. For simple systems with known forcing functions which allow use of the method of undetermined coefficients, the classical procedures of Chapter IV are usually simpler and quicker than the z transform alternative. However, as the complexity of the system becomes greater both with respect to the order of the equation and the nature of the forcing function, the relative usefulness of the z transform usually increases. The z transform has a distinct advantage over the variation-of-parameters method when generalized forcing functions are confronted.

Unfortunately, use of the z transform does not alleviate the necessity of finding the roots of characteristic equations in order to solve system difference equations. This difficulty is partially eased by use of numerical methods to extract such roots when necessary, but as is obvious from the material on the Newton–Bairstow method in Chapter IV, it is never an easy task. We have indicated that experienced designers can do a lot to perfect a system with system transforms in the z domain, but the working out of all details in terms of z is still difficult at best. In Chapter IX, however, we will see how one very important design criterion, namely that of system stability, can be checked and in some cases the general conditions for stability derived in the z domain without explicit determination of these roots. Thus if a design is inherently unstable, this fact can be determined and the laborious task of root determination avoided completely.

We now turn our attention to the topic of criteria for evaluating system performance and the effects of various types of exogenous perturbations and system parameter values in meeting these criteria.

Exercises

6.1 Use power series expansion to find the inverse transform of the following functions of z:

    a. $z/(z - 2)$ **b.** $z/(z - 2)^2$ **c.** $z^2/(z - 2)^2$
    d. $(2z^2 - 3z + 1)/(z - 2)^2$ **e.** $z/(z + 3)^2$ **f.** $z/(z + 3)^3$
    g. $(z^3 - 2z^2 + 3)/(z + 3)^3$ **h.** $2z/(z - 1)^4$ **i.** $z/(z + a)^5$

6.2 **(a)** through **(i).** Use Maclaurin series expansion to find the inverse transform of the function of z given in the corresponding part of Exercise 6.1.

6.3 **(a)** through **(i).** Use partial fraction expansion to find the inverse transform of the function of z given in the corresponding part of Exercise 6.1.

6.4 Find the inverse transform of the following functions of z:

    a. $\dfrac{z}{z^2 + z - 2}$ **b.** $\dfrac{z^2 + 3z}{z^2 + z - 2}$ **c.** $\dfrac{z^2 + 3z}{z^2 + 3z + 2}$

    d. $\dfrac{2z^2 + 4z}{2z^2 + 3z + 1}$ **e.** $\dfrac{z^2}{z^2 - 4z + 5}$ **f.** $\dfrac{2z^2}{3z^2 + 6z + 6}$

    g. $\dfrac{z^2 + 3z + 2}{z^2 - 4z + 5}$ **h.** $\dfrac{z^2}{(z - 2)^3(z - 3)}$ **i.** $\dfrac{z^2}{(z + 3)^3(z^2 + 3z + 2)}$

    j. $\dfrac{z^2}{(z + 1)^3(z^2 - 2z + 10)}$ **k.** $\dfrac{z}{(z + 1)^3(z^2 - 4)^2}$ **l.** $\dfrac{2}{z^2 - 4z + 5}$

    m. $\dfrac{3}{z^3 - z}$ **n.** $\dfrac{1}{z^3}$ **o.** $\dfrac{4}{z^4(z - 2)}$

6.5 Find the inverse transform of $X(z)$ determined in Exercise 5.6a (for the difference equation in 4.2a) given $x(0) = 1.9$ for the following forcing functions (compare answers with those of 4.11):

a. 3   **b.** 3^i **c.** $3 + 3^i$ **d.** 2^i **e.** i **f.** $\sin 2i$
g. $2^i \sin 2i$

6.6   Find the inverse transform of $X(z)$ determined in Exercise 5.6b (for the difference equation in 4.2b) given $x(0) = 0$ and $x(1) = 1$ for the following forcing functions (compare answers with those of 4.12):

a. 4   **b.** 4^i **c.** i **d.** $4i \sin i$ **e.** $i + 4$

6.7   Find the inverse transform of $X(z)$ determined in Exercise 5.6c (for the difference equation in 4.2c) given $x(0) = 0$ and $x(1) = 1$ for the following forcing functions (compare answers with those of 4.13):

a. $\sqrt{2}$ **b.** $(\sqrt{2})^i$ **c.** $\sin \pi i/4$ **d.** $(\sqrt{2})^i \sin \pi i/4$
e. $(\sqrt{2})^i + \sin \pi i/4$

6.8   Find the inverse transform of $X(z)$ determined in Exercise 5.6d (for the difference equation in 4.2d) given $x(0) = 0$ and $x(1) = 1$ for the following forcing functions (compare answers with those of 4.14):

a. 3   **b.** i **c.** 3^i **d.** $i + 3$ **e.** $i + 3^i$ **f.** $3^i + 3$
g. $\sin 3i$

6.9   Find the inverse transform of $X(z)$ determined in Exercise 5.6e (for the difference equation in 4.2e) given $x(0) = 0$ and $x(1) = 1$ for $f(i) = i^3 + i2^i$ (compare the answer with that of 4.25).

6.10   Find the inverse transform of $X(z)$ determined in Exercise 5.6f (for the difference equation in 4.2f) given $x(0) = 0$ and $x(1) = 1$ for $f(i) = i^3 + \sin i$ (compare the answer to that of 4.26).

6.11   **(a)** through **(i)**.   From the results of the corresponding part of Exercise 5.6 identify the appropriate transfer function relating $F(z)$ to $X(z)$.

6.12   **(a)** through **(g)**.   From the results of the corresponding part of Exercise 5.7 identify the appropriate transfer function relating $F(z)$ to $M(z)$.

6.13 and **6.14**   The same as 6.12 except using the results of the various parts of 5.8 and 5.9, respectively.

6.15

   a.   From the results of 5.10a, identify the transfer functions relating Demand and Safety Stock to Net Inventory. (*Note:* each forcing function requires a slightly different transfer function.)
   b.   From the results of 5.10b, identify the transfer functions relating Demand and Safety Stock to Replenishment Order Placed.
   c.   From the results of 5.10c, identify the transfer functions relating Demand and Safety Stock to Replenishment Amount on Order.

6.16 through **6.19**   From the results of Exercises 5.11, 5.12, 5.13, and 5.14, respectively, identify the various transfer functions relating each of the appropriate forcing functions to the appropriate dependent variable.

6.20 **(a)** through **(e)**.   From the results of the corresponding parts of Exercise 5.15, identify the transfer functions relating D, $E(z)$, and $P(z)$ to $M(z)$. (*Note*: each forcing function requires a slightly different transfer function.)

6.21 **a.**   (1), (2), (3) and **b.**   (1), (2), (3). From the results of the corresponding parts of Exercise 5.16, identify the transfer functions relating actual customer orders and forecast demand to order quantity. (*Note*: the two forcing functions require slightly different transfer functions.)

6.22   Find the control decision rule g_D for a rapid-response process controller with desired level of operation $D = 0$ and no measurement error such that for a perturbation $p(i) = ip$, p a constant, the forcing function solution is $s(i) = BA^i$, where $0 < A < 1$ and B is an unspecified constant.

6.23   A linear, time-invariant system has a unit-impulse response $g(i)$ such that $g(0) = 3$, $g(1) = 2$, $g(2) = 1$, and $g(i) = 0$ for $i \geq 3$. Find the output $x(i)$, $i \geq 0$, for the following forcing functions:

    a. $f(i) = \delta(i - 3)$ **b.** $f(0) = 1$, $f(1) = 2$, $f(i) = 0$ for $i \geq 2$
    c. $f(i) = (\tfrac{1}{4})^i$

6.24   A linear, time-invariant system has a unit-impulse response $g(i) = i$, $i \geq 0$. Find the output $x(i)$, $i \geq 0$, for the following forcing functions:

    a. $f(i) = \delta(i - 2)$ **b.** $f(i) = 2\,\delta(i) + 4\,\delta(i - 1)$
    c. $f(i) = 3$ for all $i \geq 0$ **d.** $f(i) = (\tfrac{1}{2})^i$

6.25   Find the output $x(i)$, $i \geq 0$, for a system with transfer function $G(z) = 3/(z + 0.1)$ for the following forcing functions:

    a. $f(i) = \delta(i)$ **b.** $f(i) = \delta(i - 4)$ **c.** $f(i) = 2$ for all $i \geq 0$
    d. $f(i) = 2^i$ **e.** $f(i) = A \sin(wi + \theta)$, $A > 0$

6.26   Find the output $x(i)$, $i \geq 0$, for a system with transfer function $G(z) = 3/(z - 0.2)(z + 0.5)$ for the following forcing functions:

    a. $f(i) = e^i$ **b.** $f(i) = \sin 3i$
    c. a generalized $f(i)$, assuming all applicable initial conditions are zero

62.7   Find the output $x(i)$, $i \geq 0$, for a system with transfer function $G(z) = 2(z + 0.3)(z - 0.4)$ for the following forcing functions:

    a. $f(i) = 1$ for all $i \geq 0$ **b.** $f(i) = \cos 5i$
    c. a generalized $f(i)$, assuming all applicable initial conditions are zero.

6.28   Use both partial fraction expansion and long division to find the unit-impulse response of the systems having the following transfer functions:

   a. $G(z) = z/(z - a)$ **b.** $G(z) = 1/(z - a)$
   c. $G(z) = z/(z - a)(z - b)$ **d.** $G(z) = 1/(z - a)(z - b)$

6.29   Find the inverse transform of $X(z)$ determined in Exercise 5.6a (for the difference equation in 4.2a) in terms of the generalized forcing function $f(i)$. Then substitute the specific forcing functions and initial conditions given in 6.5 and compare the results with those of 6.5.

6.30 through 6.34   Same as 6.29 except using the $X(z)$, specific forcing functions, and initial conditions from 6.6, 6.7, 6.8, 6.9, and 6.10, respectively.

6.35   Find the inverse transform of $X(z)$ determined in Exercise 5.6g for a generalized forcing function and set of initial conditions.

6.36 and 6.37   Same as 6.31 except using the $X(z)$ from Exercises 5.6h and 5.6i, respectively.

6.38   **(a)** through **(e)**   Find $m(i)$, $i \geq 0$, the inverse transform of $M(z)$ determined in the corresponding part of 5.15 (from the difference equations in 4.27) in terms of the forcing functions D, $\varepsilon(i)$, and $p(i)$, the various system parameters, and appropriate initial conditions.

6.39   **a.**   (1), (2), (3), and **b.**   (1), (2), (3). Find $q(i)$, $i \geq 0$, the inverse transform of $Q(z)$ determined in the corresponding part of 5.16 (from the difference equations in 4.28) in terms of the forcing functions $p(i)$ and $p(i, k)$, the various system parameters, and appropriate initial conditions.

Chapter VII | System Performance: Measures and Environmental Effects

In Chapters I through VI we have developed the basic methodology to describe discrete, linear, time-invariant systems by linear difference equations and to determine solutions to these equations. We are now in a position to discuss the application of this methodology to the analysis and design of working systems.

The solution of the system difference equation provides the basis for analysis of the performance of the system represented. Of perhaps more importance, as shown in Section 6.7, it also provides the foundation for the design of a system. In the simplest situations, the design task is the determination of the values of the parameters of the control decision rule, while in other cases designation of the form of the control decision rule, the selection of the physical properties of the process, and even the configuration of the system as a whole must be considered. Thus, in some cases we may need a means of finding the best values of the proportionality constant K and difference constant L for a rapid-response proportional-plus-difference controller. In other cases, however, we may need a way of comparing our "best" proportional-plus-difference controller with the "best" of other configurations such as proportional-plus-summation or even a general third- or fourth-order controller. In still other situations, we may even have to consider systems which are not rapid response or which feed forward or predict exogenous inputs.

Obviously, as mentioned in Section 1.1 in our discussion of systems theory, all of these evaluations and comparisons require the adoption of clearly defined criteria and, hopefully, a quantitative scale on which to measure the degree of fulfillment of each such criterion. When these are combined with the information contained in the solution to the system difference equation

184

and any applicable cost factors, all the essential ingredients are available for the determination of optimal or at least preferred designs.

A brief discussion of applicable criteria for the evaluation of discrete, linear control systems is presented in the first section of this chapter. Mention is made of economic criteria such as profit maximization and cost minimization, but the major emphasis is given to several surrogate measures directed at various aspects of system performance, many of which are traditionally used by control engineers. This is followed by an examination of the effects on system performance of several patterns of perturbations commonly experienced in control-system operation. To facilitate discussion, we will limit our attention in this chapter to the context of choosing a preferred value for the proportionality constant K of the cylinder-diameter controller introduced in Section 2.2. In Chapter VIII we continue the task of finding a preferred value for the proportionality constant of a simple proportional controller, but broaden the context to include the effects of sampling and instrumentation errors.

7.1 Control-System Performance Criteria

Economic Criteria

Ideally, any system designer would like to find *the* design which would somehow minimize the total costs of design, construction, installation, operation, and performance of the system. Attainment of this ideal, however, is almost impossible for a number of reasons.

First is the usual difficulty of gathering dependable cost data of any kind; although current developments in management information systems could assist in the areas of design, construction, installation, and operation. Costs or benefits of system performance, however, are usually much harder to measure. Often, performance is not even thought of in monetary terms. For example, what is the dollar benefit of providing a tracking radar with a given maximum tracking error? If the radar is an air-traffic-control radar at a busy airport, performance is measured in terms of the probability of an accident. Conversion of accident potential into dollar losses may be done reasonably well for the property and equipment losses that could be involved, but gets very difficult when one attempts to include the costs of human lives. A similar problem arises with air-defense radars. The tracking error is related in complicated ways to the probability of a successful intercept; but, given enough assumptions about the tactics used by the attackers and interceptors, evaluations of these probabilities are made. Department of Defense studies are full of them. However, to attach a cost to an unsuccessful intercept, even though frequently attempted by enterprising OR teams, requires even more assumptions about what the attacker is going to do after he escapes interception. Will he destroy a command post and thwart a potentially successful maneuver, or will he bomb concentrations of civilians? In either case, what is the dollar cost?

A less extreme example involves a production process. How will an appliance with a small imperfection in its porcelain finish affect the customer? How are the number, size, and location of such blemishes related to customer acceptance of the product and hence to the profits for the product line? Can sales of other products of the same company also be affected and, if so, how?

In spite of increasing acceptance of the systems approach by industrial managers, government and military officials, and others, the answers to such questions as these are often still too ponderous to consider explicitly in any but the simplest cases. A compromise approach is to consider system performance as a benefit or describe it by its effectiveness on any appropriate scale. This puts us in the realm of cost effectiveness or cost-benefit analysis, about which an extensive literature has emerged [see, for example, Hitch and McKean (1965), Quade and Boucher (1968), and Kendall (1971)]. The practical outcome of most such analyses, however, is to do one of the following: (1) maximize the system performance (effectiveness, benefit) for a given total cost of design, construction, installation, and operation, or (2) minimize the total cost to realize a given level of performance. Obviously, either approach calls for a quantitative measure of system performance such as probability of collision, probability of intercept, or some function of the distributions of the number and size of blemishes in the finish of porcelain covered products. Hence our interest in surrogate measures. After all, once such measures are established, we may eventually learn how to convert them to a dollar scale.

In some areas, however, the notion of a strict profit or cost model for system evaluation can be reasonably practical. We are all familiar with the multitude of cost models used to compute " optimal" inventory levels [see, for example, Churchman et al. (1957), Scarf et al. (1963), Naddor (1966), and Wagner (1969)]. Process controllers can also be selected on the basis of economic criteria. Consider, for example, a chemical reactor whose operation can be modeled as a generalized process controller. It is not uncommon for the efficiency, and hence the quantity of chemical produced per unit of input materials, to fall off approximately as the square of the deviation of reaction temperature $m(i)$ from the design value D. Thus, assuming a constant output flow rate of chemical, the raw material cost per period could be expressed as

$$C_M = c_0 + c_1[m(i) - D]^2, \tag{7.1.1}$$

where c_0 and c_1 are cost coefficients whose values depend on the actual materials involved. Similarly, the cost of adjusting the process setting is often proportional to the size of the adjustment. If we ignore for now the effects of measurement errors, this adjustment cost can be expressed as

$$C_A = c_2 a(i) = c_2 |g_D[D - m(0), D - m(1), \ldots, D - m(i)]|, \quad (7.1.2)$$

where c_2 is the cost of a unit adjustment. Thus, the total cost of operation for T_0 sampling periods is

$$C(T_0) = \sum_{i=0}^{T_0} \{c_0 + c_1[m(i) - D]^2 + c_2|g_D[D - m(0), \ldots, D - m(i)]|\}. \quad (7.1.3)$$

Regardless of the cost functions used, for any given control rule g_D, the expression for $m(i)$ as found from the appropriate system equation can be substituted in the cost equation so that the total cost of operation is expressed as a function of the control-rule parameters and the assumed forcing functions. Optimal values of the control-rule parameters can then be found from the resulting expression for $C(T_0)$ either by analytic or numerical means as the functions involved dictate. This process can then be repeated for as many different control-decision rules g_D as time and energy permit and the configuration with the best optimal performance chosen. Note, however, that all of the so-called optimal values of control-rule parameters found in this way depend on the specific forcing functions assumed in finding the $m(i)$. If this is the pattern that will actually be experienced in operation, the solution is then indeed optimal. In the much more likely event, however, that the actual patterns of perturbations are not completely known, it would behoove the designer to try several of the more likely possibilities and combinations thereof. The nasty thing about this procedure is that different assumed patterns of perturbations may result in different optimal values of control-rule parameters and even different rankings of control rules. At this point judgment enters, hopefully the judgment of those most familiar with the process being controlled, the environments in which it must operate, and the performance objectives to be fulfilled. Under these circumstances it would seem presumptuous to call the design selected " optimal." At best it is " good," and hopefully better than most. The word often used to describe it is " preferred design " or " preferred system," a wording we strongly " prefer."

Traditional Measures of System Performance

The most common performance measures traditionally used by control engineers involve various aspects of system response to fairly simple deterministic input functions. These measures include criteria known as stability, speed of response, peak overshoot, steady-state response, and frequency response.

Stability

Conceptually, stability is the capability of a system to recover from the effects of minor, short-lived perturbations. More precisely, a system is said to be stable if the deviation between the desired output and the actual output following an impulse perturbation approaches zero as i increases indefinitely. For an unstable system, this deviation would increase without bound as i increased. Since, as was shown in Section 6.8, the response of a system to any forcing function $f(i)$ is the convolution of $f(i)$ and the system's unit-impulse response $g(i)$, any system designed to operate for any significant length of time must be stable. Generally, therefore, stability is a prerequisite for system usefulness and must be assured while examining system performance relative to other design criteria. It is so important that we devote all of Chapter IX to stability and some tests to determine whether or not a system is stable.

Speed of Response

Speed of response is a measure of how quickly a system can readjust to a change in desired operating level or compensate for the effects of a particular perturbation. It is usually described in terms of the length of time or number of adjustments required to compensate for some percentage of the deviation caused by the change involved. Most frequently, the term is applied to responses to a step change in desired level or to a step perturbation.

Peak Overshoot

In many circumstances a system responding to a change in desired operating level or to a perturbation will overshoot its mark one or more times before finally settling down to the value being sought. The largest of these overshoots is called the peak overshoot. Should this peak be excessive, an otherwise excellent system (e.g., stable, fast response, etc.) could damage itself or ruin its product during transient periods following changes in inputs. In many cases, the magnitude of the peak overshoot increases as the speed of response is increased, so the need for some sort of trade-off between these two criteria is quite common.

Steady-State Response

The steady-state response of a system to a given forcing function is the limiting response reached by a stable system after the various adjustment transients have died out. Specifically, it is the limit approached by the system output of a stable system as i increases without bound. It is, therefore, based on the particular solution of the system difference equation. In situations where external disturbances are relatively infrequent, the system is at steady state most of the time. Thus its performance is well described by the steady-state response.

Frequency Response

The frequency response of a system describes the steady-state output of the system for a sinusoidal forcing function with unit amplitude and zero phase angle, i.e., $f(i) = \sin wi$. As illustrated by (4.3.13) in Example 2 of Section 4.3, the particular solution resulting from a sinusoidal forcing function is sinusoidal with the same frequency. Therefore, the steady-state solution will be a sinusoid with frequency equal to that of the forcing sine function but with amplitude and phase determined by the system structure and parameters. What we call the frequency response consists of plots of the amplitude and phase angle of the steady-state output. The amplitude plot is called the *amplitude frequency response* and the phase-angle plot the *phase response*. The frequency response thus completely describes the steady-state output resulting from a sinusoidal component of any given frequency which may be present in current or anticipated system inputs.

Frequency response information is extremely important not only because of likely sinusoidal perturbations which result from such factors as rotating machinery but also because many general functions of time can be described in terms of sinusoidal components by means of the Fourier series expansion for periodic functions of time and by the Fourier or Laplace transforms for more general functions of time. Thus steady-state system performance can be predicted from the system frequency response for general forcing functions. Those not familiar with these concepts are referred to the work of Guillemin (1949) or Truxal (1955).

All of the measures just defined will be illustrated in the remaining sections of this chapter in conjunction with specific forms of forcing function.

Probability-Based Measures of System Performance

Once the methodology was developed to permit ready inclusion of random inputs in system models, a whole new set of performance criteria became both available and necessary for the evaluation of system design and operation. These criteria are all based on various aspects of probability theory and involve such factors as the distribution of system output, various moments of this distribution, and combinations of probabilistic and traditional factors. For example, we might wish the output of a production process to conform as closely as possible to a specified probability distribution to provide tight fits with a known population of mating parts. A common situation is to attempt to maximize the probability of the output falling between tolerance limits agreed to by the customer or minimizing the probability of running out of inventory. In the latter case, to avoid going broke by avoiding shortages by simply carrying exhorbitant cushion stocks, other factors such as limiting total carrying costs must be included. Where uniformity of product is important such as producing matched sets of piston rings or silverware, minimizing the variance of the output is important. In some situations we might simultaneously wish to limit output variance while maximizing the rate of system response.

There are obviously many possibilities for formulation of probability-based criteria and combinations of probabilistic and traditional criteria. Some of these will be mentioned briefly in Section 7.6 of this chapter. Extensive discussion of such criteria including eight detailed examples is presented in Chapter VIII.

7.2   The Cylinder-Diameter Controller

As has been stated, the context in this chapter for our discussion and illustration of the effects on system performance of various patterns of per-turbations is the cylinder-diameter controller described in Section 2.2. The system difference equation for this controller as derived in (2.2.5) is

$$m(i + 1) - (1 - K)m(i) = KD + p(i). \tag{7.2.1}$$

To concentrate attention on the effects of the perturbations $p(i)$, we let $D = 0$. As explained in Section 6.7, this can be done with no loss of generality. Thus we have

$$m(i + 1) - (1 - K)m(i) = p(i). \tag{7.2.2}$$

Solution of (7.2.2) by either variation of parameters or the z transform yields

$$m(i) = m(0)(1 - K)^i + \sum_{n=0}^{i-1} p(n)(1 - K)^{i-1-n}. \tag{7.2.3}$$

For further simplification, $m(0)$ is assumed to be zero, so (7.2.3) becomes simply

$$m(i) = \sum_{n=0}^{i-1} p(n)(1 - K)^{i-1-n}, \tag{7.2.4}$$

or, if one prefers,

$$m(i) = \sum_{n=0}^{i-1} p(i - 1 - n)(1 - K)^n. \tag{7.2.5}$$

Before considering specific functional forms for $p(i)$, let us consider the stability of this system, the criterion we said was so highly important for con-tinued useful system operation. Note that each individual $p(i)$ can be written

$$p(i) = p(i)\,\delta(i - n) = \begin{cases} p(i) & \text{for} \quad n = i, \\ 0 & \text{elsewhere.} \end{cases} \tag{7.2.6}$$

That is, each $p(i)$ is an impulse with magnitude $p(i)$ occurring at time i. Since stability is based on the impulse response, we see that $m(i)$ as given by (7.2.4) is nothing more than the sum of the sequence of impulses $p(n)$, $n = 0, \ldots,$ $i - 1$, each one multiplied by a power of $1 - K$. Moreover, the exponent of the coefficient of $p(n)$ is an increasing function of i. Thus, if this controller is to effectively eliminate the effects of these perturbations as i increases without bound, the coefficient of each $p(n)$ must decrease in magnitude with increasing i. Obviously for this to occur,

$$|1 - K| < 1 \qquad (7.2.7)$$

or

$$0 < K < 2. \qquad (7.2.8)$$

Therefore, for the cylinder-diameter controller to be stable, the proportionality constant K must satisfy (7.2.8). Inequality (7.2.8) defines what is called the "stable range" of K. Values of K below zero or above two cause the effects of the $p(n)$ to increase with increasing i, the condition we have referred to as "instability." The special cases of $K = 0$ and $K = 2$ result in perpetuating the effects of a given $p(n)$ with constant magnitude. Some refer to this situation as "marginal stability," although many authors simply class all systems which are not actually stable as unstable.

The conditions of stability, marginal stability, and instability are illustrated for selected values of K in Table 7.2.1 for the simple case of $p(0) = p$

Table 7.2.1

Effects of Controller Proportionality Constant on System Stability[a]

K	$m(i)$				$\lim_{i \to \infty} m(i)$	System performance
	$m(1)$	$m(2)$	$m(3)$	$m(4)$		
-0.5	p	$1.5p$	$2.25p$	$3.375p$	∞	Unstable
0	p	p	p	p	p	Marginally stable
0.5	p	$0.5p$	$0.25p$	$0.125p$	0	Stable
1.0	p	0	0	0	0	Stable
1.5	p	$-0.5p$	$0.25p$	$-0.125p$	0	Stable
2.0	p	$-p$	p	$-p$	$\pm p$	Marginally stable
2.5	p	$-1.5p$	$2.25p$	$-3.375p$	$\pm \infty$	Unstable

[a] $p(0) = p$, $p(i) = 0$ for $i \geq 1$.

and $p(i) = 0$ for $i \geq 1$. Note that for $K < 1$ the effect of the perturbation has the same sign for all i, whereas for $K > 1$ the sign of the effect alternates with increasing i. Note also the increasing rapidity with which the effects of $p(0)$ are eliminated as $|K| \to 1$.

In the remaining sections of this chapter we will examine in turn the effects on the system as represented by (7.2.4) of perturbation patterns in the forms of an impulse, a step, a sinusoid, and a sequence of random variables. Criteria for the selection of the parameter K will also be considered.

7.3   Impulse Perturbation

In the format developed for system descriptions in Chapter II, each $p(i)$ is defined as an incremental change in process output caused by factors external to the process and its controller. Thus if the tool is moved by outside perturbations an amount p between the boring of the nth and $(n + 1)$st engine blocks, this would appear in the pattern of perturbations as an impulse of magnitude p at $i = n$. Relative to the position of the tool, the function appears as a step. Such an event could result from a tool slipping in its tool post or a tool change with improper positioning of the new tool. No further shifts due to outside influences are assumed. It is of course up to the controller to compensate for the shift.

Assume a perturbation of magnitude p occurs at $i = 0$, i.e.,

$$p(i) = p\ \delta(i) = \begin{cases} p & \text{for}\quad i = 0, \\ 0 & \text{elsewhere.} \end{cases} \tag{7.3.1}$$

Substitution into (7.2.4) yields

$$m(i) = p(1 - K)^{i-1}, \qquad i = 1, 2, \dots. \tag{7.3.2}$$

If desired, one could now plot $m(i)$ vs i for various values of K and pick K on the basis of the shape of the curves obtained. Such curves are not included here since Table 7.2.1 indicates the general patterns followed by $m(i)$ following such a perturbation for selected values of K. As a practical matter, if the system is to operate for long periods of time, K must be limited to the stable range $0 < K < 2$. Only in this range will the perturbation eventually be fully compensated. From (7.3.2) it is apparent that correction of impulse perturbations is most rapid for $K = 1$. Furthermore, overshoots may be avoided completely by limiting K to the range $0 < K \le 1$.

A more quantitative approach to picking a value for K can be made in terms of the speed of response of the system to an impulse perturbation. For example, suppose a design criterion were stated that the system must reduce the magnitude of any perturbation which may occur to a fraction no greater than a of its original value within n periods after the occurrence of the perturbation. Thus, for an impulse at time zero,

$$|m(n)| = |p(1 - K)^n| \le |ap|. \tag{7.3.3}$$

Canceling p and solving for K yields

$$1 - a^{1/n} < K < 1 + a^{1/n}. \tag{7.3.4}$$

Similarly, if K has been chosen, the number of periods required to reduce a given perturbation to a fraction a of its original value can be found by solving (7.3.3) for n to obtain

$$n \geq \frac{\log a}{\log(1 - K)}. \tag{7.3.5}$$

7.4 Step Perturbation

A step-type perturbation of magnitude p beginning at $i = n$ is represented by

$$p(i) = pu(i - n) = \begin{cases} 0 & \text{for} \quad i < n, \\ p & \text{for} \quad i \geq n. \end{cases} \tag{7.4.1}$$

Relative to the tool position, it appears like a ramp function in that, once started, the tool continues to be displaced an amount p during each time period thereafter. This type of function often serves as a reasonable representation of tool wear or of heat loss from a chemical reactor.

For a step perturbation of magnitude p beginning at time zero,

$$m(i) = \sum_{n=0}^{i-1} p(1 - K)^{i-1-n} = p \sum_{n=0}^{i-1} (1 - K)^{i-1-n}. \tag{7.4.2}$$

Note that in the summation as n goes from zero to $i - 1$, the exponent of $1 - K$ goes from $i - 1$ to zero. Thus, (7.4.2) can be equivalently expressed as:

$$m(i) = p \sum_{n=0}^{i-1} (1 - K)^n, \tag{7.4.3}$$

which may be summed to yield

$$m(i) = p \frac{(1 - K)^i - 1}{(1 - K) - 1} = \frac{p}{K} [1 - (1 - K)^i]. \tag{7.4.4}$$

Figure 7.4.1 shows plots of $m(i)$ for several values of K for $p = 1$. We have already assumed $m(0) = 0$, so for all K, $m(1) = 1$, the value of the step. At $i = 1$ the deviation from $D = 0$ is detected and the controller goes into action. However, the range of K for which any real benefit is realized is $1 < K < 2$. For $K < 1$, even for K in the stable range, the effect of the controller is to move $m(i)$ away from the desired value of zero. For $K = 1$, no action takes place. For $K = 2$, a continual shift from $+1$ to -1 occurs which, even though $m(i)$ gets no worse, provides no improvement and is very likely to fatigue the tool-adjustment mechanism. For $K > 2$ the system is unstable causing reversals of increasing magnitude.

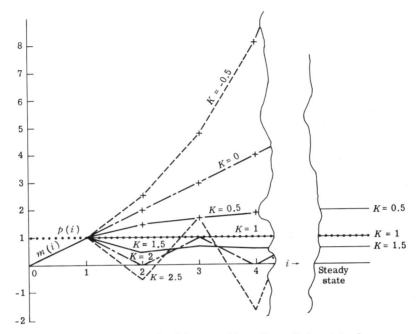

FIG. 7.4.1. Plots of $m(i)/p$ versus i for $p(i)$ a unit step at $i = 0$.

From (7.4.4) it is seen that $m(i)$ approaches p/K as i increases without bound. The value p/K is thus the steady-state output and represents the amount by which the proportional controller lags the step perturbation. Thus, a first-order system cannot ever fully compensate for a step perturbation; although the magnitude of the steady-state error is inversely proportional to K and can therefore be held to a limit just slightly exceeding $p/2$ while maintaining stable system operation. Because of this relationship, a design specification limiting steady-state error imposes a lower bound on K. It will be shown in Chapter X how a second-order system can be designed to eliminate completely the steady-state error from a step perturbation.

7.5   Sinusoidal Perturbation

Many processes consisting of rotating parts or working in environments in which physical or acoustical coupling to rotating machinery is possible are subject to sinusoidal perturbations of whatever frequency or frequencies may be present. Periodic patterns which can be approximated by sinusoids can also be caused by multihead or multistation indexing and positioning equipment or even individual differences among workmen as shifts change. Such devices could very well be present in an engine-block line. In addition, we have noted that by means of the Fourier transform, general functions of time can be expressed in terms of sinusoidal components. For all of these reasons, sinusoidal forcing functions are of great importance in system analysis and design.

To investigate this type of perturbation, let

$$p(i) = p \sin(wi + \theta), \tag{7.5.1}$$

where p is the amplitude of the sinusoid, w the angular velocity in radians per time period (or 2π times the frequency, where frequency is measured in cycles per period), and θ the phase angle in radians, i.e., the value of the argument of the sinusoid at $i = 0$. Substitution of (7.5.1) into (7.2.4) yields

$$m(i) = \sum_{n=0}^{i-1} [p \sin(wn + \theta)](1 - K)^{i-1-n}. \tag{7.5.2}$$

Little intuitive insight can be gained from this form of $m(i)$ except that the process operating level does have a sinusoidal variation with angular frequency w. However, even though we may be concerned with transient phenomena such as peak overshoot, usually the primary interest when sinusoidal inputs are involved is in steady-state performance. We therefore will confine our attention here to finding and analyzing the steady-state output, i.e., the limit of $m(i)$ as i increases without bound.

Derivation of Steady-State Output

Let m be the steady-state output, i.e.,

$$m = \lim_{i \to \infty} m(i). \tag{7.5.3}$$

To derive m by actually finding the limit of (7.5.2) is difficult at best. The usual alternative to be followed in such cases is to apply the z transform and use the final-value theorem. However, the final-value theorem does not apply where sinusoids are involved. A special approach for sinusoidal forcing functions must be followed instead.

We start by finding the z transform of $m(i)$. Had the z transform been used to find $m(i)$ in the first place, it would of course already be available. $m(i)$ is the convolution of the forcing function $p \sin(wi + \theta)$ and the weighting function $(1 - K)^i$ with a real backward translation of one time period (since the upper limit of the summation is $i - 1$ and not i). Therefore,

$$Z[m(i)] = \frac{1}{z} \cdot Z[p \sin(wi + \theta)] \cdot Z[(1 - K)^i]. \tag{7.5.4}$$

From (5.1.11),

$$Z[(1 - K)^i] = z/[z - (1 - K)]. \tag{7.5.5}$$

To find the z transform of $\sin(wi + \theta)$, we first make use of the trigonometric identity

$$\sin(wi + \theta) = \sin wi \cos \theta + \cos wi \sin \theta \tag{7.5.6}$$

to obtain

$$Z[p \sin(wi + \theta)] = p \cos \theta \cdot Z[\sin wi] + p \sin \theta \cdot Z[\cos wi]. \tag{7.5.7}$$

From (5.1.14) through (5.1.18), $Z[\sin wi]$ can be expressed as

$$Z[\sin wi] = \frac{z \sin w}{(z - e^{jw})(z - e^{-jw})}, \tag{7.5.8}$$

where $j = \sqrt{-1}$. Similarly, $Z[\cos wi]$ can be expressed as

$$Z[\cos wi] = \frac{z^2 - z \cos w}{(z - e^{jw})(z - e^{-jw})}. \tag{7.5.9}$$

In both (7.5.8) and (7.5.9) the denominator was left in factored form since partial fraction expansion is to follow. Combining (7.5.4) through (7.5.9) yields

$$Z[m(i)] = \frac{p}{z} \cdot \frac{\cos \theta(z \sin w) + \sin \theta(z^2 - z \cos w)}{(z - e^{jw})(z - e^{-jw})} \left[\frac{z}{z - (1 - K)} \right]$$

$$= \frac{p \cos \theta(z \sin w)}{(z - e^{jw})(z - e^{-jw})[z - (1 - K)]}$$

$$+ \frac{p \sin \theta(z^2 - z \cos w)}{(z - e^{jw})(z - e^{-jw})[z - (1 - K)]}. \tag{7.5.10}$$

We now employ a partial fraction expansion in which, to simplify the presentation, we will work with each term of (7.5.10) separately. Dividing the first term by z and expanding yields

$$\frac{p \cos \theta \sin w}{(z - e^{jw})(z - e^{-jw})[z - (1 - K)]} = \frac{K_+}{z - e^{jw}} + \frac{K_-}{z - e^{-jw}} + \frac{K_1}{z - (1 - K)}, \tag{7.5.11}$$

where K_+, K_-, and K_1 are found by the methods discussed in Chapter VI to be

$$K_+ = \frac{p \cos \theta}{2j[e^{jw} - (1 - K)]}, \tag{7.5.12}$$

$$K_- = \frac{p \cos \theta}{-2j[e^{-jw} - (1 - K)]}, \tag{7.5.13}$$

and

$$K_1 = \frac{p \cos \theta \sin w}{(1 - K)^2 - 2(1 - K) \cos w + 1}. \tag{7.5.14}$$

Note that K_1 is the coefficient for the term which will become $(1 - K)^i$ in the i domain and vanish as i increases without bound (assuming stable operation). Since K_1 will not enter the expression for steady-state m, there was really no need to have calculated it at all. With this in mind, we will simply use the symbol K_1 in what follows.

Substitution for K_+ and K_- in (7.5.11) gives

$$\frac{p \cos \theta}{2j[e^{jw} - (1 - K)]} \cdot \frac{z}{z - e^{jw}}$$

$$+ \frac{p \cos \theta}{-2j[e^{-jw} - (1 - K)]} \cdot \frac{z}{z - e^{-jw}} + K_1 \frac{z}{z - (1 - K)}. \quad (7.5.15)$$

Multiplying by z and taking the inverse transformation yields

$$\frac{p \cos \theta}{2j[e^{jw} - (1 - K)]} (e^{jw})^i + \frac{p \cos \theta}{-2j[e^{-jw} - (1 - K)]} (e^{-jw})^i + K_1(1 - K)^i$$

$$= \frac{p \cos \theta}{2j} \left[\frac{e^{jwi}}{e^{jw} - (1 - K)} - \frac{e^{-jwi}}{e^{-jw} - (1 - K)} \right] + K_1(1 - K)^i. \quad (7.5.16)$$

Now

$$\frac{1}{e^{jw} - (1 - K)} = \frac{1}{\cos w + j \sin w - (1 - K)} = \frac{[\cos w - (1 - K)] - j \sin w}{[\cos w - (1 - K)]^2 + \sin^2 w} \quad (7.5.17)$$

which can be expressed as

$$\frac{1}{e^{jw} - (1 - K)} = \frac{1}{[(1 - K)^2 - 2(1 - K) \cos w + 1]^{1/2}} e^{j\phi}, \quad (7.5.18)$$

where ϕ is defined as

$$\phi = \tan^{-1} \frac{-\sin w}{\cos w - (1 - K)}. \quad (7.5.19)$$

Similarly,

$$\frac{1}{e^{-jw} - (1 - K)} = \frac{1}{[(1 - K)^2 - 2(1 - K) \cos w + 1]^{1/2}} e^{-j\phi}. \quad (7.5.20)$$

Substitution of (7.5.18) and (7.5.20) into (7.5.16) gives

$$\frac{p \cos \theta}{2j[(1 - K)^2 - 2(1 - K) \cos w + 1]^{1/2}} [e^{jwi}e^{j\phi} - e^{-jwi}e^{-j\phi}] + K_1(1 - K)^i$$

$$= \frac{p \cos \theta}{[(1 - K)^2 - 2(1 - K) \cos w + 1]^{1/2}} \left[\frac{e^{j(wi + \phi)} - e^{-j(wi + \phi)}}{2j} \right] + K_1(1 - K)^i$$

$$= \frac{p \cos \theta}{[(1 - K)^2 - 2(1 - K) \cos w + 1]^{1/2}} \sin(wi + \phi) + K_1(1 - K)^i.$$

$$(7.5.21)$$

By the same procedure, the second term of (7.5.10) can be shown to be

$$\frac{p \sin \theta}{[(1 - K)^2 - 2(1 - K) \cos w + 1]^{1/2}} \cos(wi + \phi) + K_2(1 - K)^i, \quad (7.5.22)$$

where

$$K_2 = \frac{p \sin \theta \cos w}{(1 - K)^2 - 2(1 - K) \cos w + 1}, \quad (7.5.23)$$

so the two terms (7.5.21) and (7.5.22) can be combined to give:

$$m(i) = \frac{p[\sin(wi + \phi) \cos \theta + \cos(wi + \phi) \sin \theta]}{[(1 - K)^2 - 2(1 - K) \cos w + 1]^{1/2}} + C(1 - K)^i, \quad (7.5.24)$$

where $C = K_1 + K_2$. The identity in (7.5.6) permits simplification of this expression to

$$m(i) = \frac{p}{[(1 - K)^2 - 2(1 - K) \cos w + 1]^{1/2}} \sin(wi + \theta + \phi) + C(1 - K)^i.$$
$$(7.5.25)$$

Now as i increases, the first term of (7.5.25) will oscillate with angular velocity w and amplitude $p/[(1 - K)^2 - 2(1 - K) \cos w + 1]^{1/2}$, but the term involving $(1 - K)^i$ approaches zero. Thus, the steady-state output is still a function of i and is given by

$$m_i = \frac{p}{[(1 - K)^2 - 2(1 - K) \cos w + 1]^{1/2}} \sin(wi + \theta + \phi). \quad (7.5.26)$$

Note that the steady state is not really "steady" in the sense of approaching and maintaining a constant value, but instead the system performance has stabilized in the form of a sampled sinusoid with constant frequency, amplitude, and phase. The angular velocity w is the same as that of the sinusoidal forcing function. The amplitude $|m|$, represented by

$$|m| = \frac{p}{[(1 - K)^2 - 2(1 - K) \cos w + 1]^{1/2}}, \quad (7.5.27)$$

is proportional to the amplitude p of the perturbing sinusoid and to the frequency-dependent weighting factor $[(1 - K)^2 - 2(1 - K) \cos w + 1]^{-1/2}$, which we shall refer to as the *relative amplitude*. It represents the amount by which a sinusoid in the system forcing function is amplified (should it be greater than one) or attenuated (where it is less than one). The phase angle θ is the same as the phase angle of the forcing function, while ϕ, which is frequency sensitive, indicates how many radians the steady-state output sinusoid lags (for $\phi < 0$) or leads (for $\phi > 0$) the forcing function sinusoid, knowledge of which can be valuable in helping to diagnose the source of an unwanted periodic perturbation.

System Frequency Response

The relationships of the relative amplitude and phase angle ϕ with w constitute what was referred to in the beginning of this chapter as the frequency response of the system. As has been stated, the function relating relative amplitude to w is known as the amplitude frequency response and the function relating the phase angle to w is called the phase frequency response or more commonly just the phase response of the system. Both can be useful in the study of discrete, linear systems, as will be discussed below.

Amplitude Frequency Response

Figure 7.5.1 is a plot of the amplitude frequency response of the simple proportional controller. It shows how the relative amplitude

$$[(1 - K)^2 - 2(1 - K) \cos w + 1]^{-1/2}$$

varies with w over the range $0 \leq w \leq 2\pi$ for a few representative values of K. The values $K = 0$ and $K = 2$ are included to indicate system performance at

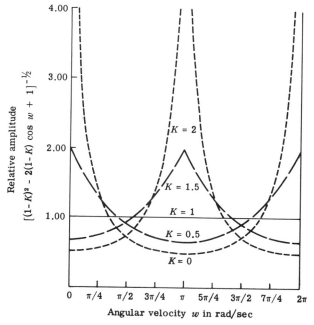

FIG. 7.5.1. Amplitude frequency response of simple proportional controller for selected values of K.

the bounds of stable operation. The patterns simply repeat for values of $w > 2\pi$.

For a proportionality constant of unity the system passes all frequencies with their amplitudes neither amplified nor attenuated as indicated by the horizontal line at a relative amplitude value of one. As K deviates from one,

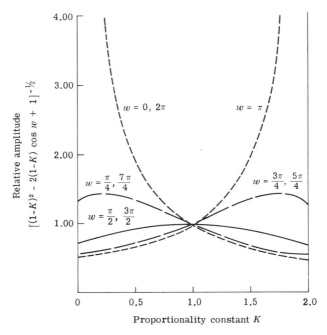

FIG. 7.5.2. Relative amplitude versus proportionality constant for simple proportional controller for selected values of angular velocity.

however, the relative amplitude is observed to become increasingly frequency sensitive until at zero and two (which are of course values not actually attainable for a stable system but represent limiting values of K) it varies from a low of 0.5 to infinity. It may also be observed that for K values in the range $0 < K < 1$, the relative amplitude peaks at a value of $1/K$ at even integer multiples of π rad $(0, 2\pi, 4\pi, \ldots)$ and reaches a minimum of $1/(2 - K)$ at odd integer multiples of π ($\pi, 3\pi, 5\pi, \ldots$). For K values in the range $1 < K < 2$, the situation is reversed. Based on these results, if a system designer suspects a significant sinusoidal perturbation of a given frequency to affect his system, he should tend to pick K values for which the relative amplitude is small for the frequency involved. If several such perturbations are anticipated, some compromise must be sought based on estimates of the input amplitudes (the p

values) for each sinusoid and the effects on the system output of each of the various frequencies. If there is no particular knowledge of what frequencies might be present, $K = 1$ provides a handy minimax solution to the design problem.

For situations in which other design criteria must also be considered in the selection of the proportionality constant, it is convenient to have a plot of the relative amplitude against K for various values of w. Even though all the information contained in such plots can be obtained from Fig. 7.5.1, they do provide a ready reference for determining the effects of a possible sinusoidal perturbation of a given frequency for any K used and could be helpful in the selection of K. Figure 7.5.2 illustrates such plots for several values of w.

From this figure we see that the relative amplitudes resulting from values of w in the first and fourth quadrants, i.e., $0 \le w < \pi/2$ and $3\pi/2 < w \le 2\pi$, approach their minimum values as K increases toward 2. These frequencies of course result in cos w being positive. On the other hand, the relative amplitudes resulting from values of w in the second and third quadrants, i.e., for $\pi/2 < w < 3\pi/2$ where the cosine is negative, approach their minimum values as K decreases toward zero. At $w = \pi/2$ and $w = 3\pi/2$ where cos $w = 0$, the relative amplitude curve is symmetrical around $K = 1$ so approaches its minimum of $\sqrt{2}/2$ at both $K = 0$ and $K = 2$. Furthermore, by examining the derivatives of $[(1 - K)^2 - 2(1 - K) \cos w + 1]^{-1/2}$ it can be found that each curve reaches a maximum of $1/|\sin w|$ at $K = 1 - \cos w$. Thus if it is important to limit the effects of any particular frequency of sinusoidal perturbation, the system designer should attempt to avoid use of proportionality constants in the neighborhood of $1 - \cos w$.

Phase Frequency Response

Should phase information be of importance in any application, phase frequency response curves could be plotted for various values of K using Eq. (7.5.19). Likewise, plots of phase angle vs K for various values of w could be constructed. In general, however, phase information is not usually of prime importance in the analysis of the long-run performance of a discrete system. As long as w is not equal to a rational number times π (a rational number is the ratio of two integers), $\sin(wi + \phi)$ will eventually cover the entire range of -1 to $+1$ as i continues to increase, regardless of the value of ϕ. Therefore, no phase plots are included here.

For the situation in which w is an integer multiple of π (a special case of rational numbers), θ assumes significant importance as may be seen from the following. Let $w = n\pi$, where n is a nonnegative integer. Then

$$\begin{aligned} \sin(wi + \theta + \phi) &= \sin(n\pi i + \theta + \phi) \\ &= \sin n\pi i \cos(\theta + \phi) + \cos n\pi i \sin(\theta + \phi). \quad (7.5.28) \end{aligned}$$

Since ni must be an integer, $\sin n\pi i = 0$ and $\cos n\pi i = \pm 1$ depending upon whether the product ni is odd ($\cos n\pi i = -1$) or even ($\cos n\pi i = +1$). Further, (7.5.19) becomes

$$\phi = \tan^{-1} \frac{-\sin nw}{\cos nw - (1 - K)} = \begin{cases} \tan^{-1} \dfrac{0}{K} = \tan^{-1} 0 = 0 & \text{for} \quad n \text{ even,} \\[3mm] \tan^{-1} \dfrac{0}{-2 + K} = \tan^{-1} \pi = 0 & \text{for} \quad n \text{ odd,} \end{cases}$$

(7.5.29)

i.e., ϕ is zero for all n. Under these conditions, therefore, $\sin(wi + \theta + \phi)$ reduces to $\pm \sin \theta$. For n even, the plus sign holds for all i; for n odd, the plus sign holds for i even and the minus sign for i odd. Thus,

$$m_i \Big|_{\substack{w = n\pi \\ n \text{ integer}}} = \begin{cases} \dfrac{p}{K} \sin \theta & \text{for} \quad n \text{ even,} \\[3mm] (-1)^i \dfrac{p}{2 - K} \sin \theta & \text{for} \quad n \text{ odd.} \end{cases}$$

(7.5.30)

These expressions show the importance of proper synchronization between the production intervals of the discrete system and the source of the periodic perturbation with angular velocity $n\pi$ to hold θ to as near zero or π as possible. If perfect synchronization can be achieved, the steady-state deviation from the desired level caused by this type of sinusoid can be held to zero regardless of K, and K can be selected on the basis of other criteria.

For the general case of w being equal to a rational number times π rad per time period, the interrelationship between θ and ϕ can be important. To illustrate, assume a perturbation

$$p(i) = p \sin(2\pi w i/3 + \theta) \tag{7.5.31}$$

on a simple proportional controller with $K = 1$. Then, from (7.5.27), $|m| = p$, and from (7.5.19), $\phi = 4\pi/3$ so that

$$m_i = p \sin\left(\frac{2\pi i}{3} + \theta + \frac{4\pi}{3}\right) = \begin{cases} p \sin\left(\theta + \dfrac{4\pi}{3}\right) & \text{for} \quad i = 3n, \\[3mm] p \sin(\theta) & \text{for} \quad i = 3n + 1, \\[3mm] p \sin\left(\theta + \dfrac{2\pi}{3}\right) & \text{for} \quad i = 3n + 2, \end{cases}$$

(7.5.32)

where n is an integer. These three outputs will continue to repeat in order, their exact values being dictated by θ. An optimal value of θ could be determined for any given criterion of system performance. Conversely, if θ were not adjustable, optimal values of K could be derived or perhaps optimal combinations of K and θ determined if both are controllable.

7.6   Random Perturbations

It is not at all uncommon for the perturbations experienced by many types of processes to be random in nature. Impurities in raw materials, the build-up and breaking away of chips on cutting tools, patterns of customer orders, and the performance of human operators all tend to impose random variations on the outputs of processes.

A random variable is a variable whose value cannot be forecast exactly, but can be described according to a probability distribution. Therefore, if the probability distribution of $p(i)$ is known or can be either logically derived or empirically estimated, we can make statements such as "the probability that $p(i)$ is equal to or less than a_i, where a_i is a real number, is b," where b is of course in the range $0 \leq b \leq 1$. A comprehensive discussion of probability distributions and various descriptors of random processes may be found in the work of Laning and Battin (1956, Chapter 3).

We now examine the effects of random perturbations on the performance of the engine-block cylinder-diameter control system. These effects of course depend on the nature of the process which generates the perturbations.

The simplified equation for the system output with $D = 0$ is

$$m(i) = \sum_{n=0}^{i-1} p(n)(1 - K)^{i-1-n} \tag{7.6.1}$$

from which we see that $m(i)$ is the weighted sum of i random variables $p(0)$ through $p(i - 1)$. Thus, $m(i)$ is a random variable.

Ideally, we would like to be able to develop an expression for the probability distribution function of $m(i)$, but, because each $p(i)$ is weighted by a different power of $1 - K$, such a derivation is impractical if not actually impossible for all distributions of the $p(i)$ except the normal distribution. For normally distributed $p(i)$, $m(i)$ will also be normal. When the $p(i)$ are approximately normally distributed, it may well be that a normal approximation of $m(i)$ will suffice. Even when the $p(i)$ are more generally distributed, for some combinations of control-system parameters (K for the example case here) and distribution functions for the $p(i)$, the central limit theorem would indicate that $m(i)$ should at least approach being normal for large i.

Whether or not the exact distribution of $m(i)$ can be determined or even approximated, it is informative to compute various moments of the distribution, in particular the mean $E[m(i)]$ and the variance $V[m(i)]$. E is used here as the expectation operator. Variance is defined as the second moment around the mean. The simplest case to model is when the $p(i)$ are independently and identically distributed, a case which is not necessarily uncommon for discrete systems. In this context, "identically" means that the process generating the $p(i)$ is stationary.

In the remainder of this section we will discuss the derivation of expressions for the mean and variance of $m(i)$ for the various situations that can arise when one considers dependence and stationarity.

$p(i)$ Mutually Independent

When the $p(i)$ are independently distributed, the mean and variance of $m(i)$ can be expressed, respectively, as

$$E[m(i)] = E\left[\sum_{n=0}^{i-1} p(n)(1 - K)^{i-1-n}\right] = \sum_{n=0}^{i-1} E[p(n)](1 - K)^{i-1-n} \quad (7.6.2)$$

and

$$V[m(i)] = V\left[\sum_{n=0}^{i-1} p(n)(1 - K)^{i-1-n}\right]$$

$$= \sum_{n=0}^{i-1} V[p(n)][(1 - K)^{i-1-n}]^2 = \sum_{n=0}^{i-1} V[p(n)](1 - K)^{2(i-1-n)}. \quad (7.6.3)$$

The derivation of (7.6.3) makes use of the theorem which states that for independently distributed random variables X_i,

$$V\left[\sum_{i=1}^{q} a_i X_i\right] = \sum_{i=1}^{q} a_i^2 V[X_i]. \quad (7.6.4)$$

In the situation where the $p(i)$ are not identically distributed, (7.6.2) and (7.6.3) are about all that can be generally stated about the mean and variance of $m(i)$ unless some pattern is specified which relates $E[p(i)]$ and $V[p(i)]$ to i. If such patterns can be identified, it may be possible to carry out one or both of the summations in (7.6.2) and (7.6.3) to obtain closed-form expressions for $E[m(i)]$ and $V[m(i)]$. If so, steady-state values can also often be obtained by taking the limit as $i \to \infty$. Thus, if $E[p(i)] = p \cdot A^i$, $A > 0$, (7.6.2) can be summed to yield

$$E[m(i)] = p \frac{A^i - (1 - K)^i}{A - 1 + K}.$$

Similarly, if $V[p(i)] = \sigma_p^2 B^i$, $B > 0$, we obtain from (7.6.3),

$$V[m(i)] = \sigma_p^2 \frac{B^i - (1 - K)^{2i}}{B - (1 - K)^2}.$$

Let us define the steady-state mean and variance as M and V, respectively, i.e.,

$$M = \lim_{i \to \infty} E[m(i)] \quad (7.6.5)$$

and

$$V = \lim_{i \to \infty} V[m(i)]. \tag{7.6.6}$$

Assuming K in the stable range, in the examples cited $M = 0$ for $A < 1$. For $A > 1$, M is unbounded. Similar results are obtained for V.

For the case in which the $p(i)$ are identically as well as independently distributed, let

$$E[p(i)] = p \qquad \text{for all } i \tag{7.6.7}$$

and

$$V[p(i)] = \sigma_p^2 \qquad \text{for all } i. \tag{7.6.8}$$

Note that (7.6.7) and (7.6.8) would result from setting A and B equal to unity in the example above. Substitution of (7.6.7) yields

$$E[m(i)] = \frac{p}{K}[1 - (1 - K)^i] \tag{7.6.9}$$

and substitution of (7.6.8) and (7.6.3) gives

$$V[m(i)] = \frac{\sigma_p^2}{2K - K^2}[1 - (1 - K)^{2i}]. \tag{7.6.10}$$

The expression for $E[m(i)]$, (7.6.9), is observed to be identical to that obtained for the step perturbation in (7.4.4). Steady-state values of $E[m(i)]$ and $V[m(i)]$ are found from (7.6.5) and (7.6.6), respectively, as

$$M = p/K \tag{7.6.11}$$

and

$$V = \sigma_p^2/(2K - K^2). \tag{7.6.12}$$

Should the $p(i)$ be independently normally distributed, $m(i)$ will also be normally distributed. Therefore, the probability density function of $m(i)$ is

$$\phi(m(i)) = \frac{1}{(2\pi V[m(i)])^{1/2}} \exp\left[-\frac{\{m(i) - E[m(i)]\}^2}{2V[m(i)]}\right], \tag{7.6.13}$$

where $E[m(i)]$ and $V[m(i)]$ are given by (7.6.9) and (7.6.10), respectively. In the steady state, the density function becomes

$$\phi(m) = \frac{1}{(2\pi V)^{1/2}} \exp\left[-\frac{(m - M)^2}{2V}\right], \tag{7.6.14}$$

where M and V are given by (7.6.11) and (7.6.12), respectively.

p(i) Interdependent

Joint Probabilities

When the $p(i)$ are not mutually independent, the mean of $m(i)$ is still given by (7.6.2), but the task of variance determination is considerably more complicated than for the independent case. In this situation we need to know not only the marginal distributions of each individual $p(i)$, but also the joint distribution of each pair of $p(i)$. This joint distribution, which is sometimes called the second probability distribution [see Laning and Battin (1956, Chapter 3)], can be represented as

$$\Phi_2(i, a_i; n, a_n) = \Pr[p(i) \le a_i, p(n) \le a_n], \qquad (7.6.15)$$

where the subscript 2 designates a "second probability." Equation (7.6.15) is the joint probability that both $p(i) \le a_i$ and $p(n) \le a_n$, where a_i and a_n are constants. If $p(i)$ is discrete, we let

$$\phi_2(i, a_i; n, a_n) = \Pr[p(i) = a_i, p(n) = a_n] \qquad (7.6.16)$$

and, for $p(i)$ continuous,

$$\phi_2(i, a_i; n, a_n) = \frac{\partial^2 \Phi_2(i, a_i; n, a_n)}{\partial p(i)\, \partial p(n)}. \qquad (7.6.17)$$

For consistency, it must follow that

$$\phi(i, a_i) = \int_{p(n) = -\infty}^{\infty} \phi_2(i, a_i; n, a_n)\, dp(n) \qquad (7.6.18)$$

and, through symmetry,

$$\phi(n, a_n) = \int_{p(i) = -\infty}^{\infty} \phi_2(i, a_i; n, a_n)\, dp(i), \qquad (7.6.19)$$

where ϕ represents the marginal probability for $p(i)$ discrete and the marginal probability density should $p(i)$ be continuous.

In situations in which the probability functions and density functions are the same for all values of the time index i, i.e., when the process generating the perturbations is stationary, the marginal probability function may be written as

$$\Phi(a) = \Pr[p(i) \le a] \qquad \text{for all } i. \qquad (7.6.20)$$

Therefore, for $p(i)$ discrete,

$$\phi(a) = \Pr[p(i) = a] \qquad \text{for all } i, \qquad (7.6.21)$$

and for $p(i)$ continuous,

$$\phi(a) = \frac{d\Phi(a)}{dp(i)} \qquad \text{for all } i. \qquad (7.6.22)$$

Also for stationary processes, the second probability distribution and density function depend only on the time difference between the occurrence of the two variables involved. That is, for

$$\tau = i - n, \tag{7.6.23}$$

we may write

$$\Phi_2(\tau) = \Pr[p(i) \le a_i; \, p(i - \tau) \le a_{i-\tau}] \qquad \text{for all} \quad i. \tag{7.6.24}$$

Note, τ may be positive or negative. Similarly, for $p(i)$ discrete,

$$\phi_2(\tau) = \Pr[p(i) = a_i; \, p(i - \tau) = a_{i-\tau}] \qquad \text{for all} \quad i, \tag{7.6.25}$$

and for $p(i)$ continuous,

$$\phi_2(\tau) = \frac{\partial^2 \Phi_2(\tau)}{\partial p(i) \, \partial p(i - \tau)} \qquad \text{for all} \quad i. \tag{7.6.26}$$

As is well known, when the $p(i)$ are mutually independent, the second probability reduces to the product of the first probabilities of the two events involved.

$p(i)$ Identically Distributed

Let us assume that the $p(i)$ are identically distributed with mean p and variance $\sigma_p{}^2$. Let us further assume that the second probability distribution is such that the first joint moment, better known as the autocorrelation function, is

$$E[p(i)p(i - \tau)] = \int_{p(i) = -\infty}^{\infty} \int_{p(i-\tau) = -\infty}^{\infty} p(i)p(i - \tau)\phi_2(\tau) \, dp(i) \, dp(i - \tau)$$

$$= \alpha_{11}(\tau) \qquad \text{for all} \quad i. \tag{7.6.27}$$

Were the $p(i)$ not identically distributed, it would be necessary to write this moment as

$$\alpha_{11}(i, i - \tau) = \int_{p(i) = -\infty}^{\infty} \int_{p(i-\tau) = -\infty}^{\infty} p(i)p(i - \tau)$$

$$\cdot \phi_2(i, p(i); i - \tau, p(i - \tau)) \, dp(i) \, dp(i - \tau). \tag{7.6.28}$$

In either case, this function figures prominently in the variance expressions for nonindependent sequences of random variables. (In passing, we note that the general joint moment r, s is defined as

$$\alpha_{rs}(\tau) = E[p(i)^r p(i - \tau)^s] \qquad \text{for all} \quad i.) \tag{7.6.29}$$

The variance is of course defined as the second moment of a distribution around its mean. This can be shown to be equal to the second moment around zero minus the square of the mean. Thus, for $m(i)$,

$$V[m(i)] = E[m(i)^2] - \{E[m(i)]\}^2. \tag{7.6.30}$$

For the stationary case being considered, $E[m(i)]$ is given by (7.6.9) which is simply squared to give the second term of $V[m(i)]$. The first term can be developed as follows:

$$E[m(i)^2] = E\left[\sum_{n=0}^{i-1} p(n)(1-K)^{i-1-n}\right]^2$$

$$= E\left\{\left[\sum_{n=0}^{i-1} p(n)(1-K)^{i-1-n}\right]\left[\sum_{q=0}^{i-1} p(q)(1-K)^{i-1-q}\right]\right\}. \quad (7.6.31)$$

Multiplication of the series of terms from each summation results in a total of i^2 terms. i of these terms correspond to products of terms with the same time index, i.e., terms for which $n = q$. The remaining $i(i-1)$ terms correspond to products of terms with different time indices, i.e., for which $n \neq q$. Furthermore, there will be two such terms for every combination of time indices. For example, a term involving $p(3)$ and $p(5)$ results both for $n = 3$, $q = 5$ and for $n = 5$, $q = 3$. Therefore, (7.6.31) can be represented by the following two summation terms:

$$E[m(i)^2] = E\left[\sum_{n=0}^{i-1} p(n)^2(1-K)^{2(i-1-n)}\right.$$

$$\left. + 2\sum_{n=1}^{i-1}\sum_{q=0}^{n-1} p(n)p(q)(1-K)^{2(i-1)-n-q}\right]$$

$$= \sum_{n=0}^{i-1} E[p(n)^2](1-K)^{2(i-1-n)}$$

$$+ 2\sum_{n=1}^{i-1}\sum_{q=0}^{n-1} E[p(n)p(q)](1-K)^{2(i-1)-n-q}. \quad (7.6.32)$$

Since the $p(i)$ are identically distributed, we are more interested in the difference between n and q than in their specific values. Therefore, we let

$$\tau = n - q \quad (7.6.33)$$

or

$$q = n - \tau. \quad (7.6.34)$$

Substitution for q in (7.6.32) yields

$$E[m(i)^2] = \sum_{n=0}^{i-1} E[p(n)^2](1-K)^{2(i-1-n)}$$

$$+ 2\sum_{n=1}^{i-1}\sum_{\tau=1}^{n} E[p(n)p(n-\tau)](1-K)^{2(i-1-n)+\tau}. \quad (7.6.35)$$

Making use of (7.6.27), the definitional formula for $\alpha_{11}(\tau)$, we get

$$E[m(i)^2] = \sum_{n=0}^{i-1} \alpha_{11}(0)(1-K)^{2(i-1-n)} + 2\sum_{n=1}^{i-1}\sum_{\tau=1}^{n} \alpha_{11}(\tau)(1-K)^{2(i-1-n)+\tau},$$

$$(7.6.36)$$

which can be shown to be equivalent to

$$E[m(i)^2] = \alpha_{11}(0) \sum_{n=0}^{i-1} (1 - K)^{2(i-1-n)} + 2 \sum_{\tau=1}^{i-1} \alpha_{11}(\tau)(1 - K)^{\tau} \sum_{n=\tau}^{i-1} (1 - K)^{2(i-1-n)}.$$

(7.6.37)

Summing over n yields

$$E[m(i)^2] = \frac{\alpha_{11}(0)}{2K - K^2} [1 - (1 - K)^{2i}]$$

$$+ 2 \sum_{\tau=1}^{i-1} \alpha_{11}(\tau)(1 - K)^{\tau} \frac{[1 - (1 - K)^{2(i-\tau)}]}{2K - K^2},$$

(7.6.38)

which is about as far as the expression can be simplified without specific knowledge of the process correlation function $\alpha_{11}(\tau)$. Equation (7.6.38) can now be substituted into (7.6.30) along with (7.6.9), the expression for $E[m(i)]$, to obtain $V[m(i)]$.

Actually, $V[m(i)]$ can be obtained much more simply than by brute-force manipulation of (7.6.30) by taking advantage of the superposition property of linear systems. Instead of considering $p(i)$ as a random variable with mean p, we instead consider $p(i)$ as consisting of two statistically independent components, one a step function of magnitude equal to the known bias, and the other an unbiased random variable with all the aforementioned variational properties except that its expected value is now zero. The response of a linear system to the sum of any number of perturbations is just the sum of the responses of the system to each one separately. Likewise, since the two components of $p(i)$ can be considered statistically independent, the variance of the system performance is the sum of the variances of the system responses to the two components individually. Since the variance of the constant step perturbation representing the bias is zero, the total system variance will be equal to that of the unbiased random component alone. Another way of looking at it is that since the variance is the second moment of the distribution around its own mean, scale changes which simply shift the mean in no way affect the variance. Thus, in the absence of other extraneous sources of random variation,

$$V[m(i)] = E\{m(i)^2 \,|\, E[m(i)] = 0\}.$$

(7.6.39)

Furthermore, since the correlation function of $p(i)$ is related to the covariance function by

$$\alpha_{11}(i, i - \tau) = \sigma_p(i, i - \tau) + E[p(i)]E[p(i - \tau)],$$

(7.6.40)

where $\sigma_p(i, i - \tau)$ is the autocovariance between $p(i)$ and $p(i - \tau)$, then for an unbiased perturbation,

$$\alpha_{11}\{i, i - \tau \,|\, E[p(i)] = E[p(i - \tau)] = 0\} = \sigma_p(i, i - \tau),$$

(7.6.41)

and, if the $p(i)$ are identically distributed,

$$\alpha_{11}\{\tau \,|\, E[p(i)] = 0\} = \sigma_p(\tau).$$

(7.6.42)

Note that for $\tau = 0$, $\sigma_p(i, i)$ represents the covariance of a given $p(i)$ with itself which is by definition the variance $V[p(i)]$. Thus for $p(i)$ unbiased,

$$\alpha_{11}\{i, i \mid E[p(i)] = 0\} = \sigma_p(i, i) = V[p(i)]. \tag{7.6.43}$$

Therefore, from (7.6.38), (7.6.39), and (7.6.42),

$$\begin{aligned}
V[m(i)] &= E\{m(i)^2 \mid E[m(i)] = 0\} \\
&= \frac{\sigma_p(0)}{2K - K^2} [1 - (1 - K)^{2i}] \\
&\quad + 2 \sum_{\tau=1}^{i-1} \sigma_p(\tau)(1 - K)^\tau \frac{[1 - (1 - K)^{2(i-\tau)}]}{2K - K^2}. \tag{7.6.44}
\end{aligned}$$

Using (7.6.44), one can find $V[m(i)]$ directly without having to bother subtracting the expected value term after finding $E[m(i)^2]$ from (7.6.38). Furthermore, covariance information can usually be computed much more readily from empirical data using commonly available computer programs than correlation function information.

Example   To illustrate the calculation of the variance of the system output for intercorrelated, unbiased, identically distributed random perturbations, assume

$$\sigma_p(\tau) = \sigma_p^2 A^{-|\tau|}, \tag{7.6.45}$$

where A is a number greater than one, indicating positive correlation that decreases as the separation in time increases. Substituting in (7.6.44) gives

$$\begin{aligned}
V[m(i)] &= \frac{\sigma_p^2}{2K - K^2} [1 - (1 - K)^{2i}] \\
&\quad + 2 \sum_{\tau=1}^{i-1} \sigma_p^2 A^{-|\tau|}(1 - K)^\tau \frac{[1 - (1 - K)^{2(i-\tau)}]}{2K - K^2} \\
&= \frac{\sigma_p^2}{2K - K^2} \left\{ [1 - (1 - K)^{2i}] + 2 \sum_{\tau=1}^{i-1} \frac{(1 - K)^\tau - (1 - K)^{2i-\tau}}{A^\tau} \right\} \\
&= \frac{\sigma_p^2}{2K - K^2} \left\{ [1 - (1 - K)^{2i}] + 2 \left[\sum_{\tau=1}^{i-1} \left(\frac{1 - K}{A} \right)^\tau \right. \right. \\
&\quad \left. \left. - (1 - K)^{2i} \sum_{\tau=1}^{i-1} [A(1 - K)]^{-\tau} \right] \right\} \\
&= \frac{\sigma_p^2}{2K - K^2} \left\{ [1 - (1 - K)^{2i}] + 2 \frac{1 - K}{A - (1 - K)} \left[1 - \left(\frac{1 - K}{A} \right)^{i-1} \right] \right. \\
&\quad \left. - 2 \frac{(1 - K)^{2i} - A^2 \left(\frac{1 - K}{A} \right)^{i+1}}{A(1 - K) - 1} \right\} \tag{7.6.46}
\end{aligned}$$

which could perhaps be further simplified or at least expressed in one of a variety of alternative forms.

Instead of pursuing $V[m(i)]$ further, let us examine the steady-state variance which is found to be, for a stable system,

$$V = \lim_{i \to \infty} V[m(i)] = \frac{\sigma_p^2}{2K - K^2} \left\{ 1 + 2 \frac{1 - K}{A - (1 - K)} \right\}. \qquad (7.6.47)$$

Note that the multiplier $\sigma_p^2/(2K - K^2)$ is the steady-state variance for the uncorrelated process, i.e., when the $p(i)$ are all mutually independent. Note also that V in the example can be either greater or less than the steady-state variance for the uncorrelated case depending on whether the term $2(1 - K)/[A - (1 - K)]$ is positive or negative; however, since this term can never be less than minus one for $A > 1$, V will always be positive as the variation must. Thus, a system designer can conceivably take advantage of interdependence among system perturbations to reduce variance in the random variation in the system output.

It may be seen from (7.6.47) that for a given K, the term

$$2(1 - K)/[A - (1 - K)]$$

is a decreasing function of A and vanishes as A becomes infinite. Thus an A value of infinity corresponds to an uncorrelated process. It may also be noted that this term equals zero for $K = 1$, so that the effects of intercorrelations on the variance of $m(i)$ can be eliminated or effectively limited relative to the uncorrelated case by selecting K values equal to or at least close to one.

The optimizers among the readers are probably already thinking in terms of setting the derivative of V with respect to K equal to zero to find the value of K which minimizes V for any given A. This is very good thinking for designers of systems where continual variation in output and the resulting control activity is costly, degrades the product, or produces undue fatigue in various moving parts. Furthermore, derivation of an expression for a K to minimize V puts the student one up on the exercises at the end of this chapter. However, one will soon find out that the stationary points of dV/dK must be found by solving a cubic equation, a possible but unpleasant task. Checking of the second derivative to determine which such points minimize V is even worse. For this reason, some simple observations regarding dV/dK might be just as useful in providing insight to the designer.

After some manipulation it is found from (7.6.47) that

$$\frac{dV}{dK} = \sigma_p^2 \frac{-K^3 - (A + 3)K^2 - (A^2 - 2A - 3)K + (A^2 - 1)}{[K^3 + (A - 3)K^2 + 2(1 - A)K]^2}. \qquad (7.6.48)$$

Since the denominator is nonnegative, the values of K for which $dV/dK = 0$ are the roots of the equation formed by setting the numerator equal to zero. Substitution of a few values of K into the numerator indicates the presence

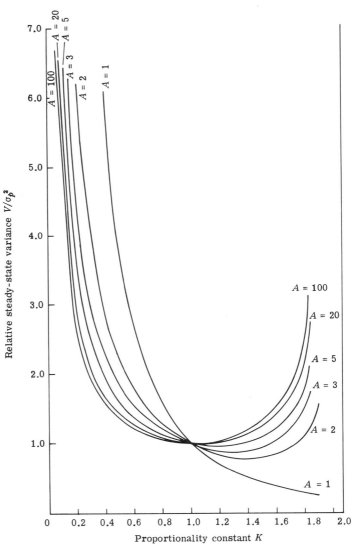

FIG. 7.6.1. Relative steady-state variance versus proportionality constant.

of one and only one root in the stable range $0 < K < 2$. Substitution of a few values of K for some selected A values into (7.6.47), the formula for V, shows that V decreases from an infinite value at $K = 0$ to σ_p^2 at $K = 1$. A minimum is reached for some K in the range $1 < K < 2$ after which V increases as K increases and becomes unbounded at $K = 2$. The K for which V is minimized increases from 1 toward 2 as A decreases toward unity (i.e., as the correlation effects become longer lived). This is borne out by the fact that for all $A > 1$, dV/dK is negative for $0 < K < 1$, indicating V is decreasing in this range, and is positive at $K = 2$, showing that V is increasing there. The switch in sign comes of course at the actual minimum point which, from the evidence stated, must lie in the range $1 < K < 2$. Figure 7.6.1 presents a few example plots of V/σ_p^2 vs K for various values of A. Note that as A increases the curve tends toward being symmetrical around $K = 1$.

7.7   Selection of K

The material in this chapter confirms what the experienced designer already knows, that the so-called best design for a system depends not only on the physical properties of the system itself but also on the environment in which the system must operate and the relative importance of various performance and output criteria. Stated in terms of the engine-block cylinder-diameter control system, what constitutes the "best" value for the proportionality constant K depends on the characteristics of the $p(i)$, the specifications on the operation of the engine block, and the costs of deviations from these specifications.

Any design effort, therefore, must begin with a careful analysis of the working environment of the system. This must be accompanied by a study of the product or output of the system and its ultimate use or disposition. An evaluation of significant costs associated with system operation and maintenance, product use and repair, and customer relations should be made. These items should then be tied together by an appropriate cost or performance model (or models) based on the solution of the system difference equation. Preferred values of K can then be obtained by manipulation of this model subject to any applicable constraints. In cases where the $p(i)$ can be fairly completely described, theoretically optimal values of K might be determined. In practice, one might select an initial value for K on the basis of an analysis of the type just described, but make final fine-tuning adjustments in K once the system is installed and operating based on empirical observations of system performance.

In the next chapter we will investigate the determination of the proportionality constant K for the first-order system in the context of the generalized discrete process controller. Various combinations of measurement error, perturbations, and performance criteria will be discussed.

Exercises

7.1

   a.   Prepare an annotated bibliography of at least five articles involving cost-effectiveness (cost-benefit) analysis. Comments should speak to the points of philosophy, technique, application, and evaluation of contribution.

   b.   Prepare a paper discussing the application of cost-effectiveness (cost-benefit) analysis to the design of control systems. Consider possible variations in applicability caused by the size, complexity, environment, and control objective of the system.

   c.   Prepare a paper presenting the application of cost-effectiveness (cost-benefit) analysis to the design of a control system of interest to a systems engineer. Specifically discuss applicable performance criteria and measures, costs and their relation to system operation, and anticipated perturbations.

7.2

   a.   List three deterministic measures of control-system performance other than those mentioned specifically in this chapter (although combinations of those mentioned are encouraged) and discuss the application of each to a specific type of control system.

   b.   Repeat part (a) using probabilistic performance measures.

   c.   Repeat part (a) using combinations of deterministic and probabilistic performance measures.

7.3

   a.   Find the z transform of (7.2.2), the system difference equation for the cylinder-diameter controller with $D = 0$, solve the result for $M(z)$, and use the final-value theorem to verify that the stable range of K is $0 < K < 2$.

   b.   Find the z transform of (7.2.1), the system difference equation for the cylinder-diameter controller explicitly including D and $p(i)$, and use the result to find the response of $m(i)$ to a unit-impulse change in D.

   c.   Use the results of part (b) to verify that the stable range of K is $0 < K < 2$.

7.4   For the cylinder-diameter controller, find the number of adjustments n required to reduce an impulse perturbation $p(i)$ to a fraction no greater than a of its original value for the following values of K and a:

   a. $K = 0.80, a = 0.10$ **b.** $K = 0.39, a = 0.22$
   c. $K = 1.61, a = 0.22$ **d.** $K = 0.20, a = 0.10$

7.5   For the cylinder-diameter controller, find the values of proportionality constant K which will reduce an impulse perturbation $p(i)$ to a fraction no larger than a of its original value within n adjustments after its occurrence for the following values n and a:

a. $n = 2$, $a = 0.10$ **b.** $n = 4$, $a = 0.22$
c. $n = 10$, $a = 0.10$ **d.** $n = 5$, $a = 0.50$

7.6 **(a)** through **(d)**.   The same as 7.4 (a) through (d), respectively, except considering the impulse to be in D instead of $p(i)$.

7.7 **(a)** through **(d)**.   The same as 7.5 (a) through (d), respectively, except considering the impulse to be in D instead of $p(i)$.

7.8 For the cylinder-diameter controller, plot the output $m(i)$ for $i = 1$ through 5 given $m(0) = D = 0$ and a unit step perturbation occurring at $i = 0$ for the following values of proportionality constant K. Also show the steady-state value of $m(i)$.

a. $K = 0.25$ **b.** $K = 0.75$ **c.** $K = 1.25$ **d.** $K = 1.75$

7.9 For the cylinder-diameter controller, plot the output $m(i)$ for $i = 1$ through 5 given $m(0) = p(i) = 0$ and $D = 2.0$ for the following values of K. Also show the steady-state value of $m(i)$.

a. $K = 0.25$ **b.** $K = 0.50$ **c.** $K = 0.75$ **d.** $K = 1.0$
e. $K = 1.25$ **f.** $K = 1.50$ **g.** $K = 1.75$

7.10 **(a)** through **(g)**.   For the cylinder-diameter controller, plot the output $m(i)$ for $i = 1$ through 5 given $D = p(i) = 0$ and $m(0) = -1.0$ for the value of K in the corresponding part of Exercise 7.9. Also show the steady-state value of $m(i)$.

7.11 The transfer function of a certain system is given by $G(z) = 3/(z - 0.5)$.
   a. Find the unit-impulse response $g(i)$ of this system.
   b. Show that the system is stable.
   c. At what value of i will the output resulting from an impulse of magnitude 9 occurring at $i = 2$ be reduced to a value equal to or less than 2?
   d. For a unit step input at $i = 0$ find the steady-state response.
   e. For an input $r(i) = 5 \cos(3w + 2)$ find the maximum magnitude of the system output and the value of i at which it occurs.
   f. For the input of part (e) find the steady-state response.

7.12 Two costs often significant in the operation of a process controller are the cost of a deviation between desired and actual output and the cost of making an adjustment. Thus, in the engine-block line the cost affected by the operation of the controller which determines the diameter of the number one cylinder can be modeled as $C(i, K) = C_D[D - m(i)] + C_C[a(i)]$, where $C(i, K)$ is the incremental cost of the ith engine block, $C_D[\cdot]$ is the functional relationship between deviation of the cylinder diameter and the resulting cost, and $C_C[\cdot]$ is the functional relationship between the adjustment made and the resulting cost. For $m(0) = D = 0$ and $p(i)$ a step of size p at $i = 0$.

find the value of the control parameter K which will minimize the steady-state cost per engine block for the following cost functions:

a. $C(i, K) = C_1[m(i)]^2 + C_2[a(i)]^2$, C_1, C_2 constants
b. $C(i, K) = C_1[m(i)]^2 + C_2|a(i)|$, C_1, C_2 constants
c. $C(i, K) = C_1|m(i)| + C_2|a(i)|$, C_1, C_2 constants

7.13   (a) through (c).   The same as 7.12 except instead of minimizing the steady-state cost per period we seek to minimize the total cost for the first three blocks.

7.14   (a) through (c).   The same as 7.12 except given $m(0) = p(i) = 0$ and $D \neq 0$.

7.15   (a) through (c).   The same as 7.13 except given $m(0) = p(i) = 0$ and $D \neq 0$.

7.16   Assume the engine-block line is subject to a sinusoidal perturbation with amplitude p and angular frequency w radians per time period. A cost $C_m[m]$ is incurred when the magnitude of the steady-state sinusoid is $|m|$ and a cost $C_C[K]$ results when the control parameter has a value of K. Determine the necessary and sufficient conditions for the cost $C(K) = C_m[m] + C_C[K]$ to be minimized for the following cost functions. Where possible, analytically determine an expression for the optimum K which minimizes $C(K)$. Where an analytic solution for K is not possible, indicate a procedure to follow to determine K for given values of w and p.

a. $C(K) = C_1 m^2 + C_2 K$, C_1, C_2 constants
b. $C(K) = C_1 m^2 + C_2 K^2$, C_1, C_2 constants
c. $C(K) = C_1|m| + C_2 K$, C_1, C_2 constants
d. $C(K) = C_1|m| + C_2 K^2$, C_1, C_2 constants

7.17   (a) through (d).   Find the optimal value of K for the appropriate part of Exercise 7.16 for $w = \pi/4$ rad and $p = 1$.

7.18   (a) through (d).   Find the optimal value of K for the appropriate part of Exercise 7.16 for $w = \pi/3$ and $p = 1$.

7.19   (a) through (d).   Find the optimal value of K for the appropriate part of Exercise 7.16 for $w = \pi/2$ and $p = 1$.

7.20   (a) through (d).   Plot the optimal value of K and the resulting minimum value of $C(K)$ as functions of w, $0 \leq w \leq 2\pi$, for the appropriate part of 7.16 assuming $p = 2$, $C_1 = 1$, and $C_2 = 3$.

7.21   (a) through (c).   The same as 7.13 except given $p(i) = p \sin(\pi i/6 + \pi/3)$.

7.22   From the appropriate derivatives of $|m|$, the amplitude of the steady-state output sinusoid of the cylinder-diameter controller as given by (7.5.27), verify that each plot of amplitude vs K for a given angular frequency w reaches a maximum of $1/\sin w$ at $K = 1 - \cos w$.

7.23   For the engine-block line with $m(0) = D = 0$ and $p(i)$ a sinusoid with angular frequency w rad per time period, plot the phase frequency response for the following values of K:

    a. 0   **b.** 0.5   **c.** 1.0   **d.** 1.5   **e.** 2.0

7.24   For the engine-block line with $m(0) = D = 0$ and $p(i)$ a sinusoid with angular frequency w rad per time period, plot the angle ϕ by which the steady-state output leads or lags the input sinusoid vs K for the following values of w:

    a. 0   **b.** $\pi/4$ **c.** $\pi/2$ **d.** $3\pi/4$ **e.** π
    f. $5\pi/4$ **g.** $3\pi/2$ **h.** $7\pi/4$ **i.** 2π

7.25   For the engine-block line with $m(0) = D = 0$, $p(i) = p \sin(2\pi/3 + \theta)$, and $K = 1$, find the value of θ which minimizes the following steady-state cost criteria (n is a large positive integer):

    a. $\displaystyle C(m_i) = \sum_{i=3n}^{3n+2} |m_i|$ **b.** $\displaystyle C(m_i) = \sum_{i=3n}^{3n+2} m_i{}^2$

    c. $\displaystyle C(m_i) = \sum_{i=3n}^{3n+2} |m_i{}^3|$

7.26   **(a)** through **(c)**.   The same as 7.25 except given $K = 0.5$.

7.27   For the engine-block line with $m(0) = D = 0$, $p(i) = p \sin(\pi i/2 + \theta)$, and $K = 1$, find the value of θ which minimizes the following steady-state cost criteria (n is a large positive integer):

    a. $\displaystyle C(m_i) = \sum_{i=4n}^{4n+3} |m_i|$ **b.** $\displaystyle C(m_i) = \sum_{i=4n}^{4n+3} m_i{}^2$

    c. $\displaystyle C(m_i) = \sum_{i=4n}^{4n+3} |m_i{}^3|$

7.28   For the engine-block line with $m(0) = D = 0$ and $p(i)$ a stationary, unbiased, independently distributed random variable with variance $\sigma_p{}^2$, find the values of proportionality constant K which will limit the steady-state variance of the process output to 4×10^{-6} in.^{2} for the following values of $\sigma_p{}^2$:

    a. 2×10^{-6} in.^{2} **b.** 4×10^{-6} in.^{2} **c.** 8×10^{-6} in.^{2}

7.29

    a. For the engine-block line with $m(0) = D = 0$ and $p(i)$ a stationary, unbiased, random variable with autocovariance function $\sigma_p(\tau)$ given by:

τ	$\sigma_p(\tau)$
0	σ_p^2
± 1	$0.5\sigma_p^2$
all other values	0

derive an expression for the variance of the process output $V[m(i)]$.

   b.   Find the steady-state output variance, assuming stable operation.

   c.   For what values of K will the steady-state variance in (b) be greater than it would have been had the perturbation been independently distributed with variance σ_p^2?

7.30   (**a**) through (**c**).   The same as 7.29 except $\sigma_p(\tau)$ given by:

τ	$\sigma_p(\tau)$
0	σ_p^2
± 1	$-0.5\sigma_p^2$
all other values	0

7.31   (**a**) through (**c**).   The same as 7.29 except $\sigma_p(\tau) = 2^{-|\tau|}\sigma_p^2$.

7.32   (**a**) through (**c**).   The same as 7.29 except $\sigma_p(\tau) = (-2)^{-|\tau|}\sigma_p^2$.

7.33   The steady-state variance of process output V for the engine-block line with $m(0) = D = 0$ and $p(i)$ a stationary, unbiased, random variable with autocovariance function $\sigma_p(\tau) = A^{-|\tau|}\sigma_p^2$, $|A| > 1$, is given by (7.6.47). Find the value of the proportionality constant K which minimizes V for:

   a. $A = 2$ **b.** $A = -2$ **c.** $A = 10$ **d.** $A = -10$
   e.   a general $A > 1$ **f.**   a general $A < -1$

7.34   (**a**) through (**d**).   The same as 7.16 except
$$p(i) = p_1\mu(i) + p_2 \sin wi.$$

7.35 through **7.37**   The same as 7.17 through 7.19, respectively, except
$$p(i) = p_1\mu(i) + p_2 \sin wi.$$

7.38   (**a**) through (**c**).   The same as 7.25 except
$$p(i) = 2p\mu(i) + p \sin(2\pi i/3 + \theta).$$

7.39   (**a**) through (**c**).   The same as 7.26 except
$$p(i) = 2p\mu(i) + p \sin(2\pi i/3 + \theta).$$

7.40   (**a**) through (**c**).   The same as 7.27 except
$$p(i) = 2p\mu(i) + p \sin(\pi i/2 + \theta).$$

| **Parameter Selection in First-Order Systems Considering Sampling and Instrumentation Errors**

In this chapter we consider the actual selection of the value of the proportionality constant K of a discrete, linear, first-order controller for several specific combinations of perturbations and system performance criteria. The context used is the simple proportional controller configuration of the generalized discrete process controller presented in Section 2.3. The discrete representation of the generalized rapid-response controller appears in Fig. 2.3.2. Recall the inclusion in this system of an additive measurement error $\varepsilon(i)$ in the determination of the output $m(i)$.

This system was chosen as the basis for this portion of our presentation because the simplicity of the system structure and process dynamics permits us to focus our attention on the topics which are new to this chapter. These include the nature and properties of sampling and instrumentation errors and the effects of each on system performance, performance criteria which take the effects of these kinds of errors into account, and various approaches to finding values of the control-system parameter K which optimize these criteria.

In the sections which follow, we first briefly review the simple proportional process controller with measurement error and then take a detailed look at the two potential components of measurement error, namely sampling and instrumentation errors and their effects on process output. This is followed by eight optimization examples, each based on a different combination of perturbations and performance criteria.

8.1   The Simple Proportional Process Controller with Measurement Error

Relative to the system shown in Fig. 2.3.2, the simple proportional controller corresponds to the unit-impulse response of the control decision rule being

$$g_D = K\,\delta(i) \tag{8.1.1}$$

and the unit-impulse response of the physical properties of the process being

$$g_P = \delta(i) \tag{8.1.2}$$

resulting in

$$s(i) = c(i) = K\rho(i) = K[D - m(i) - \varepsilon(i)]. \tag{8.1.3}$$

Thus, the system equation, given originally in (2.3.21), is

$$m(i+1) - (1-K)m(i) = KD - K\varepsilon(i) + p(i). \tag{8.1.4}$$

As has been done heretofore, we will assume $m(i)$ is measured relative to the desired output level which results in D vanishing from the system equation. This yields

$$m(i+1) - (1-K)m(i) = -K\varepsilon(i) + p(i), \tag{8.1.5}$$

the form of the system equation to be used as a basis for all further work in this chapter.

Now (8.1.5) is identical to the system equation for the cylinder-diameter controller considered earlier except the forcing function in (8.1.5) contains the additional term $-K\varepsilon(i)$. Therefore, the solutions will be of the same form with the appropriate change of forcing functions. Thus,

$$m(i) = \sum_{n=0}^{i-1} [p(n) - K\varepsilon(n)](1-K)^{i-1-n}. \tag{8.1.6}$$

The stable range of K is of course $0 < K < 2$.

8.2   Properties of Measurement Error $\varepsilon(i)$

The measurement error will in general be a random variable. Thus, all the material addressed previously in Chapter VII to the random perturbation $p(i)$ in concept applies to $\varepsilon(i)$. The effect on the system output, however, will differ from that of the random $p(i)$ because of the $-K$ coefficient of $\varepsilon(n)$ in (8.1.6). Let us examine the effects of measurement error on system output for various assumed error characteristics. This is easily accomplished by setting $p(n)$ to zero in (8.1.6) to obtain

$$m(i) = \sum_{n=0}^{i-1} [-K\varepsilon(n)](1-K)^{i-1-n} = -K\sum_{n=0}^{i-1} \varepsilon(n)(1-K)^{i-1-n}, \tag{8.2.1}$$

a procedure permitted by the superposition property of linear systems.

For the simple case of the measurement error being independently and identically distributed, let

$$E[\varepsilon(i)] = \varepsilon \qquad \text{for all } i \tag{8.2.2}$$

and

$$V[\varepsilon(i)] = \sigma_\varepsilon^2 \qquad \text{for all } i. \tag{8.2.3}$$

Then, in the manner shown in Eqs. (7.6.9) and (7.6.10), it can be demonstrated that

$$E[m(i)] = -\varepsilon[1 - (1 - K)^i] \tag{8.2.4}$$

and

$$V[m(i)] = K^2 \sum_{n=0}^{i-1} \sigma_\varepsilon^2 (1 - K)^{i-1-n} = \frac{K\sigma_\varepsilon^2}{2 - K} [1 - (1 - K)^{2i}]. \tag{8.2.5}$$

The resulting steady-state expressions are

$$M = \lim_{i \to \infty} E[m(i)] = -\varepsilon \tag{8.2.6}$$

and

$$V = \lim_{i \to \infty} V[m(i)] = K\sigma_\varepsilon^2/(2 - K). \tag{8.2.7}$$

Should the $\varepsilon(i)$ furthermore be normally distributed, $m(i)$ will be normally distributed with mean and variance given by (8.2.4) and (8.2.5), respectively, and the steady-state output will be normal with mean and variance given by (8.6.6) and (8.6.7), respectively.

For an identically distributed but interdependent measurement error, the mean of $m(i)$ is as given by (8.2.4), but the variance becomes, in the manner in which (7.6.44) was derived,

$$V[m(i)] = \frac{K\sigma_\varepsilon^2}{2 - K} [1 - (1 - K)^{2i}]$$

$$+ \frac{2K}{2 - K} \sum_{\tau=1}^{i-1} \sigma_\varepsilon(\tau)(1 - K)^\tau [1 - (1 - K)^{2(i - \tau)}], \tag{8.2.8}$$

where $\sigma_\varepsilon(\tau)$ is the covariance between two measurement errors τ periods apart and $\sigma_\varepsilon(0) = \sigma_\varepsilon^2$. A closed-form expression may be found for specific functions of $\sigma_\varepsilon(\tau)$ if the summation in (8.2.8) can be carried out. If so, an expression can also be derived for the steady-state variance V. Again, $m(i)$ will be normally distributed if the $\varepsilon(i)$ are normal, regardless of dependence or independence.

In general, a nonstationary measurement error would preclude obtaining closed-form expressions for expected value or variance of $m(i)$, unless a pattern for $E[m(i)]$ or $V[m(i)]$ with respect to i were observed which would permit the summing indicated in (8.2.1) to be carried out. Therefore, little can be said about this situation except in the case of specific examples.

8.3   Sampling

When the generalized process controller was first introduced in Chapter I the notion of periodically sampling the process output was an explicit feature of the system. We have not said any more about this since in the interim our primary example has been the engine-block line in which all blocks, i.e., 100% of the product, was inspected. Output sampling, however, is very commonly involved in the control of a large variety of processes. It is of course essential where testing is destructive. It is a practical necessity for continuous sheet processes such as paper and sheet plastic where scanning gages are used whose observation windows are only a fraction of an inch in the cross-sheet direction. Sampling is economically advantageous where high production rates combine with relatively expensive or time-consuming measurements of individual product items. The ultimate of the latter situation is the development of political strategies based on the results of public opinion polls directed at measuring the acceptance of past political actions. Obtaining the opinions of all those involved without exceeding the pollster's budget and in short enough time so that the results are timely for use in making future decisions is almost always out of the question. Thus carefully stratified random samples are used.

Since individual items of product usually differ from one another in somewhat unpredictable ways, errors are present wherever sampling is used. These errors are based on the inherent individual differences among the items in the output population and hence among those selected for the sample and have nothing to do with errors due to the instrumentation used to measure the items. The error due to these individual differences we call *sampling error*; error due to the instrumentation (including any human error in reading the instruments) we call *instrumentation error*. Whenever sampling is used, both types of errors can be expected to some degree. If either is felt or found to be significant, it must be included in the system equation for analysis or design of the system. The means for doing so will now be presented.

Each estimate of process output is now to be based on the average of the observed values of one or more individual product units. Specifically, the ith sample is that set of units selected according to some sampling scheme from the process at or as near as possible to instant i. It is the group of items upon which estimates of the process operating level $m(i)$ at time i are based. Let N be the number of units in the sample, i.e., the "sample size" (assumed here as independent of i), and $\gamma(i, q)$ the actual value of the qth unit in the sample drawn at time i; $q = 1, \ldots, N$. When individual differences among items or units of product exist, the process output $m(i)$ can be thought of as an average or expected output at time i. If the entire sample can be selected from items produced close enough to time i that it can be safely assumed that all items come from essentially the same production population, then we may state that

$$E[\gamma(i, q)] = m(i), \qquad q = 1, \ldots, N. \tag{8.3.1}$$

Let us further define the individual difference between $\gamma(i, q)$ and $m(i)$, or the error in the qth unit of sample i, as

$$y(i, q) = \gamma(i, q) - m(i), \tag{8.3.2}$$

whose expected value must, of course, be zero. The assumption that all items in the sample come from the same population, i.e., have the same distribution, allows us to represent the variance of $y(i, q)$ as

$$V[y(i, q)] = \sigma_y{}^2(i), \qquad q = 1, \ldots, N. \tag{8.3.3}$$

Should the distribution of individual items around $m(i)$ be stationary from sampling period to sampling period, we let $\sigma_{y(i)}^2 = \sigma_y{}^2$ for all i, which means that

$$V[y(i, q)] = \sigma_y{}^2 \qquad \text{for all } i, q. \tag{8.3.4}$$

In addition, let $\eta(i, q)$ be the instrumentation error associated with the measurement of item q in sample i. It is reasonable to assume that the instrumentation error will also be stationary within the space of drawing and measuring a given sample. Therefore, we define

$$E[\eta(i, q)] = H(i), \qquad q = 1, \ldots, N, \tag{8.3.5}$$

and

$$V[\eta(i, q)] = \sigma_{\eta(i)}^2, \qquad q = 1, \ldots, N. \tag{8.3.6}$$

Whether or not this error can be considered stationary from sampling period to sampling period depends on the instrumentation and the operating conditions. Frequent set-point calibration checks of meters and gages could change $H(i)$ without affecting $\sigma_{\eta(i)}^2$. Some environmental factors could change $\sigma_{\eta(i)}^2$ with no effect on $H(i)$. Other factors could affect both. However, when long-term stationarity of instrumentation errors can be assumed, we let $H(i) = H$ for all i and $\sigma_{\eta(i)}^2 = \sigma_\eta{}^2$ for all i, which gives

$$E[\eta(i, q)] = H \qquad \text{for all } i, q \tag{8.3.7}$$

and

$$V[\eta(i, q)] = \sigma_\eta{}^2 \qquad \text{for all } i, q. \tag{8.3.8}$$

The overall measurement error $\varepsilon(i, q)$ associated with item q in sample i will be affected by both individual-difference variance and instrumentation error. For an additive instrumentation error,

$$\varepsilon(i, q) = y(i, q) + \eta(i, q). \tag{8.3.9}$$

Generally, especially for additive errors, the instrumentation error will be independent of the value being measured. Therefore,

$$E[\varepsilon(i, q)] = E[y(i, q)] + E[\eta(i, q)] = 0 + H(i) = H(i) \qquad (8.3.10)$$

and

$$V[\varepsilon(i, q)] = V[y(i, q)] + V[\eta(i, q)] = \sigma^2_{y(i)} + \sigma^2_{\eta(i)}, \qquad q = 1, \ldots, N. \quad (8.3.11)$$

Now the measurement error $\varepsilon(i)$ will be the error in the average of the items in the sample, i.e.,

$$\varepsilon(i) = (1/N) \sum_{q=1}^{N} \varepsilon(i, q) = (1/N) \sum_{q=1}^{N} y(i, q) + (1/N) \sum_{q=1}^{N} \eta(i, q). \quad (8.3.12)$$

Therefore,

$$E[\varepsilon(i)] = (1/N) \sum_{q=1}^{N} H(i) = H(i) \qquad (8.3.13)$$

and

$$V[\varepsilon(i)] = (1/N^2) \sum_{q=1}^{N} \sigma^2_{y(i)} + (1/N^2) \sum_{q=1}^{N} \sigma^2_{\eta(i)} = (1/N)\sigma^2_{y(i)} + (1/N)\sigma^2_{\eta(i)}. \quad (8.3.14)$$

If the $y(i, q)$ and $\eta(i, q)$ are normally distributed, $\varepsilon(i)$ will of course be normal also. However, even if the $y(i, q)$ and $\eta(i, q)$ are only approximately normal or even uniform, $\varepsilon(i)$ will still tend toward normality as N increases because of the central limit theorem.

The effect of these errors on $m(i)$ can be shown by substituting (8.3.12) for $\varepsilon(i)$ in (8.2.1), the expression for $m(i)$ in terms of the $\varepsilon(i)$ alone. As already mentioned, if either component of $\varepsilon(i)$ is not stationary across sampling periods, closed-form expressions for the mean and variance of $m(i)$ are not usually obtainable unless patterns are noted which permit the appropriate summations to be carried out. If, however, both components are stationary across sampling periods so that (8.3.4), (8.3.7), and (8.3.8) hold, the mean and variance of $\varepsilon(i)$ can be expressed, respectively, as

$$E[\varepsilon(i)] = H \qquad (8.3.15)$$

and

$$V[\varepsilon(i)] = (1/N)\sigma_y^2 + (1/N)\sigma_\eta^2 \qquad (8.3.16)$$

for all i. Comparison with (8.2.2) and (8.2.3) indicates that

$$E[\varepsilon(i)] = \varepsilon = H \qquad (8.3.17)$$

and

$$V[\varepsilon(i)] = \sigma_\varepsilon^2 = (1/N)\sigma_y^2 + (1/N)\sigma_\eta^2, \qquad (8.3.18)$$

from which

$$E[m(i)] = -H[1 - (1 - K)^i] \qquad (8.3.19)$$

and

$$V[m(i)] = \frac{K}{2-K}\left[\frac{1}{N}\sigma_y^2 + \frac{1}{N}\sigma_\eta^2\right][1-(1-K)^{2i}]. \qquad (8.3.20)$$

In the steady state, the mean and variance are

$$M = -H \qquad (8.3.21)$$

and

$$V = \frac{K}{2-K}\left[\frac{1}{N}\sigma_y^2 + \frac{1}{N}\sigma_\eta^2\right]. \qquad (8.3.22)$$

8.4 Instrumentation

Instrumentation errors can affect measurements in various ways, the most common of which is to add a random variable to the true value being measured. Some devices, such as the nucleonic detectors used in the tobacco and paper industries, tend to multiply the true value by a random variable. In other devices the effects are even more complicated. We showed in the previous section the modeling of the additive instrumentation error, and will confine our coverage in the material which follows to this type.

Instrumentation errors may be completely independent of one another or highly correlated depending on the instrument, the stableness of the environment, and the time between measurements. Some electrical and electronic instruments tend to drift slightly with changes in line voltage or variations in the performance of their own components. Some instrumentation is temperature or humidity sensitive; some is affected by ambient noise or radiation levels. Instrument readings by human operators can be affected by such factors as parallax and personal number preferences. If the line voltage in a plant is subject to variations caused by changing power demands in the plant, the errors produced in the readings of meters sensitive to such variations will be highly correlated within periods wherein the voltage remains fixed but will tend to be uncorrelated over longer periods of time. Frequent recalibration and automatic line-voltage compensation can lessen the effects of these variations, but in many applications they remain significant and must be represented explicitly in models of the system.

A type of instrumentation error becoming increasingly important with the advent of digital and hybrid analog–digital information handling devices such as minicomputers in the feedback loops of control systems is quantization error. This is the error caused by rounding off or classifying into discrete intervals values of inherently continuous phenomena. Widrow (1956) has shown that such errors can be treated for most analysis purposes as independently uniformly distributed with mean zero and variance $d^2/12$, where

d is the width of the classification interval, i.e., the range of the random error variable. Thus, if the engine-block cylinder-diameter readings were rounded to the nearest 0.0001 in., the resulting error could be considered uniform and unbiased with a variance of 8.33×10^{-10} in.^{2} (or a standard deviation of 2.88×10^{-5} in.). Whether or not this is significant enough to be included explicitly in the system model depends on the quality desired in the engine blocks and the relative effect of such an error on the system output. Actually the principal use of information of this sort would be to design the information system so that a sufficient number of digits or bits would be included to assure that the effects of the resulting quantization errors are negligible.

8.5   Distribution of Individual Product Units

Since, in the situation being considered here, individual product units vary around the process output $m(i)$ and $m(i)$ in turn varies even in the absence of perturbations $p(i)$ due to the feedback of previous measuring errors, it is meaningful to examine the total variation of the individual units. As will be demonstrated later in this chapter, numerous evaluation criteria can be based on this distribution. The total variation is of course caused by the combined effects of changes in $m(i)$ and the variation of the units around $m(i)$. Assuming the individual differences among product units to be mutually independent so that the $y(i, q)$, $q = 1, \ldots, N$, are independent of all past measurement errors, then the total variation in $y(i, q)$ around $E[m(i)]$ will have a variance $\sigma_T^2(i)$:

$$\sigma_T^2(i) = V[m(i)] + \sigma_{y(i)}^2, \tag{8.5.1}$$

i.e., we simply add the variance of the distribution of $y(i, q)$ around $m(i)$ to the variance of $m(i)$ itself. For stationary $y(i, q)$, $\sigma_T^2(i)$ becomes

$$\sigma_T^2(i) = V[m(i)] + \sigma_y^2 \tag{8.5.2}$$

which in the steady state is given by

$$V_T = \lim_{i \to \infty} \sigma_T^2(i) = V + \sigma_y^2. \tag{8.5.3}$$

Of course, either (8.2.5) or (8.2.8) can be substituted in (8.5.1) or (8.5.2) as appropriate.

We will now consider several examples which illustrate how the material presented thus far can be used to select values of the proportionality constant K for various combinations of $\varepsilon(i)$, $p(i)$, and performance criteria. The examples also illustrate the use of several different optimization approaches for finding optimal values of K in specific situations. It should be emphasized that in no way is it implied that the techniques used in any given example are necessarily the best ones for that case. They simply provide the reader with an indication of what is available and how they are applied.

8.6   Maximum Speed of Response with Bounded Steady-State Process-Output Variance

Conditions

$p(i)$:   Occasional impulse, pattern unknown, magnitude possibly severe

$\varepsilon(i)$:   Mutually independent, identically distributed with mean zero and variance σ_ε^2, independent of $p(i)$

Criterion   React as quickly as possible to $p(i)$ without steady-state variance V exceeding an upper limit V_{max}.

Since the perturbations are occasional and their time and magnitude patterns unknown, explicit treatment of $p(i)$ in determining steady-state conditions is not feasible. Our main interest with regard to the steady state is in the effect of continually feeding back the measurement error into the process. Therefore, given an unbiased, independent, identically distributed measurement error, the steady-state variance can be expressed as in (8.2.7). Thus, with an upper bound of V_{max},

$$V = K\sigma_\varepsilon^2/(2 - K) \le V_{max}. \qquad (8.6.1)$$

V is of course an increasing function of K in the stable range of $0 < K < 2$, so (8.6 1) imposes an upper bound on K. Solving (8.6.1) for K in terms of σ_ε^2 and V_{max} yields

$$K \le K_{max} = 2V_{max}/(\sigma_\varepsilon^2 + V_{max}). \qquad (8.6.2)$$

Now let us consider the part of the performance criterion dealing with response of the system to a perturbation when one does occur. In Chapter VII it was pointed out (and illustrated in Table 7.2.1) that the rapidity of the response to an impulse increases as K approaches one. This is further borne out by (7.3.2) in Section 7.3. Thus, to react to an impulse as quickly as possible, we would like to set $K = 1$ and will indeed do so providing this value of K does not cause a V which exceeds V_{max}. Should this occur, we would then make K as close to one as we can without violating the constraint on V.

In summary, we select K for this case as follows:

1.   Calculate K_{max} from (8.6.2).
2.   If $K_{max} \le 1$,   use $K = K_{max}$;
    $K_{max} > 1$,   use $K = 1$.

Under no circumstances would we ever use a $K > 1$ in the situation involved here. Note from (8.6.2) that $K_{max} \le 1$ and will thus be binding in selecting K when $V_{max} \le \sigma_\varepsilon^2$. For $V_{max} > \sigma_\varepsilon^2$, $K_{max} > 1$ and $K = 1$. Use of this selection rule and the results just mentioned are illustrated in Table 8.6.1 for the control of the diameter of a gear blank. The maximum allowable variance V_{max} in m, the steady-state process output, is 0.0002 in.^{2}. Various assumed values of the variance of the measurement error σ_ε^2 are shown.

Table 8.6.1

Selection of K for $V_{max} = 0.0002$ in.² and Various Values of σ_ε^2

σ_ε^2(in.²)	K_{max}	K
0.0000	2.000	1.000
0.0001	1.333	1.000
0.0002	1.000	1.000
0.0003	0.800	0.800
0.0005	0.571	0.571
0.0010	0.333	0.333

A more general relationship can be developed in terms of the ratio r of σ_ε^2 to V_{max}. That is, if we define r as

$$r = \sigma_\varepsilon^2 / V_{max} \tag{8.6.3}$$

and substitute $r V_{max}$ for σ_ε^2 in (8.6.2), the expression for K_{max} can be written

$$K_{max} = 2 V_{max}/(r V_{max} + V_{max}) = 2/(r + 1). \tag{8.6.4}$$

Figure 8.6.1 is a plot of K_{max} and K versus r.

A nontrivial design task is the determination of V_{max}. In many cases, it may be based directly on stated customer requirements, control-equipment

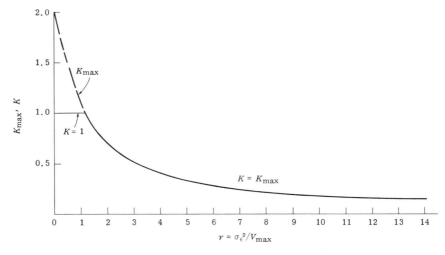

FIG. 8.6.1 K_{max} and K as functions of $\sigma_\varepsilon^2/V_{max}$.

guarantees, a knowledge of competitors' capabilities, or simply on a feeling of what the traffic will bear. In other cases, however, customer requirements may be expressed in the form of tolerance limits on the steady-state output m which should not be exceeded more than some small portion of the time.

To illustrate, let m_U and m_L represent upper and lower tolerance limits, respectively, on m, and let β be the maximum allowable probability of m falling outside the range $m_L \leq m \leq m_U$, i.e.,

$$\Pr[m_L \leq m \leq m_U] \geq 1 - \beta. \tag{8.6.5}$$

Note that with m defined as the deviation between actual and desired outputs, normally $m_L < 0$ and $m_U > 0$ for $D = 0$. The relationships among m_L, m_U, β, and V_{max}, and therefore K_{max}, depend on the distribution of m. If careful calibration and reading of instruments can be assumed, we would expect $\varepsilon(i)$ to be unbiased and hence $M = 0$. Further, if, as is commonly the case, the tolerances are symmetrical (i.e., $m_L = -m_U$) and the distribution of m is symmetrical, we can express the constraint on m in terms of either m_L or m_U as follows:

$$\Pr[m < m_L] = \Pr[m > m_U] \leq \beta/2. \tag{8.6.6}$$

Suppose now m is normally distributed as it would be if the $\varepsilon(i)$ were normal. The number of standard normal deviates $t_{\beta/2}$ from $M = 0$ which have a probability of $\beta/2$ of being exceeded can be found from Table 8.6.2, a table of the cumulative normal distribution. Then from

$$t_{\beta/2}\sqrt{V} \leq m_U, \tag{8.6.7}$$

we can find

$$V_{max} = (m_U/t_{\beta/2})^2. \tag{8.6.8}$$

This calculation would obviously be made more complicated by nonsymmetrical limits, a biased measurement error, or a nonnormal (especially nonsymmetrical) measurement error.

If m is normally distributed, $m_U = -m_L = 0.001$ in., and $\beta = 0.01$, we would search Table 8.6.2 for the entry $1 - \beta/2 = 0.9950$ which occurs for $t = 2.576$. Then, from (8.6.8),

$$V_{max} = (0.001/2.576)^2 = 0.151 \times 10^{-6} \text{ in.}^2.$$

8.7 Maximum Speed of Response with Bounded Steady-State Variance of Individual Product Units

Conditions

$p(i)$: Occasional impulse, pattern unknown, magnitude possibly severe

$\varepsilon(i)$: Mutually independent, identically distributed with mean zero and variance σ_ε^2, independent of $p(i)$

Table 8.6.2

Cumulative Probabilities of the Normal Probability Distribution[ab]

t	.00	.01	.02	.03	.04	.05	.06	.07	.08	.09
.0	.5000	.5040	.5080	.5120	.5160	.5199	.5239	.5279	.5319	.5359
.1	.5398	.5438	.5478	.5517	.5557	.5596	.5636	.5675	.5714	.5753
.2	.5793	.5832	.5871	.5910	.5948	.5987	.6026	.6064	.6103	.6141
.3	.6179	.6217	.6255	.6293	.6331	.6368	.6406	.6443	.6480	.6517
.4	.6554	.6591	.6628	.6664	.6700	.6736	.6772	.6808	.6844	.6879
.5	.6915	.6950	.6985	.7019	.7054	.7088	.7123	.7157	.7190	.7224
.6	.7257	.7291	.7324	.7357	.7389	.7422	.7454	.7486	.7517	.7549
.7	.7580	.7611	.7642	.7673	.7704	.7734	.7764	.7794	.7823	.7852
.8	.7881	.7910	.7939	.7967	.7995	.8023	.8051	.8078	.8106	.8133
.9	.8159	.8186	.8212	.8238	.8264	.8289	.8315	.8340	.8365	.8389
1.0	.8413	.8438	.8461	.8485	.8508	.8531	.8554	.8577	.8599	.8621
1.1	.8643	.8665	.8686	.8708	.8729	.8749	.8770	.8790	.8810	.8830
1.2	.8849	.8869	.8888	.8907	.8925	.8944	.8962	.8980	.8997	.9015
1.3	.9032	.9049	.9066	.9082	.9099	.9115	.9131	.9147	.9162	.9177
1.4	.9192	.9207	.9222	.9236	.9251	.9265	.9279	.9292	.9306	.9319
1.5	.9332	.9345	.9357	.9370	.9382	.9394	.9406	.9418	.9429	.9441
1.6	.9452	.9463	.9474	.9484	.9495	.9505	.9515	.9525	.9535	.9545
1.7	.9554	.9564	.9573	.9582	.9591	.9599	.9608	.9616	.9625	.9633
1.8	.9641	.9649	.9656	.9664	.9671	.9678	.9686	.9693	.9699	.9706
1.9	.9713	.9719	.9726	.9732	.9738	.9744	.9750	.9756	.9761	.9767
2.0	.9772	.9778	.9783	.9788	.9793	.9798	.9803	.9808	.9812	.9817
2.1	.9821	.9826	.9830	.9834	.9838	.9842	.9846	.9850	.9854	.9857
2.2	.9861	.9864	.9868	.9871	.9875	.9878	.9881	.9884	.9887	.9890
2.3	.9893	.9896	.9898	.9901	.9904	.9906	.9909	.9911	.9913	.9916
2.4	.9918	.9920	.9922	.9925	.9927	.9929	.9931	.9932	.9934	.9936
2.5	.9938	.9940	.9941	.9943	.9945	.9946	.9948	.9949	.9951	.9952
2.6	.9953	.9955	.9956	.9957	.9959	.9960	.9961	.9962	.9963	.9964
2.7	.9965	.9966	.9967	.9968	.9969	.9970	.9971	.9972	.9973	.9974
2.8	.9974	.9975	.9976	.9977	.9977	.9978	.9979	.9979	.9980	.9981
2.9	.9981	.9982	.9982	.9983	.9984	.9984	.9985	.9985	.9986	.9986
3.0	.9987	.9987	.9987	.9988	.9988	.9989	.9989	.9989	.9990	.9990
3.1	.9990	.9991	.9991	.9991	.9992	.9992	.9992	.9992	.9993	.9993
3.2	.9993	.9993	.9994	.9994	.9994	.9994	.9994	.9995	.9995	.9995
3.3	.9995	.9995	.9995	.9996	.9996	.9996	.9996	.9996	.9996	.9997
3.4	.9997	.9997	.9997	.9997	.9997	.9997	.9997	.9997	.9997	.9998

t	1.282	1.645	1.960	2.326	2.576	3.090	3.291	3.891	4.417
$F(t)$	.90	.95	.975	.99	.995	.999	.9995	.99995	.999995
$2[1 - F(t)]$	.20	.10	.05	.02	.01	.002	.001	.0001	.00001

[a] Areas under the normal curve from $-\infty$ to t.

[b] From A. M. Mood, *Introduction to the Theory of Statistics*, p. 423. Copyright 1950 by McGraw-Hill, Inc. Used with permission of McGraw-Hill Book Co.

Criterion   React as quickly as possible to $p(i)$ without total steady-state variance of individual items V_T exceeding an upper limit $V_{T\,max}$.

This case is the same as the example in Section 8.6 except here the criterion involves individual product items $y(i, q)$, instead of their expected value $m(i)$. We, therefore, need to know the variance of the individual items around their mean σ_y^2. The relationship among σ_ε^2, σ_y^2, the instrumentation error variance σ_η^2, and the sample size N is given by (8.3.18). V is expressed in terms of σ_y^2, σ_η^2, and N in (8.3.22). Substitution of (8.3.22) into (8.5.3), the expression for V_T, yields

$$V_T = \frac{K}{2 - K}\left[\frac{1}{N}\,\sigma_y^2 + \frac{1}{N}\,\sigma_\eta^2\right] + \sigma_y^2 \leq V_{T\,max}. \qquad (8.7.1)$$

V_T is an increasing function of K, so $V_{T\,max}$ imposes an upper bound K_{max} on K. K_{max} is found by setting $V_T = V_{T\,max}$ in (8.7.1) and solving for K. This yields

$$K_{max} = \frac{2[V_{T\,max} - \sigma_y^2]}{V_{T\,max} + \sigma_\eta^2/N - (N - 1)\sigma_y^2/N}. \qquad (8.7.2)$$

Determination of $V_{T\,max}$ can follow any of the procedures mentioned in the previous example with regard to V_{max}. For the case of specified tolerances on $y(i, q)$ which may not be exceeded by more than some small portion of the product items, let y_U and y_L be the upper and lower tolerances, respectively, on $y(i, q)$ for all i, q, and let β again represent the maximum allowable probability of y, the steady-state value of $y(i, q)$, falling outside the range $y_L \leq y \leq y_U$. For the case of symmetrical tolerances, $y_L = -y_U$, and η and y normally distributed, $V_{T\,max}$ can easily be found from

$$V_{T\,max} = (y_U/t_{\beta/2})^2, \qquad (8.7.3)$$

where $t_{\beta/2}$ is as previously defined.

Both common sense and a look at (8.7.1) lead us to conclude that compatibility problems can arise if one attempts to set $V_{T\,max}$ too low relative to the natural variance of individual product items around their mean. If $\sigma_y^2 > V_{T\,max}$, then even if $m = M = 0$, the product variance exceeds the constraint. This is easily seen by subtracting σ_y^2 from both sides of the inequality in (8.7.1), which gives

$$V = \frac{K}{2 - K}\left[\frac{1}{N}\,\sigma_y^2 + \frac{1}{N}\,\sigma_\eta^2\right] \leq V_{T\,max} - \sigma_\varepsilon^2 < 0, \qquad (8.7.4)$$

meaning V would have to be negative for the constraint to be met, an obviously impossible situation. A check of compatibility between σ_y^2 and $V_{T\,max}$ should always be made before calculations begin, for should $\sigma_y^2 > V_{T\,max}$ the problem is no longer one of finding values of control-system parameters; it is a problem either of inadequate process design which is permitting too much item-to-item variation in process output or of unrealistic specification of performance criteria. Should one overlook such a check, calculation of K_{max} from (8.7.2) will provide a quick reminder in the form of a value outside the stable range $0 < K_{max} < 2$. K_{max} will be negative for $\sigma_y^2 > V_{T max}$ but with a positive denominator in (8.7.2). For σ_y^2 large enough for the denominator of (8.7.2) also to be negative, K_{max} will be greater than two.

For a compatible set of values, the constant K will of course be taken as one should $K_{max} > 1$ and made equal to K_{max} should $K_{max} \leq 1$. Note that (8.7.2) can be used to study the potential benefits of increasing sample size, investing in more rigid production machinery to reduce σ_y^2, or improving instrumentation to hold down σ_η^2. Cost analyses based on a combination of process-design, operation, and instrumentation costs can be made using relationships such as (8.7.2) and the resulting expressions for process output.

To illustrate the use of (8.7.2) in selecting K, consider the control of the diameter of the gear blank mentioned previously. Let

$$\sigma_y^2 = 0.005 \text{ in.}^2, \qquad \sigma_\eta^2 = 0.001 \text{ in.}^2,$$
$$N = 5, \qquad \beta = 0.05, \qquad y_L = -y_U.$$

For $\beta = 0.05$, $t_{\beta/2} = 1.96$ and $V_{max} = (y_U/1.96)^2$. Substitution of these values into (8.7.2) gives, after some manipulation,

$$K_{max} = (2y_U^2 - 0.038416)/(y_U^2 - 0.014598).$$

This yields the values of K_{max} and K for various values of y_U shown in Table 8.7.1.

Table 8.7.1

Values of K_{max} and K^a

y_U (in.)	K_{max}	K
$y_U < 0.12082$	>2	Incompatible
$0.12082 < y_U < 0.13859$	<0	Incompatible
0.14000	0.15674	0.15674
0.14300	0.42420	0.42420
0.14600	0.62757	0.62757
0.14900	0.78732	0.78732
0.15200	0.91606	0.91606
0.15433	1.00000	1.00000
$y_U > 0.15433$	>1.00000	1.00000

[a] Values for $\sigma_y^2 = 0.005$ in.^{2}, $\sigma_\eta^2 = 0.001$ in.^{2}, $N = 5$, $\beta = 0.05$, $y_L = -y_U$, and various values of y_U.

8.8   Maximum Speed of Response with Random Perturbation and Bounded Steady-State Process-Output Variance

Conditions

$p(i)$: Occasional impulse, pattern unknown, magnitude possibly severe, plus an independently distributed stationary random perturbation with mean zero and variance $\sigma_p{}^2$

$\varepsilon(i)$: Mutually independent, identically distributed with mean zero and variance $\sigma_\varepsilon{}^2$, independent of $p(i)$

Criterion   React as quickly as possible to the impulse component of $p(i)$ without steady-state variance V exceeding an upper limit.

The random perturbation was discussed at length in Section 7.6. It was shown that for the independent, stationary case the effects on the steady-state mean and variance of process output m were p/K, where p is the expected value of the random variable, and $\sigma_p{}^2/(2K - K^2)$, respectively. For the unbiased random perturbation of course $p = 0$ so there is no effect on M.

In the example here, since $\varepsilon(i)$ and $p(i)$ are mutually independent, their individual effects on V can be added. Thus by combining (7.6.12) and (8.2.7) we get

$$V = \frac{\sigma_p{}^2}{2K - K^2} + \frac{K\sigma_\varepsilon{}^2}{2 - K} = \frac{\sigma_p{}^2 + K^2\sigma_\varepsilon{}^2}{K(2 - K)}. \qquad (8.8.1)$$

With the specified limit $V_{\max}$ on V,

$$(\sigma_p{}^2 + K^2\sigma_\varepsilon{}^2)/K(2 - K) \leq V_{\max}, \qquad (8.8.2)$$

from which

$$\sigma_p{}^2 + \sigma_\varepsilon{}^2 K^2 \leq 2V_{\max} K - V_{\max} K^2$$

or

$$(\sigma_\varepsilon{}^2 + V_{\max})K^2 - 2V_{\max} K + \sigma_p{}^2 \leq 0. \qquad (8.8.3)$$

Let $g(K)$ represent the left-hand side of (8.8.3). Then,

$$g'(K) = \frac{dg}{dK} = 2(\sigma_\varepsilon{}^2 + V_{\max})K - 2V_{\max} \qquad (8.8.4)$$

and

$$g''(K) = \frac{d^2g}{dK^2} = 2(\sigma_\varepsilon{}^2 + V_{\max}). \qquad (8.8.5)$$

Now $g''(K) > 0$ for all K, so $g(K)$ is everywhere strictly convex. Thus, the stationary point found by setting $g'(K)$ to zero will be the global minimum of $g(K)$. This is found from (8.8.4) to occur at

$$\hat{K} = V_{\max}/(\sigma_\varepsilon{}^2 + V_{\max}), \qquad (8.8.6)$$

which must fall in the range $0 < \hat{K} < 1$, a subset of the stable range. The global minimum of $g(K)$ is found by substituting (8.8.6) for K in $g(K)$ to be

$$g(\hat{K}) = \min_{K} g(K) = \sigma_p^2 - [V_{\max}^2/(\sigma_\varepsilon^2 + V_{\max})]. \qquad (8.8.7)$$

Since $g(K) \leq 0$ for $V \leq V_{\max}$, it is obvious from (8.8.7) that if

$$\sigma_p^2 > V_{\max}^2/(\sigma_\varepsilon^2 + V_{\max}) \qquad (8.8.8)$$

so that the minimum of $g(K)$ is positive, the constraint on V cannot be met for any value of K. Therefore,

$$\sigma_p^2 < V_{\max}^2/(\sigma_\varepsilon^2 + V_{\max}) \qquad (8.8.9)$$

is the compatibility condition on σ_p^2 which must be met if the design criteria are to be satisfied.

It can be shown by simple substitution that

$$g(0) = \sigma_p^2 \qquad (8.8.10)$$

and

$$g(2) = 4\sigma_\varepsilon^2 + \sigma_p^2, \qquad (8.8.11)$$

showing that $g(K)$ is positive at the limits of the stable range of K. Therefore, if the compatibility condition (8.8.9) is met, $g(K)$ will be negative over a range somewhere between 0 and 2 bounded by the roots of

$$g(K) = (\sigma_\varepsilon^2 + V_{\max})K^2 - 2V_{\max} K + \sigma_p^2 = 0. \qquad (8.8.12)$$

These roots are found to be

$$K_{\max} = \frac{V_{\max}}{\sigma_\varepsilon^2 + V_{\max}} + \left[\frac{V_{\max}^2 - \sigma_p^2(\sigma_\varepsilon^2 + V_{\max})}{(\sigma_\varepsilon^2 + V_{\max})^2} \right]^{1/2}$$

$$= \hat{K} + \left[\hat{K}^2 - \frac{\sigma_p^2}{\sigma_\varepsilon^2 + V_{\max}} \right]^{1/2} \qquad (8.8.13)$$

and

$$K_{\min} = \frac{V_{\max}}{\sigma_\varepsilon^2 + V_{\max}} - \left[\frac{V_{\max}^2 - \sigma_p^2(\sigma_\varepsilon^2 + V_{\max})}{(\sigma_\varepsilon^2 + V_{\max})^2} \right]^{1/2}$$

$$= \hat{K} - \left[\hat{K}^2 - \frac{\sigma_p^2}{\sigma_\varepsilon^2 + V_{\max}} \right]^{1/2}. \qquad (8.8.14)$$

Note that if the compatibility condition is not met, both roots are complex and $g(K)$ is nowhere negative.

We of course want to make K as close to one as possible and indeed to make $K = 1$ if $K_{min} < 1 < K_{max}$. From (8.8.14) it is observed that $K_{min} < \hat{K}$, which in turn is less than one. Therefore, the range of K for which $V \leq V_{max}$ will always extend below $K = 1$, and K_{min} can never force K to lie above one. Therefore, the rule for selection of K remains essentially the same as in the first two examples, namely:

1. Calculate K_{max} from (8.8.13).
2. If $K_{max} \leq 1$, use $K = K_{max}$,
   $K_{max} > 1$, use $K = 1$,
   K_{max} is complex, increase V_{max} or decrease σ_p^2 or σ_ε^2 until (8.8.9) is satisfied.

8.9 Maximum Speed of Response with Random Perturbation and Bounded Steady-State Variance of Individual Product Units

Conditions

$p(i)$: Occasional impulse, pattern unknown, magnitude possibly severe, plus an independently distributed stationary random perturbation with mean zero and variance σ_p^2

$\varepsilon(i)$: Mutually independent, identically distributed with mean zero and variance σ_ε^2, independent of $p(i)$

Criterion   React as quickly as possible to the impulse component of $p(i)$ without total steady-state variance of individual items V_T exceeding an upper limit $V_{T\,max}$.

This case combines the features of the previous two examples since it includes both the variation of individual product items and the effects of a random perturbation component as well as instrumentation error. Combining the appropriate expressions previously developed, we get for the total steady-state variance V_T,

$$V_T = \frac{\sigma_p^2}{2K - K^2} + \frac{K}{2 - K}\left[\frac{1}{N}\sigma_y^2 + \frac{1}{N}\sigma_\eta^2\right] + \sigma_y^2 \leq V_{T\,max}. \qquad (8.9.1)$$

It is left as an exercise for the reader to show the following:

a.   There is a compatibility condition on σ_y^2 given by

$$\sigma_y^2 \leq V_{T\,max}. \qquad (8.9.2)$$

b.   There is a compatibility condition on σ_p^2 given by

$$\sigma_p^2 \leq \frac{(V_{T\,max} - \sigma_y^2)^2}{V_{T\,max} + \sigma_\eta^2/N - (N-1)\sigma_y^2/N}. \qquad (8.9.3)$$

c. Given that both compatibility conditions are met, V_T is minimized for $K = \hat{K}$ given by

$$\hat{K} = \frac{V_{T\,max} - \sigma_y^2}{V_{T\,max} + \sigma_\eta^2/N - (N-1)\sigma_y^2/N} < 1. \tag{8.9.4}$$

d. $V_T \leq V_{T\,max}$ for K in the range $K_{min} \leq K \leq K_{max}$, where

$$K_{max} = \hat{K} + \left[\hat{K}^2 - \frac{\sigma_p^2}{V_{T\,max} + \sigma_\eta^2/N - (N-1)\sigma_y^2/N} \right]^{1/2} \tag{8.9.5}$$

and

$$K_{min} = \hat{K} - \left[\hat{K}^2 - \frac{\sigma_p^2}{V_{T\,max} + \sigma_\eta^2/N - (N-1)\sigma_y^2/N} \right]^{1/2}. \tag{8.9.6}$$

e. The rule for the selection of K is:

1. Calculate K_{max} from (8.9.5).
2. If $K_{max} \leq 1$, use $K = K_{max}$,
   $K_{max} > 1$, use $K = 1$,
   K_{max} is complex, increase $V_{T\,max}$ or decrease σ_p^2, σ_η^2, or σ_y^2 until (8.9.2) and (8.9.3) are satisfied.

In all the examples considered so far, very little has been assumed to be known about the perturbation $p(i)$, except perhaps the mean and variance of its distribution. Although this is a common situation for a system designer to be in, there are situations in which some additional information may be available. Let us examine a few cases in which a step perturbation of known magnitude is present.

8.10   Minimum Steady-State Mean-Square Deviation of Process Output

Conditions

$p(i)$:   Step function of magnitude p starting at $i = 0$
$\varepsilon(i)$:   Mutually independent, identically distributed with mean zero and variance σ_ε^2

Criterion   Minimize the mean-square deviation of the steady-state output from $D = 0$.

The minimum-mean-square-deviation criterion is in very common use in theoretical systems work, primarily because of some fortuitous mathematical properties. Nevertheless, it also represents the operation of many real processes with a high degree of realism. For example, the usefulness or salability of a product could very well fall off as the square of the deviation between its actual length, weight, surface finish, etc., and the specified value.

The efficiency of some chemical reactions also decreases approximately as the square of the deviation of the reaction temperature, pressure, or other operating conditions and the design-center values.

For the situation specified for this example, the equation for $m(i)$, (8.1.6), becomes

$$m(i) = \sum_{n=0}^{i-1} [p(n) - K\varepsilon(n)](1 - K)^{i-1-n}$$

$$= \sum_{n=0}^{i-1} [p - K\varepsilon(n)](1 - K)^{i-1-n}$$

$$= \frac{p}{K} [1 - (1 - K)^i] - K \sum_{n=0}^{i-1} \varepsilon(n)(1 - K)^{i-1-n}. \qquad (8.10.1)$$

The expected value and variance of $m(i)$ are

$$E[m(i)] = (p/K)[1 - (1 - K)^i] \qquad (8.10.2)$$

and

$$V[m(i)] = [K\sigma_\varepsilon^2/(2 - K)][1 - (1 - K)^{2i}], \qquad (8.10.3)$$

respectively, from which the steady-state values are found to be

$$M = p/K \qquad (8.10.4)$$

and

$$V = K\sigma_\varepsilon^2/(2 - K). \qquad (8.10.5)$$

The mean-square deviation around $D = 0$, or equivalently the second moment of the steady-state output Q, is

$$Q = V + M^2 = \frac{K\sigma_\varepsilon^2}{2 - K} + \frac{p^2}{K^2} = \frac{\sigma_\varepsilon^2 K^3 - p^2 K + 2p^2}{K^2(2 - K)}. \qquad (8.10.6)$$

Our design task is to find the value of K in the stable range which minimizes Q. The derivative of Q with respect to K is

$$Q' = \frac{dQ}{dK} = \sigma_\varepsilon^2 \frac{2}{(2 - K)^2} - \frac{2p^2}{K^3} = \frac{2\sigma_\varepsilon^2 K^3 - 2p^2 K^2 + 8p^2 K - 8p^2}{K^3(2 - K)^2}. \qquad (8.10.7)$$

For Q' to equal zero, the numerator must vanish, i.e., the cubic equation

$$\sigma_\varepsilon^2 K^3 - p^2 K^2 + 4p^2 K - 4p^2 = 0 \qquad (8.10.8)$$

must be satisfied. The second derivative of Q,

$$Q'' = \frac{d^2 Q}{dK^2} = \frac{4\sigma_\varepsilon^2}{(2 - K)^3} + \frac{6p^2}{K^4}, \qquad (8.10.9)$$

is positive for all K in the range $0 < K < 2$ (actually over the range $-\infty < K$ to an upper bound greater than 2). Therefore, Q is everywhere convex in the stable range. Hence, there can be no more than one stationary point in this region, and, if there is one, it is the global minimum of Q in this region. Note that the numerator of Q' equals $-4p^2$ at $K = 0$ and $8\sigma_\varepsilon^2$ at $K = 2$. Thus the numerator changes sign somewhere in the region $0 < K < 2$ while the denominator remains positive. Therefore, Q' must equal zero for some $K = \hat{K}$ in the region for any combination of values of σ_ε^2 and p, and $\hat{K}$ thus corresponds to the global minimum of Q in the stable range. $\hat{K}$ can be found by applying the cubic formula to (8.10.8) or by numerical methods for specified values of σ_ε^2 and p. Where numerical methods are used or where analysis of the effects of various combinations of σ_ε^2 and p is desired, it is easier to work with the ratio r_p of σ_ε^2 and p^2. That is, define

$$r_p = \sigma_\varepsilon^2/p^2 \tag{8.10.10}$$

and substitute $r_p p^2$ for σ_ε^2 in (8.10.8) to obtain, after dividing out p^2,

$$r_p K^3 - K^2 + 4K - 4 = 0. \tag{8.10.11}$$

Plots of Q_{min} and K against r_p could be quite useful as a basis for quantitative determination of the benefits to be obtained in system performance for any specific reduction in either p or σ_ε^2. It would be possible, for example, to determine the cost and benefit of reducing the pull of a steel-mill roll or of investing in more sophisticated measuring devices with lower σ_ε^2.

8.11   Minimum Steady-State Mean-Square Deviation of Individual Product Units

Conditions

$p(i)$:   Step function of magnitude p starting at $i = 0$
$\varepsilon(i)$:   Mutually independent, identically distributed with mean zero and variance σ_ε^2, independent of $p(i)$

Criterion   Minimize the total steady-state mean-square deviation of individual units $y(i, q)$ from $D = 0$.

Assuming the variation of individual units around their mean to be unbiased with variance σ_y^2, the total steady-state mean-square deviation Q_T is given by

$$Q_T = V_T + M^2 = \frac{K}{2 - K}\left[\frac{1}{N}\sigma_y^2 + \frac{1}{N}\sigma_\eta^2\right] + \sigma_y^2 + \frac{p^2}{K^2}. \tag{8.11.1}$$

However, the term in brackets is simply σ_ε^2, so the expression for Q_T can be reduced to

$$Q_T = \frac{K\sigma_\varepsilon^2}{2 - K} + \sigma_y^2 + \frac{p^2}{K^2} = V + \sigma_y^2 + M^2 = Q + \sigma_y^2. \quad (8.11.2)$$

Since σ_y^2 is independent of K, Q_T will be minimized for any given ratio of σ_ε^2 to p^2 by the same $\hat{K}$ as Q is. Thus, the results of the previous example can be used here also. The value of $Q_{T\min}$, however, will exceed the comparable value of $Q_{\min}$ by σ_y^2, a fact which could have an important bearing on system design since σ_y^2 is also a component of σ_ε^2.

8.12 Maximum Steady-State Probability of Acceptable Process Output

Conditions

$p(i)$: Step function of magnitude p starting at $i = 0$
$\varepsilon(i)$: Mutually independent, identically normally distributed with mean zero and variance σ_ε^2, independent of $p(i)$

Criterion Maximize the probability of steady-state output m lying between tolerance limits m_L and m_U.

The situation involved here is essentially the same as that in the example in Section 8.10, except our criterion involves maximizing the portion of time the steady-state output m is in the range $m_L \leq m \leq m_U$. Since probabilities are involved, it is necessary to specify a distribution for all random variables involved. For convenience of illustration, a normal distribution is assumed. Trying to maximize the probability of m lying between stated tolerances is appropriate where sales contracts specify an acceptable range of material characteristics which must be possessed by the output if payment is to be made.

It has been stated earlier, both in our discussion of random perturbations and in describing the properties and effects of measurement error, that if all random variables are normally distributed, the steady-state output will be normally distributed with mean M and variance V. Therefore, the probability of m lying between the limits m_L and m_U is

$$\Gamma = \Pr[m_L \leq m \leq m_U] = \int_{m=m_L}^{m_U} \frac{1}{(2\pi V)^{1/2}} \exp\left[-\frac{(m - M)^2}{2V}\right] dm, \quad (8.12.1)$$

where, for the conditions of this example, M and V are given by (8.10.4) and (8.10.5), respectively.

The first step in determining the value of K which maximizes Γ is to differentiate Γ with respect to K. Differentiating the integral yields

$$\Gamma' = \frac{d\Gamma}{dK} = \int_{m=m_L}^{m_U} \frac{d}{dK} \left[\frac{1}{(2\pi V)^{1/2}} \exp\left[-\frac{(m-M)^2}{2V} \right] \right] dm$$

$$+ \frac{1}{(2\pi V)^{1/2}} \exp\left[-\frac{(m_U - M)^2}{2V} \right] \frac{dm_U}{dK}$$

$$- \frac{1}{(2\pi V)^{1/2}} \exp\left[-\frac{(m_L - M)^2}{2V} \right] \frac{dm_L}{dK}. \qquad (8.12.2)$$

Since neither m_U nor m_L is a function of K, $dm_U/dK = dm_L/dK = 0$ and the last two terms of (8.12.2) vanish. Carrying out the differentiation of the integrand in the first term yields, after some manipulation,

$$\Gamma' = \int_{m=m_L}^{m_U} \frac{1}{(2\pi V)^{1/2}} \exp\left[-\frac{(m-M)^2}{2V} \right] \left\{ \frac{1}{2V} \frac{dV}{dK} \frac{(m-M)^2}{V} \right.$$

$$\left. + \frac{1}{\sqrt{V}} \frac{dM}{dK} \frac{m-M}{\sqrt{V}} - \frac{1}{2V} \frac{dV}{dK} \right\} dm, \qquad (8.12.3)$$

a form chosen to enhance performance of the integration. We now let

$$r = \frac{dM}{dK}, \qquad (8.12.4)$$

$$s = \frac{dV}{dK}, \qquad (8.12.5)$$

and

$$t = (m - M)/\sqrt{V}. \qquad (8.12.6)$$

Then

$$m = \sqrt{V} t + M \qquad (8.12.7)$$

from which

$$dm = \sqrt{V} \, dt. \qquad (8.12.8)$$

Thus, when $m = m_U$,

$$t = (m_U - M)/\sqrt{V} = t_U, \qquad (8.12.9)$$

and when $m = m_L$,

$$t = (m_L - M)/\sqrt{V} = t_L. \qquad (8.12.10)$$

Substituting (8.12.4) through (8.12.10) in (8.12.3) yields

$$\Gamma' = \int_{t=t_L}^{t_U} \frac{1}{(2\pi V)^{1/2}} \exp\left(-\frac{t^2}{2}\right) \left\{\frac{s}{2V} t^2 + \frac{r}{\sqrt{V}} t - \frac{s}{2V}\right\} \sqrt{V}\, dt$$

$$= \frac{s}{(2\pi)^{1/2} \cdot 2V} \int_{t=t_L}^{t_U} t^2 \exp\left(-\frac{t^2}{2}\right) dt + \frac{r}{(2\pi V)^{1/2}} \int_{t=t_L}^{t_U} t \exp\left(-\frac{t^2}{2}\right) dt$$

$$- \frac{s}{(2\pi)^{1/2} \cdot 2V} \int_{t=t_L}^{t_U} \exp\left(-\frac{t^2}{2}\right) dt. \tag{8.12.11}$$

To integrate the second integral in (8.12.11), we note that

$$\frac{d}{dt} \exp\left(-\frac{t^2}{2}\right) = \exp\left(-\frac{t^2}{2}\right) - \frac{1}{2} 2t = -t \exp\left(-\frac{t^2}{2}\right), \tag{8.12.12}$$

so

$$\frac{r}{(2\pi V)^{1/2}} \int_{t=t_L}^{t_U} t \exp\left(-\frac{t^2}{2}\right) dt = \frac{r}{(2\pi V)^{1/2}} \left[-\exp\left(-\frac{t^2}{2}\right)\right]\Bigg|_{t=t_L}^{t_U}$$

$$= \frac{r}{(2\pi V)^{1/2}} \left[\exp\left(-\frac{t_L^2}{2}\right) - \exp\left(-\frac{t_U^2}{2}\right)\right]. \tag{8.12.13}$$

The first integral is integrated by parts letting $u = t$ and $dv = te^{-t^2/2}\, dt$ to obtain

$$\frac{s}{(2\pi)^{1/2} \cdot 2V} \int_{t=t_L}^{t_U} t^2 \exp\left(-\frac{t^2}{2}\right) dt$$

$$= \frac{s}{(2\pi)^{1/2} \cdot 2V} \left[t_L \exp\left(-\frac{t_L^2}{2}\right) - t_U \exp\left(-\frac{t_U^2}{2}\right)\right]$$

$$+ \frac{s}{(2\pi)^{1/2} \cdot 2V} \int_{t=t_L}^{t_U} \exp\left(-\frac{t^2}{2}\right) dt. \tag{8.12.14}$$

Note that the second term of (8.12.14) is of the same form but of opposite sign as the last term of (8.12.11), so the two terms cancel. Substitution of (8.12.13) and (8.12.14) and cancellation of the terms noted yields

$$\Gamma' = \frac{s}{(2\pi)^{1/2} \cdot 2V} \left[t_L \exp\left(-\frac{t_L^2}{2}\right) - t_U \exp\left(-\frac{t_U^2}{2}\right)\right]$$

$$+ \frac{r}{(2\pi V)^{1/2}} \left[\exp\left(-\frac{t_L^2}{2}\right) - \exp\left(-\frac{t_U^2}{2}\right)\right]$$

$$= \exp\left(-\frac{t_L^2}{2}\right) \left[\frac{s}{(2\pi)^{1/2} \cdot 2V} t_L + \frac{r}{(2\pi V)^{1/2}}\right]$$

$$- \exp\left(-\frac{t_U^2}{2}\right) \left[\frac{s}{(2\pi)^{1/2} \cdot 2V} t_U + \frac{r}{(2\pi V)^{1/2}}\right]. \tag{8.12.15}$$

A necessary condition for Γ to be maximized is of course that $\Gamma' = 0$ which from (8.12.15) requires that

$$\exp\left(-\frac{t_L^2}{2}\right)\left[\frac{s}{(2\pi)^{1/2} \cdot 2V} t_L + \frac{r}{(2\pi V)^{1/2}}\right]$$

$$= \exp\left(-\frac{t_U^2}{2}\right)\left[\frac{s}{(2\pi)^{1/2} \cdot 2V} t_U + \frac{r}{(2\pi V)^{1/2}}\right]. \qquad (8.12.16)$$

Now

$$r = \frac{dM}{dK} = \frac{d}{dK}\left(\frac{p}{K}\right) = -\frac{p}{K^2} \qquad (8.12.17)$$

and

$$s = \frac{dV}{dK} = \frac{d}{dK}\left(\frac{K\sigma_\varepsilon^2}{2 - K}\right) = \frac{2\sigma_\varepsilon^2}{(2 - K)^2}. \qquad (8.12.18)$$

Substitution for r, s, t_U, and t_L in (8.12.16) yields

$$\exp\left[-\frac{(m_L - M)^2}{2V}\right]\left[\frac{2\sigma_\varepsilon^2/(2 - K)^2}{(2\pi)^{1/2} \cdot 2V} \cdot \frac{(m_L - M)}{\sqrt{V}} + \frac{-p/K^2}{(2\pi V)^{1/2}}\right]$$

$$= \exp\left[-\frac{(m_U - M)^2}{2V}\right]\left[\frac{2\sigma_\varepsilon^2/(2 - K)^2}{(2\pi)^{1/2} \cdot 2V} \cdot \frac{(m_U - M)}{\sqrt{V}} + \frac{-p/K^2}{(2\pi V)^{1/2}}\right] \qquad (8.12.19)$$

or

$$\exp\left[-\frac{(m_L - M)^2}{2V}\right]\left[\frac{\sigma_\varepsilon^2(m_L - M)}{(2 - K)^2 V} - \frac{p}{K^2}\right]$$

$$= \exp\left[-\frac{(m_U - M)^2}{2V}\right]\left[\frac{\sigma_\varepsilon^2(m_U - M)}{(2 - K)^2 V} - \frac{p}{K^2}\right]. \qquad (8.12.20)$$

Further substitution of (8.10.4) and (8.10.5) for M and V, respectively, would provide an expression consisting only of the decision variable K and the quantities p and σ_ε^2. Solving this expression explicitly for K, however, is another matter. Numerical methods of some variety are a necessity for solution of this problem. Therefore, instead of blanket substitution for M and V throughout (8.12.20), we substitute only inside the brackets to obtain

$$\exp\left[-\frac{(m_L - M)^2}{2V}\right]\left[\frac{\sigma_\varepsilon^2(m_L + p/K)}{(2 - K)^2 K\sigma_\varepsilon^2/(2 - K)} - \frac{p}{K}\right]$$

$$= \exp\left[-\frac{(m_U - M)^2}{2V}\right]\left[\frac{\sigma_\varepsilon^2(m_U + p/K)}{(2 - K)^2 K\sigma_\varepsilon^2/(2 - K)} - \frac{p}{K}\right], \qquad (8.12.21)$$

which, after considerable manipulation, can be written as

$$K = 3p \frac{\exp[-(m_U^2 - 2m_U M)/2V] - \exp[-(m_L^2 - 2m_L M)/2V]}{(m_U + p)\exp[-(m_U^2 - 2m_U M)/2V]}$$
$$- (m_L + p)\exp[-(m_L^2 - 2m_L M)/2V].$$

(8.12.22)

Even though K appears alone on the left-hand side of (8.12.22), this expression is not an explicit solution for K since both M and V which are present on the right-hand side are functions of K. However, (8.12.22) can be used iteratively to find the desired value of K, $\hat{K}$; i.e., an initial assumed value of K can be substituted into the right-hand side and a new value of K computed. This new value can then in turn be substituted in the right-hand side and another value of K calculated. This procedure is continued until two successive values of K computed in this manner are sufficiently close together to meet whatever criterion for precision may be applicable.

That the value of $\hat{K}$ thus determined corresponds to a maximum of Γ could be shown by examination of the second derivative of Γ with respect to K. An easier procedure, however, is the following. First note that Γ, being a probability function, is everywhere nonnegative as long as the variance V is nonnegative (as the variance of a real process must be). It is seen that V as given by (8.10.5) will be nonnegative for K in the stable range $0 < K < 2$. As K approaches zero, $M = p/K$ increases without bound and $V = K\sigma_\varepsilon^2/(2 - K)$ approaches zero. Thus the distribution of m becomes of infinitesimal variation located infinitely far from any finite band of acceptability $m_L \leq m \leq m_U$. Thus Γ approaches zero as K is reduced from $\hat{K}$ toward zero. Similarly, as K is increased from $\hat{K}$ toward two, V increases without bound and the distribution of m assumes infinite width. Thus, the probability of m lying in any finite range such as $m_L \leq m \leq m_U$ approaches zero. Therefore, any value of K in the stable range which satisfied (8.12.22) must correspond to a maximum of Γ.

To illustrate the use of (8.12.22) as an iteration formula to find $\hat{K}$, consider the situation in which $m_L = -2$, $m_U = 2$, and $\sigma_\varepsilon^2 = p = 1$. Substitution of these values in (8.12.22) yields

$$K = 3 \frac{e^{2M/V} - e^{-2M/V}}{3e^{2M/V} + e^{-2M/V}},$$

(8.12.23)

which for computing purposes is more conveniently expressed as

$$K = 3 \frac{e^{4M/V} - 1}{3e^{4M/V} + 1}.$$

(8.12.24)

Now

$$\frac{4M}{V} = \frac{4p/K}{K\sigma_\varepsilon^2/(2 - K)} = \frac{4(2 - K)}{K^2}$$

(8.12.25)

for $\sigma_\varepsilon{}^2 = p = 1$. For an initial guess $K_0 = 1$, $4M/V = 4$, which yields

$$K_1 = 3\,\frac{e^4 - 1}{3e^4 + 1} = 3\,\frac{54.59815003 - 1}{3 \cdot 54.59815003 + 1} = 0.9757273380.$$

Successive values of K's are listed in Table 8.12.1. Although 10 digits are

Table 8.12.1

Successive Values of K^a.

Iteration	K	$4M/V$
0	1	4
1	0.9757273380	4.303468479
2	0.9820521235	4.221982744
3	0.9805358039	4.241359219
4	0.9809075382	4.236599751
5	0.9808168884	4.237759824
6	0.9808390228	4.237476532
7	0.9808336199	4.237545679
8	0.9808349388	4.237528800
9	0.9808346169	

^{a} Values for $m_L = -2$, $m_U = 2$, $\sigma_\varepsilon{}^2 = p = 1$.

carried in the table simply because they were available from the calculator used to solve this problem, rarely would more than 3 or 4 be required in practice. Notice that convergence to 3 digits was obtained at the fourth iteration and 4 digits at the sixth. The 9 iterations shown produced convergence through 6 digits. These results are of course heavily dependent on the initial guess K_0. In this example convergence was oscillatory, i.e., the successive values of K alternated being higher and then lower than the value to which they were converging. With different parameter values, smooth convergence could conceivably occur.

The simple form of the iterative equation obtained for this example will occur whenever the specified tolerance limits are symmetrical, i.e., when

$$m_L = -m_U. \tag{8.12.26}$$

Substitution of (8.12.26) in (8.12.22) yields

$$K = 3p\,\frac{\exp[-(m_U{}^2 - 2m_U M)/2V] - \exp[-(m_U{}^2 + 2m_U M)/2V]}{(m_U + p)\exp[-(m_U{}^2 - 2m_U M)/2V]} \\ -(-m_U + p)\exp[-(m_U{}^2 + 2m_U M)/2V].$$

$$\tag{8.12.27}$$

Multiplication of numerator and denominator by $\exp(m_U^2/2V)$ results in

$$K = 3p \frac{e^{m_U M/V} - e^{-m_U M/V}}{(m_U + p)e^{m_U M/V} - (-m_U + p)e^{-m_U M/V}}$$

$$= 3p \frac{e^{2m_U M/V} - 1}{(m_U + p)e^{2m_U M/V} - (-m_U + p)}. \qquad (8.12.28)$$

Since symmetrical limits are common, (8.12.28) should be quite useful.

To find the maximum value of the criterion function Γ, we first note that (8.12.1) can be manipulated as follows:

$$\Gamma = \int_{m=m_L}^{m_U} \frac{1}{(2\pi V)^{1/2}} \exp\left[-\frac{(m - M)^2}{2V}\right] dm$$

$$= \int_{t=(m_L - M)/\sqrt{V}}^{(m_U - M)/\sqrt{V}} \frac{1}{(2\pi)^{1/2}} \exp\left(-\frac{t^2}{2}\right) dt$$

$$= \int_{t=-\infty}^{(m_U - M)/\sqrt{V}} \frac{1}{(2\pi)^{1/2}} \exp\left(-\frac{t^2}{2}\right) dt - \int_{t=-\infty}^{(m_L - M)/\sqrt{V}} \frac{1}{(2\pi)^{1/2}} \exp\left(-\frac{t^2}{2}\right) dt.$$

$$(8.12.29)$$

After finding $\hat{K}$, we compute the associated M and V and from these determine

$$t_U = (m_U - M)/\sqrt{V} \qquad (8.12.30)$$

and

$$t_L = (m_L - M)/\sqrt{V}. \qquad (8.12.31)$$

We then enter Table 8.6.2 for t_U and t_L to find $\Pr[t \leq t_U]$ and $\Pr[t \leq t_L]$, which correspond respectively to the first and second integrals of (8.12.29). Γ_{max} is then found by subtracting $\Pr[t \leq t_L]$ from $\Pr[t \leq t_U]$. For the illustrative problem, $\hat{K}$ was found to be 0.9808 to four places. Therefore,

$$M = p/\hat{K} = 1/0.9808 = 1.0196$$

and

$$V = \hat{K}\sigma_\varepsilon^2/(2 - \hat{K}) = 0.9808 \cdot 1/(2 - 0.9808) = 0.9623$$

from which

$$t_U = (2 - 1.0196)/0.9810 = 0.9994$$

and

$$t_L = (-2 - 1.0196)/0.9810 = -3.0781.$$

From Table 8.6.2 we find that

$$\Pr[t \leq 0.999] = 0.8411$$

and

$$\Pr[t \leq -3.0781] = 1 - \Pr[t \leq 3.0781] = 1 - 0.9989 = 0.0011$$

which gives

$$\Gamma_{max} = 0.8411 - 0.0011 = 0.8400.$$

8.13   Maximum Steady-State Probability of Acceptable Individual Product Units

Conditions

$p(i)$:   Step function of magnitude p starting at $i = 0$

$\varepsilon(i)$:   Mutually independent, identically normally distributed with mean zero and variance σ_ε^2, independent of $p(i)$

Criterion   Maximize the steady-state probability of individual items $y(i, q)$ lying between tolerance limits y_L and y_U.

As in the previous example, since probabilities are involved, it is necessary to specify a distribution for all random variables. Again for convenience, independent, stationary normal distributions are assumed for both $\eta(i, q)$ and $y(i, q)$. Maximizing the probability of individual product items lying within tolerance limits is a very common criterion.

For $\eta(i, q)$ and $y(i, q)$ normally distributed, the steady-state distribution of individual items y will be normal with mean M and variance V_T. Therefore, the probability of an item lying between limits y_L and y_U is

$$\Gamma_T = \Pr[y_L \leq y \leq y_U]$$

$$= \int_{y = y_L}^{y_U} \frac{1}{(2\pi V_T)^{1/2}} \exp\left[-\frac{(y - M)^2}{2V_T}\right] dy. \qquad (8.13.1)$$

The form of (8.13.1) is the same as that of (8.12.1) for Γ in the previous example except that here we have V_T in place of V. Therefore, we handle differentiation with respect to K and the solving of $\Gamma_T{}' = 0$ in the same manner as was used for Γ. Furthermore,

$$\frac{dV_T}{dK} = \frac{d}{dK}(V + \sigma_\varepsilon^2) = \frac{dV}{dK} = s = \frac{2\sigma_\varepsilon^2}{(2 - K)^2}, \qquad (8.13.2)$$

so the derivation is exactly the same as the one in Section 8.12 until it is necessary to substitute for V_T, i.e., through Eq. (8.12.20). For our current example, the counterpart of (8.12.20) is

$$\exp\left[-\frac{(y_L - M)^2}{2V_T}\right]\left[\frac{\sigma_\varepsilon^2(y_L - M)}{(2 - K)^2 V_T} - \frac{p}{K^2}\right]$$

$$= \exp\left[-\frac{(y_U - M)^2}{2V_T}\right]\left[\frac{\sigma_\varepsilon^2(y_U - M)}{(2 - K)^2 V_T} - \frac{p}{K^2}\right]. \qquad (8.13.3)$$

As in the previous example, even though substitution of (8.10.4) and (8.5.3) for M and V_T, respectively, would provide an expression consisting only of the decision variable K and the quantities p, σ_ε^2, and σ_y^2 (or $p, \sigma_\eta^2, \sigma_y^2$, and N if one wished further to substitute (8.3.18) for σ_ε^2), chances of explicit analytical solution are nil. Since numerical methods of some sort are required for solution, we could simply leave the designer with (8.13.3), his knowledge of numerical methods, and his ingenuity. However, some further modification of this expression might prove helpful. One possibility is to proceed as in Section 8.12 and substitute for M and V_T in the brackets only and then isolate a K on the left-hand side. One possible form of the result of doing this is

$$K = 2\sigma_y^2 p(2 - K)$$

$$\cdot \frac{\exp[-(y_U^2 - 2y_U M)/2V_T] - \exp[-(y_L^2 - 2y_L M)/2V_T]}{[\sigma_\varepsilon^2(Ky_U - p) - p(2 - K)(\sigma_\varepsilon^2 - \sigma_y^2)]\exp[-(y_U^2 - 2y_U M)2V_T]}{\quad - [\sigma_\varepsilon^2(Ky_L - p) - p(2 - K)(\sigma_\varepsilon^2 - \sigma_y^2)]\exp[-(y_L^2 - 2y_L M)/2V_T]}$$

$$(8.13.4)$$

which unfortunately cannot be used iteratively since it has been found, by sad experience, to diverge. It can, however, be used as an alternative form of $\Gamma' = 0$ upon which to base a one-dimensional search technique or simply to guide and check trial-and-error solutions.

To illustrate the use of (8.13.4) in the latter role consider the situation in which $m_L = -2$, $m_U = 2$, and $\sigma_\varepsilon^2 = \sigma_y^2 = p = 1$. With these values, (8.13.4) reduces to

$$K = \frac{2(2 - K)(e^{4M/V_T} - 1)}{2K(e^{4M/V_T} + 1) - (e^{4M/V_T} - 1)}.$$

$$(8.13.5)$$

A few cycles of the iteration method beginning with $K_0 = 1$ yield $K_1 = 1.23$ and $K_2 = 0.45$, signs of rapid divergence. Therefore, trial and error is used as follows. A trial K value of 1.05 substituted in the right-hand side of (8.13.5) yields 0.988, a result smaller than the trial value. A trial K value of 1.02, however, yields 1.127, a result larger than the trial value. Thus the optimal value of K lies somewhere between 1.02 and 1.05. A trial value of 1.03 yields 1.078, so the optimal value must lie between 1.03 and 1.05. The trial value 1.04 yields 1.042, which means the optimal value is just slightly higher than 1.04. Further evaluation using such trial values as 1.041, 1.042, etc., could pin down the optimal value as close as is needed, although the engineer might be perfectly satisfied with 1.04.

8.14 Other Possibilities

Obviously many other examples could be included here. There are innumerable combinations of perturbations, instrumentation errors, sampling errors, and criteria for effective process operation, many of which are quite common in many types of real systems and would provide bases for interesting and worthwhile studies. Sinusoidal perturbations can be used to describe periodic phenomena ranging from shift-to-shift variation in product quality to seasonal fluctuations in demand on inventory and production control systems. These have not been included in the examples in this section, although their basic properties were discussed earlier and they will be included in our chapters on higher-order control systems. Interdependent random variations also have not been treated explicitly. Our criteria have not explicitly included costs of operation, system adjustment, and customer satisfaction. These and other possibilities could indeed have been included. However, our purpose in these examples was simply to examine the general properties of various possible $p(i)$ and $\varepsilon(i)$ and to illustrate how a system designer might relate these to system performance criteria to choose a system parameter. Some of the other items will be covered later.

It is hoped that what has been included has indicated what is involved in system design, even when restricted to one very simple configuration. It is also hoped that some ideas for approaches to challenging design problems may have been generated, some of which might be useful in the context of the second- and higher-order systems to be discussed in Chapters X and XI, where system configuration as well as selection of parameter values may be involved.

Since stability determination is much more complicated for second- and higher-order systems than for the simple first-order systems discussed in Chapters VII and VIII, we now turn our attention to a general discussion of stability. In Chapter IX, we study the properties which must be possessed by the roots of the characteristic equation for a system to be stable and introduce some tests for determining system stability directly from the characteristic equation without the necessity of finding its roots.

Exercises

8.1   For the process controller whose output is described by (8.2.1), i.e., with $D = m(0) = p(i) = 0$, find the steady-state variance V given a stationary, independently distributed measurement error $\varepsilon(i)$ with $\sigma_\varepsilon^2 = 1$ for the following values of K:

    a. 0.25   **b.** 0.50   **c.** 0.75   **d.** 1.00   **e.** 1.25   **f.** 1.50   **g.** 1.75

8.2

    a.   For the process controller whose output is described by (8.2.1), i.e., with $D = m(0) = p(i) = 0$, and a stationary measurement error $\varepsilon(i)$ with autocovariance function $\sigma(\tau)$ given by:

τ	$\sigma_\varepsilon(\tau)$
0	σ_ε^2
± 1	$0.5\sigma_\varepsilon^2$
all other values	0

derive an expression for the variance of the process output $V[m(i)]$.

b. Find the steady-state output variance, assuming stable operation.

c. For what values of K will the steady-state variance in (b) be greater than it would have been had the $\varepsilon(i)$ been independently distributed?

8.3 **(a)** through **(c)**. The same as 8.2 except $\sigma_\varepsilon(\tau)$ given by:

τ	$\sigma_\varepsilon(\tau)$
0	σ_ε^2
± 1	$-0.5\sigma_\varepsilon^2$
all other values	0

8.4 **(a)** through **(c)**. The same as 8.2 except $\sigma_\varepsilon(\tau) = 2^{-|\tau|}\sigma_\varepsilon^2$.

8.5 **(a)** through **(c)**. The same as 8.2 except $\sigma_\varepsilon(\tau) = (-2)^{-|\tau|}\sigma_\varepsilon^2$.

8.6 The process controller whose output is described by (8.2.1), i.e., with $D = m(0) = p(i) = 0$, operates such that individual product items at time i are identically and independently distributed around the expected process output $m(i)$ with variance $\sigma_y^2(i)$. Given that a sample of N items is taken at each sampling instant i, find the steady-state mean M and variance V of process output for the following situations:

a. Negligible instrumentation error $\eta(i, q)$ for all i, q.

b. $\eta(i, q)$ unbiased and identically and independently distributed with variance σ_η^2 for all i, q.

c. $\eta(i, q)$ identically and independently distributed with expected value H and variance σ_η^2 for all i, q.

d. $\eta(i, q) = H(i)$ for all items in the ith sample (i.e., $\sigma_\eta^2(i, q) = 0$ for all i, q), where $H(i)$ is identically and independently distributed from sample to sample with mean H and variance σ_H^2.

e. $\eta(i, q)$ identically and independently distributed within the ith sample with mean $H(i)$ and variance σ_η^2 and $H(i)$ identically and independently distributed from sample to sample with mean H and variance σ_H^2.

f. The same as part (d) except $H(i)$ has a sample-to-sample autocovariance function $\sigma_H(\tau)$ given by

τ	$\sigma_H(\tau)$
0	$\sigma_H{}^2$
± 1	$0.5\sigma_H{}^2$
all other values	0

g. The same as part (d) except $H(i)$ has a sample-to-sample auto-covariance function $\sigma_H(\tau) = A^{-|\tau|}\sigma_H{}^2$, $A > 1$.

h. The same as part (e) except $H(i)$ has a sample-to-sample auto-covariance function $\sigma_H(\tau)$ as given in part (f).

i. The same as part (e) except $H(i)$ has a sample-to-sample auto-covariance function $\sigma_H(\tau) = A^{-|\tau|}\sigma_H{}^2$, $A > 1$.

j. $\eta(i, q)$ identically distributed within each sample with mean H for all i, q and autocovariance function $\sigma_\eta(\tau)$ given by

τ	$\sigma_\eta(\tau)$
0	$\sigma_\eta{}^2$
± 1	$0.5\sigma_\eta{}^2$
all other values	0

for all i. Sample-to-sample variations are mutually independent.

k. The same as part (j) except $\sigma_\eta(\tau) = A^{-|\tau|}\sigma_\eta{}^2$, $A > 1$.

l. $\eta(i, q)$ identically distributed within each sample with mean $H(i)$ tor all q and autocovariance function as given in part (j). $H(i)$ is identically and independently distributed from sample to sample with mean H and variance $\sigma_H{}^2$.

m. $\eta(i, q)$ identically distributed within each sample with mean $H(i)$ for all q and autocovariance function $\sigma_\eta(\tau) = A^{-|\tau|}\sigma_\eta{}^2$, $A > 1$. $H(i)$ is identically distributed from sample to sample with mean H and autocovariance function $\sigma_H(\tau) = B^{-|\tau|}\sigma_H{}^2$.

8.7 (a) through **(m).** For the same conditions which apply to the corresponding part of Exercise 8.6, find the steady-state mean M_T and variance V_T of individual product items around $D = 0$.

8.8 For the process controller whose output is described by (8.2.1), i.e., with $D = m(0) = p(i) = 0$, and $\varepsilon(i)$ a stationary, unbiased, independently distributed random variable with variance $\sigma_\varepsilon{}^2$, find the values of proportionality constant K which will limit the steady-state variance of the process output to 4×10^{-6} in.^{2} for the following values of $\sigma_\varepsilon{}^2$:

a. 2×10^{-6} in.^{2} **b.** 4×10^{-6} in.^{2} **c.** 8×10^{-6} in.^{2}

8.9 (a) through (c). The same as 8.8 except $\varepsilon(i)$ has autocovariance function $\sigma_\varepsilon(\tau)$ as given in 8.2(a).

8.10 (a) through (c). The same as 8.8 except $\varepsilon(i)$ has autocovariance function $\sigma_\varepsilon(\tau) = A^{-|\tau|}\sigma_\varepsilon^2$, $A > 1$.

8.11 through **8.14**   For the situations described in 8.2 through 8.5, respectively, find the value of K which minimizes the steady-state variance of process output V.

8.15   Given that x is a standard normal variable, i.e., x is normally distributed with mean zero and variance one, find the probability of occurrence of each of the following:

a. $x \le 2.5$ b. $x \le 1.35$ c. $x \ge -1.02$ d. $x \ge 1.333$
e. $0.851 \le x \le 2.097$ f. $-1.876 \le x \le -1.376$
g. $-0.703 \le x \le 1.792$

8.16   Given that x is a normally distributed random variable with expected value μ and variance σ^2, find the probability of occurrence of each of the following:

a. $x \le 2.5$ for $\mu = 0.5$, $\sigma = 2.0$
b. $x \le -2.82$ for $\mu = -3.0$, $\sigma = 0.3$
c. $-1.00 \le x \le 1.00$ for $\mu = -0.5$, $\sigma = 1.5$
d. $0 \le x \le 3.00$ for $\mu = 1.5$, $\sigma = 0.75$
e. $-3.00 \le x \le 0$ for $\mu = 1.5$, $\sigma = 2.25$

8.17   Given that x is a standard normal variable, i.e., x is normally distributed with mean zero and variance one, find the limits X_L and X_U such that the following probability statements are satisfied:

a. $\Pr[x \le X_U] = 0.6500$ b. $\Pr[x \le X_U] = 0.1239$
c. $\Pr[X_L \le x] = 0.2135$ d. $\Pr[X_L \le x] = 0.9000$
e. $\Pr[X_L \le x \le X_U] = 0.500$ given $X_U - X_L = 1.0$
f. $\Pr[X_L \le x \le X_U] = 0.1000$ given $X_U - X_L = 0.5$

8.18   Given that x is a normally distributed random variable with expected value μ and variance σ^2, find the limits X_L and X_U such that the following probability statements are satisfied:

a. $\Pr[x \le X_U] = 0.7000$ for $\mu = -2.00$, $\sigma = 1.5$
b. $\Pr[x \le X_U] = 0.3000$ for $\mu = -2.00$, $\sigma = 1.5$
c. $\Pr[X_L \le x] = 0.0352$ for $\mu = 17.291$, $\sigma = 5.42$
d. $\Pr[X_L \le x] = 0.9555$ for $\mu = 5.7$, $\sigma = 0.0003$
e. $\Pr[X_L \le x \le X_U] = 0.7500$ for $\mu = 5.7$, $\sigma = 2.0$, and $X_U - X_L = 2.2$

8.19   The ideal operating temperature for a chemical reactor is 500°F. Company policy requires that actual temperature must fall between specified tolerance limits at least 95% of the time. Temperatures are read at evenly spaced time intervals and adjustments made according to a proportional control rule. If the temperature sensing device has an unbiased, stationary, independently and normally distributed instrumentation error with variance 225 deg^2, what fraction of any measured deviation from the ideal temperature should be used as a basis for adjustment to minimize response time while adhering to company policy given the following tolerances:

a.   460 to 540°F   **b.**   470 to 530°F   **c.**   480 to 520°F

8.20   A simple discrete proportional controller is used to control the length of spacer pins at 3.000 cm. The standard deviation of these lengths is constant at 0.005 cm. Each adjustment is based on the average of a sample of 5 pins. Assuming instrumentation error is negligible, find the value of the controller proportionality constant which will limit the fraction of defective pins when $m(i) = D = 3.000$ cm to no more than two percent, yet will correct for shifts in process mean as fast as possible given the following tolerance limits on individual pins:

a.   2.980–3.020 cm   **b.**   2.988–3.012 cm   **c.**   2.989–3.011 cm

8.21   Soft drink bottles are required by law to contain at least the volume of liquid listed on the label of the bottle. The automatic filling device is therefore checked by periodically measuring the liquid content of a sample of bottles and making adjustments according to some control rule. This sampling and adjustment can also be done automatically. Suppose a line turning out 12 oz bottles has a bottle-to-bottle variation which is identically, independently, and normally distributed with variance 0.050 oz^2 and the instrumentation error is unbiased and is also identically, independently, and normally distributed with variance 0.010 oz^2. If a proportional control rule and samples of 10 bottles are to be used, find the value of the proportionality constant so that the system will respond as quickly as possible to perturbations when they occur but will limit the probability of the liquid in a bottle falling below 12 oz when the process is set at $D = 12.5$ oz to the following:

a.   0.020   **b.**   0.015   **c.**   0.010

8.22   An automatic lathe is used to turn a shaft to a diameter of 1.000 in. Tolerance limits on the shafts are ± 0.0010 and rejects, whether too big or too small, cost $0.10 apiece. The production rate is 100 shafts per hour. Assume the diameters are normally distributed with constant standard deviation of 0.0004 in.

a. What is the expected hourly cost of defective shafts when this process is operating properly (i.e., with $m(i) = 1.000$ and $\sigma = 0.0004$)? This amount is the "normal" cost of defectives.

b. What expected hourly cost of defectives will result if $m(i)$ shifts to 1.0004 in. and remains uncorrected for one hour?

c. A simple proportional control device is attached to this lathe which automatically measures each shaft and adjusts the setting of the cutting tool by an amount K times the measured deviation between the shaft diameter and 1.0000 in. Assume instrumentation errors are negligible and that each adjustment is complete before the next shaft is cut. Find the value of K which will correct shifts in process mean as fast as possible but will insure that no more than 5% defective product will be produced when the process is working properly (i.e., when no shifts due to assignable causes occur).

d. What is the expected hourly cost of defectives if the controller of part (c) is used?

e. What expected cost of defectives will be incurred during the hour following a shift in $m(i)$ to 1.0004 in. if the controller of part (c) is used?

f. What is the expected hourly cost of defectives if the proportionality constant K is set at 1.0?

g. What expected cost of defectives will be incurred during the hour following a shift in $m(i)$ to 1.0004 in. if the controller with $K = 1.0$ is used?

8.23 An automatic screw machine turns out tapered connecting pins at the rate of 300 per hour. The bogie value for pin length is 1.000 in. with engineering tolerance limits of ± 0.003 in. Defective items cost $.05 each. Past data show the pin lengths to be normally distributed with a constant standard deviation of 0.001 in.

a. What is the expected hourly cost of defective pins when the machine is operating properly (i.e., with $m(i) = 1.000$ and $\sigma_y = 0.001$)?

b. What is the expected hourly cost if on the average of once every half hour the stop which controls pin length slips by 0.001 in. and remains uncorrected? Assume the "slips" occur at the beginning of the hour and again one-half hour later.

c. A simple proportional control device is used to control the position of the stop. Each pin is measured and the stop adjusted a fixed portion K of the deviation between the measurement and 1.000 in. Assume measurement errors are negligible and that each adjustment is complete before the next pin is cut. Find the value of K which will correct shifts in mean as fast as possible but will limit the fraction of defective product produced when the stop is properly set (i.e., $m(i) = 1.000$ in.) to:

1. $p = 0.05$ **2.** $p = 0.005$ **3.** $p = 0.001$

d.   What is the expected hourly cost of defectives if each of the controllers of part (c) is used and the process is operating properly ($m(i) = 1.000$)?

e.   What is the expected hourly cost of defectives if each of the controllers of part (c) is used and shifts occur in the stop setting as described in part (b)? (For simplicity assume expected adjustments.)

8.24   A simple discrete proportional controller is used to control the length of spacer pins at 2.000 cm. The distribution of the lengths of individual pins around the process mean is normal with a constant standard deviation of 0.005 cm. Production specifications require that 95% of the pins be within ± 0.011 cm of 2.000 cm when $m(i) = D = 2.000$ cm. The measuring instrument involved has a constant error $H(i)$ for all items within the same sample (i.e., $V[\eta(i, q)] = 0$) but $H(i)$ is independently normally distributed from period to period with mean zero and standard deviation 0.001 cm. What must the sample size be to allow the above specifications to be met with a proportionality constant of $K = 1$?

8.25   The same as 8.24 except that $K = 0.5$.

8.26   **(a)** through **(c)**.   It is desired to modify the spacer-pin production process discussed in Exercise 8.20 so that a control constant of unity can be used while still holding the steady-state fraction defective to a maximum of 2%. A study of the process shows that the standard deviation of the pin lengths around the process mean can be reduced at a cost per sampling period given by $C_y = 10[(0.005 - \sigma_y)/\sigma_y]$ dollars, where σ_y is the resulting new value of pin-length standard deviation and $\sigma_y \leq 0.005$ (i.e., no "profit" results from increasing σ_y). Inspection costs $1.00 per item in the sample. For the tolerance limits given in the corresponding part of 8.20 find the minimum-cost procedure to accomplish the desired objective.

8.27   **(a)** through **(c)**.   What would the actual fraction defective be in the corresponding part of Exercise 8.20 if an unknown constant instrumentation error of 0.001 cm were present in each measurement?

8.28   **(a)** through **(c)**.   Assume the chemical reactor discussed in Exercise 8.19 is subject to a random, unbiased, stationary, independently and normally distributed perturbation $p(i)$ with variance $\sigma_p^2 = 100$ deg^2, while all other conditions remain as in 8.19. Furthermore, $p(i)$ and $\varepsilon(i)$ are independent. Solve the corresponding part of 8.19 given this modified set of conditions.

8.29   A first-order rapid-response process controller with $D = m(0) = 0$ has an unbiased, stationary, independently and normally distributed measurement error $\varepsilon(i)$ with variance σ_ε^2. It is also subject to a perturbation such that

$$p(i) = \begin{cases} A & \text{with probability } P, \\ 0 & \text{with probability } 1 - P, \end{cases}$$

for all i (i.e., $p(i)$ is binomially distributed). $\varepsilon(i)$ and $p(i)$ are mutually independent. Assuming stable operation, derive closed-form expressions for the steady-state mean and variance of process output.

8.30 (a) through (e). Verify the corresponding item in Section 8.9 which was left as an exercise to the reader.

8.31 (a) through (c). Assume that the spacer-pin production process described in Exercise 8.20 is subject to a random, unbiased, stationary, independently and normally distributed perturbation $p(i)$ with standard deviation $\sigma_p = 0.001$ cm, while all other conditions remain as in 8.20. Furthermore, $p(i)$ and $\varepsilon(i)$ are independent. Solve the corresponding part of 8.20 for this modified set of conditions.

8.32 (a) through (c). The same as Exercise 8.26 except with the random perturbation described in 8.31 present.

8.33 (a) through (c). The same as Exercise 8.32 except that the effects of the random perturbation can be eliminated at a cost of $20 per period.

8.34 A rapid-response system is to be controlled by a simple proportional controller with control constant K. The system is subject to a step perturbation of magnitude p. An unbiased measuring error $\varepsilon(i)$ with variance σ_ε^2 is present with each measurement of process output. The $\varepsilon(i)$ are mutually independent. For the following relationships between σ_ε and p, find the value of K which minimizes the second moment Q of the steady-state distribution of the process output around the desired level of output and the value of Q_{min} as a function of p:

   a. $\sigma_\varepsilon = \sqrt{2}\,p$ **b.** $\sigma_\varepsilon = p$ **c.** $\sigma_\varepsilon = (\sqrt{2}/2)p$

8.35 (a) through (c). In the control system of Exercise 8.34, assume that $\sigma_\varepsilon = \sigma_y/2$, where σ_y is the standard deviation of the distribution of individual items around the mean (the items are identically and independently distributed) and 4 the current sample size. It has been determined that $Q = 1$ is the largest second moment of $m(i)$ around $D = 0$ that can be tolerated. Studies show that σ_y can be reduced from its current value $\bar\sigma_y$ at a cost per sampling period of $C_y = 10[(\bar\sigma_y - \sigma_y)/\sigma_y]$ dollars for $\sigma_y \leq \bar\sigma_y$ (i.e., no "profit" results from increasing σ_y) and that p can be reduced from its current value of $p = 1$ at a cost per period of $C_p = 5[(1 - p)/p]$. Inspection costs $1.00 per item in the corresponding part of 8.34; find the minimum-cost procedure to accomplish the desired objective.

8.36 (a) through (c). The same as Exercise 8.35 except it is now desired to reduce Q_T, the second moment of individual items around $D = 0$, to a value no larger than one.

8.37   (a) through (c).   For the system described in Exercise 8.34, find the value of K which maximizes the probability Γ of the steady-state process mean being between the limits m_U and m_L and find the value of Γ_{max} given $m_U = 2$, $m_L = -2$, and $p = 2$. Assume $\varepsilon(i)$ is normally distributed.

8.38 through 8.41   The same as 8.37 except for the following values of m_U, m_L, and p:

    8.38 $m_U = 2$, $m_L = -2$, $p = 1$

    8.39 $m_U = 1$, $m_L = -1$, $p = 2$

    8.40 $m_U = 2$, $m_L = -1$, $p = 2$

    8.41 $m_U = 3$, $m_L = 1$, $p = 2$

8.42   (a) through (c).   In the control system described in Exercise 8.34 assume that $\sigma_\varepsilon = \sigma_y/2$, where σ_y is the standard deviation of the distribution of individual items around the mean (the items are identically, independently, and normally distributed) and the sample size is 4. Find the value of K which maximizes the steady-state probability Γ_T of individual items lying between the limits y_U and y_L and find the value of $\Gamma_{T\,max}$ given $y_U = 2$, $y_L = -2$, and $p = 2$.

8.43 through 8.46   The same as 8.42 except for the following values of y_U, y_L, and p:

    8.43 $y_U = 2$, $y_L = -2$, $p = 1$

    8.44 $y_U = 1$, $y_L = -1$, $p = 2$

    8.45 $y_U = 2$, $y_L = -1$, $p = 2$

    8.46 $y_U = 3$, $y_L = 1$, $p = 2$

Chapter IX | **System Stability**

As illustrated in Chapter VII the long range usefulness of a system depends on the stability of the system, where a stable system is one which has the ability to recover from the effects of impulse-type inputs. For this reason, system stability must be considered while determining system performance with respect to other design criteria.

Since stability determinations are so important, it behooves both the system analyst and designer to have ready means of determining whether a given system is stable or not. Of even greater advantage would be the ability to determine the ranges of system parameters over which stable system operation is obtained, for this would concentrate the search for preferred parameter values to those in the stable range. This was effectively illustrated in Chapters VII and VIII for the rapid-response, first-order process controller where consideration was given to values of the proportionality constant K only in the range $0 < K < 2$. In this simple case, the stable range was readily apparent and easily verified.

As will be demonstrated in Section 9.2, system stability is directly related to the magnitude of the roots of the characteristic equation of the system difference equation. From Section 4.5, however, it is apparent that the difficulty in obtaining these roots increases rapidly with the order of the equation. Thus for systems with complicated control rules, complex system dynamics, feedback delays, or feedforward or prediction of inputs, there is a real need for stability tests which can be applied directly to the characteristic equation in polynomial form. This is particularly true where solutions are needed in the form of functions of system parameters so that stable ranges can be determined, since for equations of more than second-order root-determination is almost always limited to numerical techniques.

Fortunately, a number of stability tests are available which require as input information only the polynomial form of the system difference equation or the polynomial form of the z transform of the system response. Furthermore, several of these tests provide much additional information as well as the absolute determination of stability versus instability. We will examine two of the more useful of these tests in Section 9.3 after first presenting some basic definitions in Section 9.1 and relating system stability to the magnitude of the roots of the characteristic equation in Section 9.2.

9.1   General Definitions

It was stated above that a system is stable if the transients caused when the system is initially put into operation or perturbed by short-lived external inputs eventually die out. Specifically, a linear system is stable if the homogeneous solution of the system difference equation goes to zero as the independent variable increases without bound, i.e.,

$$\lim_{i \to \infty} x(i)^{(H)} = 0. \qquad (9.1.1)$$

It is important to note that stability has nothing to do with the ability of a system to follow or compensate for a particular forcing function; i.e., the limit approached by the system error could be infinite even for a stable system. This was illustrated in Section 7.4.

We can now define $x(i)^{(H)}$ to be the transient solution to a linear difference equation, i.e., the transient behavior of a system (thus it could be completely eliminated by the proper initial conditions); $x(i)^{(P)}$ to be the steady-state solution to a linear difference equation, i.e., the limit approached by system behavior after the transient dies out (thus it is meaningful only if the system is stable).

9.2   Criterion for Stability of Discrete Linear Systems

It has been demonstrated in Chapter IV that the nature of $x(i)^{(H)}$ is determined by the roots of the characteristic equation. In that chapter, four categories of roots were discussed. In order to develop an all-inclusive stability criterion, all four categories of roots must be investigated. We now consider each category in turn. The material which follows, however, is described in terms of the symbolism developed in conjunction with the z transform method of solution described in Section 6.5 for the partial fraction expansion procedure where the similarity between the characteristic equation and $B(z) = 0$ was noted. Since the roots of the two equations are identical, stability can be discussed in terms of either one.

Single Real Roots

Each such root $z_r = c$ forms a term in $f(i)$ of the form

$$Kc^i.$$

As shown by the simple example in Section 7.2, for a system to be stable, $\lim_{i \to \infty} Kc^i$ must be zero. For this to occur,

$$|c| < 1.$$

For c positive, decay will be a smooth sequence of steps, but for c negative, successive values of the term will have opposite signs but decreasing magnitudes. In the special case in which $|c| = 1$, $\lim_{i \to \infty} |Kc^i| = |K|$ ($+K$ for $c = 1$ and alternating $+$ and $-$ for $c = -1$). This is the case of marginal stability.

Single Pairs of Conjugate Complex or Imaginary Roots

Each single pair of complex conjugate or imaginary roots $z_{r+} = a + jb$ and $z_{r-} = a - jb$ forms a pair of terms in $f(i)$ of the form

$$K_{r+} z_{r+}^i + K_{r-} z_{r-}^i,$$

where the coefficients are of the form

$$K_{r+} = s + jd \quad \text{and} \quad K_{r-} = s - jd.$$

As shown previously, the pair of roots may be alternatively represented in $f(i)$ by the expression (see Eq. (6.5.40))

$$2R^i(s \cos \theta i - d \sin \theta i),$$

where

$R =$ the magnitude of the root $= [(\text{Re } z_r)^2 + (\text{Im } z_r)^2]^{1/2} = (a^2 + b^2)^{1/2}$

and

$$\theta = \tan^{-1}(\text{Im } z_r / \text{Re } z_r) = \tan^{-1}(b/a).$$

For the system to be stable, $\lim_{i \to \infty} R^i \to 0$. For this to occur, $R < 1$. Again, there is the special case of $R = 1$ which produces sustained "oscillations" of angular frequency $2\pi/\theta T$, where T is the sampling interval. This is also a form of marginal stability.

Real Roots of Multiplicity p

Each such root of multiplicity p forms a sequence of p terms in $f(i)$ of the form

$$\sum_{q=1}^{p} \frac{K_{rq}}{z_r^{q-1}} \cdot \frac{i!}{(i - q + 1)!(q - 1)!} z_r^i.$$

For a system to be stable, the limit of each such term as i increases without bound must equal zero. It can be shown that for $|z_r| < 1$ this limit is zero; however, $|z_r| = 1$ leads to definite instability because of the presence of the coefficient

$$i!/(i - q + 1)!(q - 1)!$$

which increases without bound as i increases without bound. Obviously, for $|z_r| > 1$ the term also increases without bound. Therefore, the condition for stability is that $|z_r| < 1$.

Conjugate Complex or Imaginary Roots of Multiplicity p

Each conjugate complex or imaginary pair of roots of multiplicity p forms a sequence of $2p$ terms in $f(i)$ of the form

$$R^i \left\{ \sum_{q=1}^{k} \frac{i!}{R^{q-1}(i - q + 1)!(q - 1)!} \left[s_{rq} \cos \theta i - d_{rq} \sin \theta i \right] \right\}.$$

For a system to be stable, the limit of each term as it increases without bound must be zero. For this to occur, $R = |z_r| < 1$. As for the root of multiplicity p, the case of $|z_r| = 1$ leads to definite instability because of the presence of the coefficient which itself increases without bound as i increases.

General Rule   For a system to be stable, the magnitude (absolute value, index) of all roots of the characteristic equation must be less than unity, i.e., they must lie within the unit circle in the complex z plane. In some cases, a single real root or single pairs of conjugate complex or imaginary roots may lie on the unit circle (marginal stability).

Thus, all roots of $B(z) = 0$ in the z domain must be such that $|z| < 1$. (In some cases a single real root or single pairs of complex or imaginary roots may be such that $|z| = 1$.)

9.3   Tests for Stability

Although the tests to be discussed here can be used either in conjunction with the characteristic equation of the linear difference equation which describes the system or with the z transform of the system response, the discussion which follows is presented in the context of the latter, i.e., the z transform of the system response.

Since the denominator $B(z)$ of the z transform of the system response as given in (6.5.1) is identical to the left-hand side of the characteristic equation of the system difference equation, we actually need to concern ourselves only with the equation $B(z) = 0$. Furthermore, what is said here can be applied directly to the system characteristic equation.

Stability tests can be broadly categorized into two groups. The first consists of those which are applied to $B(z)$ directly. The other group includes those tests which first require a transformation of the z transform before they can be applied. We will describe in detail one of each type of test.

Tests Involving $B(z) = 0$ Directly

Two well-known stability tests which work directly with $B(z) = 0$ are the Schur–Cohn (Schur, 1917; Cohn, 1922; Marden, 1966) and Jury (1962; Jury and Blanchard, 1961) tests. These tests in general reveal only absolute stability vs instability, although for fairly simple low-order systems some indication of the sensitivity of system stability to changes in system parameters can be obtained. The Schur–Cohn test can be applied to polynomial equations with either real or complex coefficients. The Jury test is based on the same mathematical relationships as the Schur–Cohn, but is designed exclusively for polynomial equations with real coefficients. For this reason, it is much more efficient and much easier to apply to the polynomial with real coefficients. Since the characteristic equations corresponding to physically realizable systems will always have real coefficients, the Jury test is far more appropriate than the Schur–Cohn for our purposes here. Both tests are summarized by Shinners (1964).

A third test due to Bishop (1960) is similar to the Nyquist approach for continuous systems. This test, called the B plane stability test, though more cumbersome in its application than the others, permits evaluation of the sensitivity of system stability to changes in both system parameter values and system configuration in addition to an absolute determination of stability or instability. Such information can be valuable in cases where elements are subject to change with time or where components manufactured in production lots display significant individual differences, and can provide the system designer with insights concerning both the transient and steady-state performance of the system. The test is based on the fact that, since, for a system to be stable, all roots of the characteristic equation must be inside the unit circle in the complex z plane, the locus of $B(z)$ in the complex B plane must encircle the B plane origin exactly n times for an nth-order characteristic equation as z traverses the unit circle once. This test can obviously be run only for numerical coefficients in the characteristic equation. Thus the establishment of the stable ranges for system parameters, such as $0 < K < 2$ for the proportionality constant in the simple proportional controller, can be accomplished only as a numerical search, an extremely difficult and time-consuming approach which gets rapidly worse as the order of the system increases. This is further complicated by the fact that, as the order of the characteristic equation increases, the number of numerical evaluations around the unit circle in the z plane required to determine the number of encirclements of the B plane origin by $B(z)$ also increases rapidly.

All things considered, the Jury test seems by far the most appropriate of the three tests, and will, therefore, be discussed in detail.

The Jury Stability Test

The application of the Jury test requires the formation of a table whose entries are based on the coefficients of $B(z)$. Given the general form

$$B(z) = b_0 z^n + b_1 z^{n-1} + \cdots + b_k z^{n-k} + \cdots + b_{n-1} z + b_n, \qquad (9.3.1)$$

the corresponding Jury table is given in Table 9.3.1. The first two rows consist of the coefficients in $B(z)$ arranged in ascending order of powers of z in row 1 and in reverse order in row 2. All even-numbered rows are simply the reverse of the immediately preceding odd-numbered row. The elements for rows 3 through $2n - 3$ are calculated from the following determinants:

$$c_k = \begin{vmatrix} b_n & b_{n-1-k} \\ b_0 & b_{k+1} \end{vmatrix}, \qquad k = 0, \ldots, n-1, \qquad (9.3.2)$$

$$d_k = \begin{vmatrix} c_{n-1} & c_{n-2-k} \\ c_0 & c_{k+1} \end{vmatrix}, \qquad k = 0, \ldots, n-2. \qquad (9.3.3)$$

This pattern continues until finally

$$q_k = \begin{vmatrix} p_3 & p_{2-k} \\ p_0 & p_{k+1} \end{vmatrix}, \qquad k = 0, 1, 2, \qquad (9.3.4)$$

that is, until the $(2n - 3)$rd row is reached which will contain exactly three elements. Note that for second-order systems, $2n - 3 = 1$, and the table will

Table 9.3.1

Jury Table for General Form of $B(z)$

Row	z^0	z^1	z^2	z^3	z^k	z^{n-2}	z^{n-1}	z^n
1	b_n	b_{n-1}	b_{n-2}	b_{n-3}	$\cdots b_{n-k}$	$\cdots b_2$	b_1	b_0
2	b_0	b_1	b_2	b_3	$\cdots b_k$	$\cdots b_{n-2}$	b_{n-1}	b_n
3	c_{n-1}	c_{n-2}	c_{n-3}	c_{n-4}	$\cdots c_{n-k-1}$	$\cdots c_1$	c_0	
4	c_0	c_1	c_2	c_3	$\cdots c_k$	$\cdots c_{n-2}$	c_{n-1}	
5	d_{n-2}	d_{n-3}	d_{n-4}	d_{n-5}	$\cdots d_{n-k-2}$	$\cdots d_0$		
6	d_0	d_1	d_2	d_3	$\cdots d_k$	$\cdots d_{n-2}$		
$\vdots$								
$2n-5$	p_3	p_2	p_1	p_0				
$2n-4$	p_0	p_1	p_2	p_3				
$2n-3$	q_2	q_1	q_0					

consist only of row 1, the coefficients of $B(z)$ in ascending order of the power of z. For third-order systems, three rows will be present, and so forth.

By the Jury stability criterion, a system is stable if *all* of the following conditions hold:

1. $B(z = 1) > 0$, i.e., when 1 is substituted for z in $B(z)$, $B(z)$ takes on a positive value.

2. $B(z = -1) \begin{cases} > 0 & \text{for } n \text{ even,} \\ < 0 & \text{for } n \text{ odd.} \end{cases}$

3. The coefficients in the Jury table meet the following $n - 1$ constraints:

$$|b_n| < |b_0|,$$
$$|c_{n-1}| > |c_0|,$$
$$|d_{n-2}| > |d_0|,$$
$$\vdots$$
$$|q_2| > |q_0|.$$

Note that the constraint involving b_n and b_0 has the opposite inequality than the other $n - 2$ constraints. Note also that all determinants which must be calculated to form the Jury table come from 2×2 matrices, which greatly simplifies the calculations involved compared to those needed for the Schur–Cohn test.

As a matter of practicality, should either of the first two conditions in the Jury criterion be violated, there is no need to construct the Jury table at all. The system is definitely unstable. Further, if $z = 1$ is a root of $B(z) = 0$, then $B(z = 1)$ will equal zero; and if $z = -1$ is a root of $B(z) = 0$, then $B(z = -1)$ will equal zero. Thus a ready test for at least these types of marginal stability is provided.

Let us now apply the Jury test to a few examples.

Example 1 Let
$$B(z) = z^3 - 0.20z^2 - 0.25z + 0.05. \qquad (9.3.5)$$

In factored form this becomes

$$B(z) = (z - 0.50)(z + 0.50)(z - 0.20), \qquad (9.3.6)$$

which means the roots of $B(z) = 0$ are 0.50, -0.50, and 0.20 and that the system represented by $B(z)$ is stable. For illustrative purposes, however, assume we are either unable or unwilling to find these roots and apply the Jury stability test. Here $b_0 = 1.00$, $b_1 = -0.20$, $b_2 = -0.25$, and b_3 (which is b_n for this third-order equation) $= 0.05$.

First we examine

$$B(z = 1) = 1 - 0.20 - 0.25 + 0.05 = 0.60 > 0$$

and

$$B(z = -1) = -1 - 0.20 + 0.25 + 0.05 = -0.90 < 0.$$

Since this is an odd-order polynomial, it meets the first two conditions of the Jury criterion. We now construct the Jury table as follows:

Row	z^0	z^1	z^2	z^3
1	0.05	−0.25	−0.20	1.00
2	1.00	−0.20	−0.25	0.05
3	−0.9975	0.1875	0.2400	

The third-row entries were computed from

$$c_2 = q_2 = \begin{vmatrix} 0.05 & 1.00 \\ 1.00 & 0.05 \end{vmatrix} = -0.9975,$$

$$c_1 = q_1 = \begin{vmatrix} 0.05 & -0.20 \\ 1.00 & -0.25 \end{vmatrix} = 0.1875,$$

$$c_0 = q_0 = \begin{vmatrix} 0.05 & -0.25 \\ 1.00 & -0.20 \end{vmatrix} = 0.2400.$$

In this example, c_j and q_j are equivalent since there are only three rows in the table. Note that the calculation for q_1 could have been omitted since q_1 does not enter into the stability determination in any way.

Applying the third set of stability conditions we obtain:

$$|b_n| = 0.05, \qquad |b_0| = 1.00; \qquad \text{therefore} \qquad |b_n| < |b_0|$$

and

$$|c_{n-1}| = 0.9975, \qquad |c_0| = 0.2400; \qquad \text{therefore} \qquad |c_{n-1}| > |c_0|.$$

Thus both conditions on the coefficients of $B(z)$ are also met, so we conclude that the system is stable.

Example 2   Let us now change the third factor of (9.3.6) to 2 from its original value of 0.20. This, of course, implies an unstable system. In polynomial form $B(z)$ is now

$$B(z) = z^3 - 2.00z^2 - 0.25z + 0.50. \tag{9.3.7}$$

We now examine

$$B(z = 1) = 1 - 2 - 0.25 + 0.50 = -0.75 < 0,$$

which violates Jury's first condition for stability. Thus we already have identified this system as unstable so no further analysis is necessary for absolute stability considerations alone. Not all unstable systems can be discovered this easily, however, so one should always plan to carry the analysis completely through all three steps, even though in some cases, such as this, an indication of instability may be detected prior to completion of the analysis.

As a matter of interest, continuation of the Jury test for this example yields:

$$B(z = -1) = -1 - 2 + 0.25 + 0.50 = -2.25 < 0$$

which does not show instability,

$$|b_n| = 0.50, \qquad |b_0| = 1.00 \qquad \text{giving} \qquad |b_n| < |b_0|,$$

which also does not indicate instability, and

$$|c_{n-1}| = 0.75, \qquad |c_0| = 0.75 \qquad \text{giving} \qquad |c_{n-1}| = |c_0|,$$

which does violate the final condition for stability.

Example 3   As a final example let us examine a marginally stable system. To represent such a system, we change the third factor of (9.3.6) to unity, keeping the other two roots at their original values of 0.50 and -0.50. This gives for the polynomial form of $B(z)$,

$$B(z) = z^3 - z^2 - 0.25z + 0.25. \qquad (9.3.8)$$

First we look at

$$B(z = 1) = 1 - 1 - 0.25 + 0.25 = 0.$$

This indicates that $z = 1$ is a root of $B(z) = 0$ so that the system is at best marginally stable. If a marginally stable system is not satisfactory, the test can be concluded at this point and the system discarded. If, on the other hand, a marginally stable system is satisfactory for whatever the intended application might be, the remainder of the test must be run to determine whether the system is actually unstable. Continuing we obtain:

$$B(z = -1) = -1 - 1 + 0.25 + 0.25 = -1.50 < 0,$$
$$|b_n| = 0.25, \qquad |b_0| = 1.00 \qquad \text{so} \qquad |b_n| < |b_0|,$$

and

$$|c_{n-1}| = 0.9375, \qquad |c_0| = 0 \qquad \text{so} \qquad |c_{n-1}| > |c_0|.$$

Thus all other conditions for stability are met and the system is marginally stable.

The form of the Jury test just presented was formulated primarily for hand calculation. An alternate form which permits more efficient computer utilization is presened by Jury (1962).

Root-Coefficient Relationships

Before leaving the subject of stability tests based directly on $B(z)$, it might prove worthwhile to point out that the relationships which exist between the roots and the coefficients of polynomial equations may offer some extremely gross indications of system instability. Given

$$B(z) = b_0 z^n + b_1 z^{n-1} + \cdots + b_j z^{n-j} + \cdots + b_{n-1}z + b_n = 0$$

with roots z_1 through z_n, the following relationships hold:

$$b_1/b_0 = -\sum_{k=1}^{n} z_k,$$

$$b_2/b_0 = \sum_{j=1}^{n} \sum_{k=1}^{n} z_j z_k,$$

$$b_3/b_0 = -\sum_{j=1}^{n} \sum_{k=1}^{n} \sum_{p=1}^{n} z_j z_k z_p,$$

$$\vdots$$

$$b_n/b_0 = (-1)^n \prod_{k=1}^{n} z_k.$$

For all roots to be less than one in magnitude it is necessary, but not sufficient, that

$$|b_1/b_0| < n, \qquad |b_2/b_0| < n!/(n-2)!\,2! = n(n-1)/2,$$

and in general,

$$|b_j/b_0| < n!/(n-j)!\,j\,!$$

Note that for b_n/b_0, this condition is simply

$$|b_n/b_0| < 1,$$

an item that can be checked by inspection with a glance at $B(z)$.

Tests Requiring Prior Transformation of B(z)

Those familiar with continuous linear systems will recall that the criterion for stable operation for such a system is that all roots of the characteristic equation of the system differential equation lie in the left (negative) half of the complex plane. This criterion of course holds whether one is dealing with the characteristic equation itself or the Laplace transform thereof and the transform variable s. The Nyquist stability criterion, based on the number of encirclements of the origin in the complex plane by the locus of values of the characteristic equation as the transform variable s traverses the imaginary axis in the complex s plane, has led to the development of a number of stability tests for continuous systems which in effect are based on clever ways of detecting roots of the characteristic equation in the right (positive) half of the complex plane. Among these are the Bode diagram, the polar plot of the open-loop transfer function, and the polar plot of the inverse open-loop transfer function, all of which not only indicate stability or the lack thereof but also provide a numerical basis for sensitivity analysis of system stability with respect to the values of the various system parameters. The root-locus method is also a valuable tool for such purposes. A popular test for continuous-system stability only is the Routh–Hurwitz test.

These and any other stability test based on the detection of roots of a polynomial equation in the right-half complex plane can be applied to the stability analysis of discrete linear systems if a transformation of $B(z) = 0$ can be found which fulfills the following requirements for the transform variable λ:

1.   The outside of the unit circle in the complex z plane is transformed into the right half of the complex λ plane.

2.   The unit circle in the z plane is transformed into the imaginary axis of the λ plane.

3.   The inside of the unit circle in the z plane is transformed into the left half of the λ plane.

Figure 9.3.1 illustrates the effect of such a transformation.

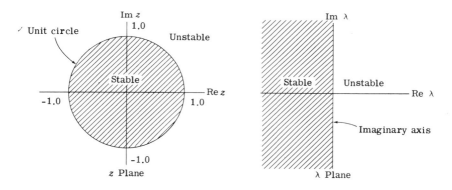

FIG. 9.3.1.   Bilinear transformation of the complex B plane into the λ plane.

The Bilinear Transformation

The bilinear transformation can be shown to meet the requirements just outlined. For this transformation we let

$$z = (\lambda + 1)/(\lambda - 1). \tag{9.3.9}$$

When solved for λ this gives

$$\lambda = (z + 1)/(z - 1), \tag{9.3.10}$$

which shows that the transformation is symmetrical.

To illustrate the transformation, consider the point $z = 1 + j1$. Applying the transformation formula yields

$$\lambda = \frac{z + 1}{z - 1} = \frac{2 + j1}{j1} = 1 - j2.$$

Note that z lies outside the unit circle in the z plane and that the resulting value of λ is in the right half of the λ plane. On the other hand, the point $z = 0.5 + j(0.5)$, which is inside the unit circle, transforms to

$$\lambda = \frac{1.5 + j(0.5)}{-0.5 + j(0.5)} = \frac{(1.5 + j(0.5))(-0.5 - j(0.5))}{(-0.5 + j(0.5))(-0.5 - j(0.5))} = \frac{-0.5 - j1}{0.5} = -1 - j2,$$

which is in the left half of the λ plane.

The transformations of several points of interest on the unit circle in the z plane are presented in Table 9.3.2. Note that all λ values are pure imaginary numbers, hence they lie on the imaginary axis of the λ plane. Note also that the point $z = 1 + j0$ corresponds to either limit of the λ plane imaginary axis, depending from which direction it is approached. In the table we have simply listed $z = 1 + j0$ as corresponding to $\lambda = -j\infty$ and $z = 1 - j0$ as corresponding to $+j\infty$ to represent the limits approached by λ as z approaches the real axis on the unit circle from above and below, respectively.

The transformation of the equation $B(z) = 0$ is illustrated using the equation of our previously considered Example 1 (see Eq. (9.3.5)):

$$B(z) = z^3 - 0.20z^2 - 0.25z + 0.05 = 0.$$

Substituting $(\lambda + 1)/(\lambda - 1)$ for z yields

$$\left(\frac{\lambda + 1}{\lambda - 1}\right)^3 - 0.20\left(\frac{\lambda + 1}{\lambda - 1}\right)^2 - 0.25\left(\frac{\lambda + 1}{\lambda - 1}\right) + 0.05 = 0.$$

We now clear the fractions by multiplying by the highest power of $\lambda - 1$ present, which in this case is $(\lambda - 1)^3$. This gives

$$(\lambda + 1)^3 - 0.20(\lambda + 1)^2(\lambda - 1) - 0.25(\lambda + 1)(\lambda - 1)^2 + 0.05(\lambda - 1)^3 = 0$$

Table 9.3.2

Bilinear Transformations of Points on
the Unit Circle

z	λ
$1 + j0$	$-j\infty$
$\sqrt{2}/2 + j\sqrt{2}/2$	$-j(2.414)$
$0 + j1$	$-j(1.0)$
$-\sqrt{2}/2 + j\sqrt{2}/2$	$-j(0.414)$
$-1 + j0$	0
$-\sqrt{2}/2 - j\sqrt{2}/2$	$j(0.414)$
$0 - j1$	$j(1.0)$
$\sqrt{2}/2 - j\sqrt{2}/2$	$j(2.414)$
$1 - j0$	$j\infty$

or

$$(\lambda^3 + 3\lambda^2 + 3\lambda + 1) - 0.20(\lambda^3 + \lambda^2 - \lambda - 1)$$
$$- 0.25(\lambda^3 - \lambda^2 - \lambda + 1) + 0.05(\lambda^3 - 3\lambda^2 + 3\lambda - 1) = 0.$$

Grouping terms in like powers of λ yields the following transformed equation:

$$\hat{B}(\lambda) = 0.60\lambda^3 + 2.90\lambda^2 + 3.60\lambda + 0.90 = 0.$$

Tests Applicable to the λ Domain

As has been stated, among the tests which can be applied once $B(z)$ has been transformed into the λ domain are the well-known continuous-system tests such as the Bode, Nyquist polar plots, root-locus, and Routh–Hurwitz. For continuous systems the first three provide a wealth of additional information regarding sensitivity of stability to changes in system parameters and several quantitative measures of other aspects of system performance. The λ plane, even though similar to the "s plane" of continuous-system analysis for stability considerations, is not quantitatively equivalent to the s plane in any other way. Thus, the use of such methods to obtain any information for the discrete system other than a determination of absolute stability would take considerable effort of a basic research nature.

For this reason, the most generally useful stability test for discrete linear systems is the Routh–Hurwitz test. It is designed to indicate stability or lack thereof and nothing more, although some notion of sensitivity to system parameter values can be obtained in some cases. It will, however, indicate exactly how many roots lie in the right half of the λ plane and provide means of evaluating roots which lie on the imaginary axis (plus some other special types of roots).

The Routh–Hurwitz Stability Test

For a system to be stable, all of the following conditions must be met by $\hat{B}(\lambda)$, the λ domain transform of $B(z)$:

1. All powers of λ in descending order from the highest to the constant term in the transformed characteristic equation must exist. This assures against multiple zero or imaginary roots. If only the constant term is missing, a single zero root is present, which may or may not be acceptable (the case of marginal stability). If it is acceptable, one may proceed to check for other roots which may lie in the right half of the λ plane by dividing $B(z)$ by z and applying the Routh–Hurwitz test to the quotient polynomial.

2. The coefficients of all powers of λ must have the same algebraic sign. This assures against positive real roots from Descartes' rule of signs.

3. There may be no changes of sign in the left-hand column of the following Routh array computed as indicated from the coefficients of

$$\hat{B}(\lambda) = B_0 \lambda^n + B_1 \lambda^{n-1} + \cdots + B_{-1}\lambda + B = 0 \qquad (9.3.11)$$

(Each change which occurs indicates a root in the right-hand λ plane):

$$
\begin{array}{cccc}
B_0 & B_2 & B_4 & B_6 & \cdots \\
B_1 & B_3 & B_5 & B_7 & \cdots \\
C_0 & C_2 & C_4 & C_6 & \cdots \\
C_1 & C_3 & C_5 & C_7 & \cdots \\
D_0 & D_2 & D_4 & D_6 & \cdots \\
D_1 & D_3 & D_5 & D_7 & \cdots \\
\vdots & \vdots & \vdots & \vdots &
\end{array}
$$

where

$$
C_0 = \frac{B_1 B_2 - B_0 B_3}{B_1}, \qquad C_2 = \frac{B_1 B_4 - B_0 B_5}{B_1}, \ldots
$$

$$
C_1 = \frac{C_0 B_3 - B_1 C_2}{C_0}, \qquad C_3 = \frac{C_0 B_5 - B_1 C_4}{C_0}, \ldots
$$

$$
D_0 = \frac{C_1 C_2 - C_0 C_3}{C_1}, \qquad D_2 = \frac{C_1 C_4 - C_0 C_5}{C_1}, \ldots
$$

$$\vdots \qquad\qquad\qquad \vdots$$

The array continues to the right and downward as long as values last. "Missing" terms are taken as zero whenever a term is called for which does not exist, so each pair of rows has one less term than the preceding pair (pairings depend on whether n is odd or even). There will always be precisely $n + 1$ rows.

To illustrate, consider the equation in λ:

$$\hat{B}(\lambda) = \lambda^6 + 9\lambda^5 + 36\lambda^4 + 86\lambda^3 + 125\lambda^2 + 97\lambda + 30 = 0. \quad (9.3.12)$$

Note that all powers of λ are present from the highest, λ^6, to the constant term and that all terms are positive. Thus no indication of instability is present so far, so we proceed to construct the Routh array.

The calculations for the third row, i.e., for C_0, C_2, and C_4, are as follows:

$$C_0 = \frac{9 \cdot 36 - 1 \cdot 86}{9} = 26.44,$$

$$C_2 = \frac{9 \cdot 125 - 1 \cdot 97}{9} = 114.33,$$

and

$$C_4 = \frac{9 \cdot 30 - 1 \cdot 0}{9} = 30.$$

Other entries are similarly computed by the appropriate formula. The resulting Routh array is:

1	36	125	30
9	86	97	
26.44	114.33	30	
47.10	87.00		
65.70	30		
65.40			
30			

Since there are no changes of sign, all three criteria are met and the system represented is stable.

When zeros occur in the first column of the array, special procedures must be adopted. Leading zeros, i.e., zeros which occur in the first column of rows containing other nonzero elements, are easily handled by replacing them with a small positive constant ε completing the array, and taking the limits of the terms involving ε as ε is made to approach zero. Consider, for example,

$$\hat{B}(\lambda) = \lambda^5 + \lambda^4 + 3\lambda^3 + 3\lambda^2 + \lambda + 2 = 0. \tag{9.3.13}$$

Using the prescribed procedures, the first three rows of the Routh array are:

1	3	1
1	3	2
0	-1	

The leading zero in the third row is replaced by ε and the procedure continued as follows:

1	3	1
1	3	2
ε	-1	
$(3\varepsilon + 1)/\varepsilon$	2	
$(-2\varepsilon^2 - 3\varepsilon - 1)/(3\varepsilon + 1)$		
2		

As ε approaches zero the first term in rows three and four is positive but the first term of row five is negative. Thus, there are two changes of sign, so two roots lie in the right-hand λ plane and the system is unstable.

Another situation involving zero elements in the Routh array occurs when two roots of $\hat{B}(\lambda) = 0$ are negatives of each other. This produces an all-zero row in the array. At best a system which produces this result will be marginally stable having a pair of imaginary roots which obviously must be negatives of each other. Should any root having a nonzero real part be the negative of another root, one of the two would have to be in the right-hand λ plane and the system would be unstable. Normally, therefore, an all-zero row would result in rejection of the system. In applications in which marginally stable systems can be used, however, it is desirable to be able to continue the test to determine whether the system is marginally stable or unstable. The procedure involves forming a subsidiary equation from the coefficients in the row just above the all-zero row and using this equation as the basis for continuing the analysis. If k pairs of roots which are negatives of one another are present in $\hat{B}(\lambda) = 0$, the all-zero row will be the first row reached which normally would contain k elements. Thus, the row above will contain $k + 1$ elements. The subsidiary equation is a polynomial equation consisting of only even power terms of λ. The first element in the row just above the all-zero row becomes the coefficient of λ^{2k} in the subsidiary equation, the second element becomes the coefficient of λ^{2k-2}, the third element, if any, becomes the coefficient of λ^{2k-4}, and so forth until the last element becomes the constant term.

The Routh array is completed by differentiating the subsidiary equation with respect to λ and using the resulting coefficients to form a row to replace the all-zero row. The array is completed from this point by the usual procedures. In addition, the roots of the subsidiary equation are the pairs of roots which are negatives of each other. Should the Routh array show no changes of sign and $\hat{B}(\lambda)$ contain all its terms with positive coefficients, then it will be necessary to find the roots of the subsidiary equation, or at least to determine their location, to find out whether the system involved is marginally stable or unstable.

Should other results, such as a change of sign in the first column of the Routh array, indicate instability, there is no need, from a stability point of view at least, in finding the roots of the subsidiary equation. If no other indications of instability exist, however, the location of the roots of the subsidiary equation can be determined by any applicable stability test geared to finding roots in the right-hand plane other than the Routh–Hurwitz.

To illustrate this situation, consider the equation

$$\hat{B}(\lambda) = \lambda^4 + 3\lambda^3 + 6\lambda^2 + 12\lambda + 8 = 0$$

which happens to have roots -1, -2, and $\pm j2$. Thus it is a marginally stable system with one pair of imaginary roots. The first four rows of the Routh array are:

$$\begin{array}{ccc} 1 & 6 & 8 \\ 3 & 12 & \\ 2 & 8 & \\ 0 & & \end{array}$$

Since the fourth row consists of only one element, we could substitute a positive ε and complete the table, which would involve merely bringing the 8 down to form the last row. Since no changes of sign appear, there can be no roots of $\hat{B}(\lambda) = 0$ in the right-hand plane. However, this analysis would be incomplete and improper conclusions could be drawn as a result. It must be noted instead that, even though the row is one element long, the entire row is zero. Thus we must follow the subsidiary equation procedure to discover the nature of the pair of roots which are negatives of each other.

From row three, we find the subsidiary equation to be

$$2\lambda^2 + 8 = 0.$$

Differentiating the left-hand side gives 4, so 4 is used to replace the 0 in row four of the array. Having made this substitution, we get for the complete array:

$$\begin{array}{ccc} 1 & 6 & 8 \\ 3 & 12 & \\ 2 & 8 & \\ 4 & & \\ 8 & & \end{array}$$

This shows no changes of sign so we must find the roots of the subsidiary equation to determine whether the system is marginally stable or unstable. These roots are easily found to be $\pm j2$. Thus no roots lie in the right half λ plane, no multiple roots lie on the imaginary axis, one pair of imaginary roots exists, so the system is marginally stable.

Consider a more complicated example. Here

$$\hat{B}(\lambda) = \lambda^7 + 6\lambda^6 + 11\lambda^5 + 6\lambda^4 + 4\lambda^3 + 24\lambda^2 + 44\lambda + 24 = 0$$

which happens to have roots -1, -2, -3, $-1 + j1$, $-1 - j1$, $1 + j1$, and $1 - j1$. Thus two roots lie in the right half plane and two pairs of roots exist which are negatives of each other. Since all terms are present with positive coefficients in $\hat{B}(\lambda) = 0$, it is necessary to construct the Routh array to determine system stability. The first five rows are:

$$\begin{array}{cccc} 1 & 11 & 4 & 44 \\ 6 & 6 & 24 & 24 \\ 10 & 0 & 40 & \\ 6 & 0 & 24 & \\ 0 & 0 & & \end{array}$$

Note that we do not worry about the internal zero in rows three and four. Since the fifth row contains two zero elements, we know that there are two pairs of roots which are negatives of one another.

The subsidiary equation, based on row four, is:

$$6\lambda^4 + 0\lambda^2 + 24 = 0.$$

Differentiating the left-hand side yields

$$24\lambda^3 + 0\lambda,$$

which, when used to form row five, gives for the first six rows of the array:

$$
\begin{array}{cccc}
1 & 11 & 4 & 44 \\
6 & 6 & 24 & 24 \\
10 & 0 & 40 & \\
6 & 0 & 24 & \\
24 & 0 & & \\
0 & 24 & &
\end{array}
$$

Again we are confronted with a leading zero; but, since the other element in the row is nonzero, the ε procedure is followed. This gives for rows six through eight:

$$
\begin{array}{cc}
\varepsilon & 24 \\
-576/\varepsilon & \\
24 &
\end{array}
$$

Thus we see that the first element of row seven is negative. This produces two changes of sign in the first column, correctly diagnosing two roots in the right-hand λ plane.

Since the system has been found to be definitely unstable the analysis could stop at this point. However, it may prove interesting to find the roots of the subsidiary equation

$$
6\lambda^4 + 24 = 0.
$$

From this we get

$$
\lambda^4 = -4 \qquad \text{or} \qquad \lambda^2 = \pm j2.
$$

Writing the expression for λ^2 in polar form gives

$$
\lambda^2 = 2e^{j\pi/2}.
$$

Taking the square root of λ^2 gives

$$
\pm \sqrt{2}\,e^{\pm j\pi/4} = \left\{
\begin{array}{l}
1 + j1 \\
-1 - j1 \\
1 - j1 \\
-1 + j1
\end{array}
\right.
$$

showing the two pairs of complex roots which are negatives of each other and confirming the instability already determined from the Routh array.

Note that since violation of any one of these three Routh–Hurwitz stability conditions indicates that the system is unstable, it is computationally advantageous to perform the easiest test first. Only if no violation occurs would it be necessary to run any succeeding test. By the same reasoning, one would choose to run, if necessary, the second easiest test second. With the Routh–Hurwitz tests, both (1) and (2) can be performed by inspection while test (3) involves a considerable amount of computational effort. For this reason, one should always check out conditions (1) and (2) first (in either order or together) and then build the array to test condition (3) only if no indication of instability results from tests (1) and (2).

With this material on stability, we are now in a position to direct our attention to the analysis and design of second- and higher-order systems.

Exercises

9.1 Indicate whether the discrete systems represented by the following characteristic equations are stable, unstable, or marginally stable. Support your answer.

a. $B(z) = (z + 0.9)(z - 0.8) = 0$ b. $B(z) = (z + 1)(z - 0.8) = 0$
c. $B(z) = (z + 1)(z - 1) = 0$ d. $B(z) = (z + 1)^2(z - 0.8) = 0$
e. $B(z) = (z + 1)(z - 0.8)^2 = 0$ f. $B(z) = (z + 0.9)(z - 2) = 0$
g. $B(z) = (z - 0.8 - j(0.3))(z - 0.8 + j(0.3)) = 0$
h. $B(z) = (z - 0.8 - j(0.3))^2(z - 0.8 + j(0.3))^2 = 0$
i. $B(z) = (z - 0.9)(z - 0.8 - j(0.3))(z - 0.8 + j(0.3)) = 0$
j. $B(z) = (z - 1)(z - 0.8 - j(0.3))(z - 0.8 + j(0.3)) = 0$
k. $B(z) = (z - 1.5)(z - 0.8 - j(0.3))(z - 0.8 + j(0.3)) = 0$
l. $B(z) = (z - 0.8 - j(0.6))(z - 0.8 + j(0.6)) = 0$
m. $B(z) = (z + 0.8 - j(0.6))(z + 0.8 + j(0.6)) = 0$
n. $B(z) = (z - 0.8 + j(0.6))^2(z - 0.8 - j(0.6))^2 = 0$
o. $B(z) = (z - 1)(z - 0.8 - j(0.6))(z - 0.8 + j(0.6)) = 0$
p. $B(z) = (z - 0.8 - j(0.3))(z - 0.8 + j(0.3))$
     $\cdot (z - 0.8 - j(0.6))(z - 0.8 + j(0.6)) = 0$
q. $B(z) = z^3(z - 0.9) = 0$
r. $B(z) = (z - 0.8 - j(0.9))(z - 0.8 + j(0.9)) = 0$
s. $B(z) = z(z + 0.5)(z - 0.8 - j(0.6))(z - 0.8 + j(0.6)) = 0$
t. $B(z) = (z - 0.8 - j(0.6))(z - 0.8 + j(0.6))$
     $\cdot (z + 0.8 - j(0.9))(z + 0.8 + j(0.9)) = 0$

9.2 Indicate whether the discrete systems represented by the following characteristic equations are stable, unstable, or marginally stable. Support your answer.

a. $B(z) = z^2 + z - 0.5 = 0$ b. $B(z) = z^2 + z + 0.5 = 0$
c. $B(z) = z^2 - z - 0.75 = 0$ d. $B(z) = z^2 - z + 0.75 = 0$
e. $B(z) = z^2 + 2z + 1 = 0$ f. $B(z) = z^2 - 0.5z - 0.5 = 0$
g. $B(z) = z^2 + 1.5z + 0.5 = 0$ h. $B(z) = z^2 - 0.25 = 0$
i. $B(z) = z^2 - 1 = 0$ j. $B(z) = 5z^2 - 2z + 2 = 0$
k. $B(z) = 2z^2 - 2z + 5 = 0$

9.3 Use the Jury test to determine whether the discrete systems represented by the following characteristic equations are stable, unstable, or marginally stable:

a. $B(z) = 5z^2 - 2z + 2 = 0$ b. $B(z) = 2z^2 - 2z + 5 = 0$
c. $B(z) = z^3 - 0.4z^2 - 0.25z + 0.1 = 0$
d. $B(z) = z^3 - 2.1z^2 + 1.8z - 0.5 = 0$
e. $B(z) = z^3 - 0.5z^2 + 0.4z - 0.2 = 0$
f. $B(z) = z^3 + 1.6z^2 - 0.02z - 0.42 = 0$
g. $B(z) = z^4 - 1.4z^3 + 0.15z^2 + 0.35z - 0.1 = 0$
h. $B(z) = z^4 + 0.5z^3 - 0.1z^2 + 0.2z - 0.2 = 0$
i. $B(z) = z^4 - 0.3z^3 - 0.15z^2 - 0.075z - 0.1 = 0$
j. $B(z) = z^4 - 5z^3 + 4.5z^2 + 0.2z - 0.3 = 0$
k. $B(z) = z^4 + 3z^3 + z^2 + 0.5z + 2 = 0$
l. $B(z) = z^7 - 0.8z^6 + 0.4z^5 - 0.32z^4 - 0.0625z^3 + 0.05z^2$
    $\quad - 0.025z + 0.02 = 0$

9.4 Use the Jury test to determine what relationships must exist among the coefficients a, b, c, and d for the system represented by the following characteristic equations to be stable:

a. $B(z) = az^2 + bz + c = 0$ b. $B(z) = z^3 - z^2 + az + b = 0$
c. $B(z) = z^3 + az^2 + bz + c = 0$
d. $B(z) = z^4 + az^3 + bz^2 + cz + d = 0$

9.5 Transform the following equations in z into the λ domain by means of the bilinear transformation:

a. $B(z) = z^3 + 2z^2 - 3z + 1 = 0$
b. $B(z) = z^4 + 3z^3 - 4z^2 - 2z + 1 = 0$

9.6 (a) through (l).   Transform the equation in z from the corresponding part of Exercise 9.3 into the λ domain by means of the bilinear transformation.

9.7 (a) through (d).   Transform the equation in z from the corresponding part of Exercise 9.4 into the λ domain by means of the bilinear transformation.

9.8 Use the Routh–Hurwitz test to determine whether the discrete systems represented by the following transformed (by the bilinear transformation) characteristic equations are stable, unstable, or marginally stable.

a. $\hat{B}(\lambda) = \lambda^9 + \lambda^8 + 3.2\lambda^7 + 7.29\lambda^6 + 5\lambda^5 + 4.93\lambda^4 + 7.1\lambda^2$
    $\quad + 2\lambda + 17 = 0$
b. $\hat{B}(\lambda) = \lambda^8 + 2\lambda^7 + 3\lambda^6 + 4\lambda^5 - 4\lambda^4 + 2\lambda^2 + 7\lambda + 3 = 0$
c. $\hat{B}(\lambda) = \lambda^5 + 16\lambda^4 + 82\lambda^3 + 268\lambda^2 + 520\lambda + 400 = 0$
d. $\hat{B}(\lambda) = \lambda^6 + 10\lambda^5 + 50\lambda^4 + 60\lambda^3 + 45\lambda^2 + 52\lambda + 12 = 0$
e. $\hat{B}(\lambda) = \lambda^6 + 3\lambda^5 + 7\lambda^4 + 15\lambda^3 + 14\lambda^2 + 12\lambda + 8 = 0$
f. $\hat{B}(\lambda) = \lambda^6 + 9\lambda^5 + 36\lambda^4 + 86\lambda^3 + 125\lambda^2 + 97\lambda + 30 = 0$
g. $\hat{B}(\lambda) = \lambda^6 + \lambda^5 + 5\lambda^4 + 3\lambda^3 + 9\lambda^2 + 5\lambda + 6 = 0$
h. $\hat{B}(\lambda) = \lambda^5 + 2\lambda^4 + 8\lambda^3 + 16\lambda^2 + 16\lambda + 32 = 0$

i. $\hat{B}(\lambda) = \lambda^6 + 12\lambda^5 + 60\lambda^4 + 160\lambda^3 + 239\lambda^2 + 188\lambda + 60 = 0$
j. $\hat{B}(\lambda) = \lambda^5 + 3\lambda^4 + 2\lambda^3 + 6\lambda^2 + \lambda + 3 = 0$

9.9 (a) through (l). From the results of the transformation performed in the corresponding part of Exercise 9.6, determine by means of the Routh–Hurwitz test whether the systems represented are stable, unstable, or marginally stable. Results may be compared with those of 9.3.

9.10 (a) through (d). From the results of the transformation performed in the corresponding part of Exercise 9.7, determine by means of the Routh–Hurwitz test the relationships which must exist among the parameters a, b, c, and d for the systems represented to be stable. Results may be compared with those of 9.4.

9.11 Use the Routh–Hurwitz test to determine what relationships must exist among the coefficients A, B, C, and D for the systems represented by the following transformed (by the bilinear transformation) characteristic equations to be stable:

a. $\hat{B}(\lambda) = A\lambda^2 + B\lambda + C = 0$ **b.** $\hat{B}(\lambda) = \lambda^3 + A\lambda^2 + B\lambda + C = 0$
c. $\hat{B}(\lambda) = \lambda^4 + A\lambda^3 + B\lambda^2 + C\lambda + D = 0$

9.12 Use the Jury criteria to determine the sets of values of the control-system parameters which result in stable system operation for the following process controllers. Use the referenced system difference equation as a starting point.

a. The first-order process controller represented by (2.3.21).
b. The proportional-plus-difference controller represented by (2.3.30) when N is set equal to zero.
c. The proportional-plus-summation controller represented by (2.3.30) when L is set equal to zero.
d. The proportional-plus-difference-plus-summation controller represented by (2.3.30).
e. The generalized second-order process controller represented by (2.3.32).
f. The generalized third-order process controller represented by (2.3.34) by setting $n = 3$.
g. The generalized fourth-order process controller represented by (2.3.34) by setting $n = 4$.

Note: z transforms for parts (a), (b), (c), and (e) were found in Exercise 5.15 and various transfer functions identified in Exercise 6.20. If either of these exercises have been done, the results should be helpful here.

9.13 (a) through (g). The same as 9.12 except using the bilinear transformation and the Routh–Hurwitz criteria.

Chapter X | Second-Order Systems

In Chapters VII and VIII a number of basic concepts were introduced in the context of the first-order system. These included (1) the effects on system performance of sampling and measurement errors and of several types of perturbation, (2) criteria for evaluating system performance, and (3) approaches to the selection of the system parameter K in several example operating situations. The relative simplicity of the first-order system facilitated the introduction of these topics. On the other hand, the coverage was of necessity fairly limited. A more complete view of the effects of various perturbations and measurement errors on the operation and performance of control systems can be obtained by examining higher-order systems, many of which can be demonstrated in the context of the second-order system.

In this chapter, we examine the second-order system in detail, primarily utilizing the general second-order, rapid-response process controller. Expressions for its response to several common types of perturbation and measurement error are derived and, where appropriate, comparisons are made to those of the first-order, rapid-response process controller. Finally, a few specific second-order control decision rules are discussed.

10.1 The Second-Order System

A second-order system is, by definition, a system which can be represented by a second-order difference equation. Therefore, a second-order, linear, time-invariant system is represented by a second-order, linear difference equation with constant coefficients. In general terms, this equation is, from (2.1.2),

$$a_0 \, \Delta^2 x(i) + a_1 \, \Delta x(i) + a_2 \, x(i) = f(i) \qquad (10.1.1)$$

in difference-operator form or, from (2.1.3),

$$b_0 x(i + 2) + b_1 x(i + 1) + b_2 x(i) = f(i) \qquad (10.1.2)$$

in expanded form. The characteristic equation is of course

$$b_0 h^2 + b_1 h + b_2 = 0, \qquad (10.1.3)$$

the roots of which are

$$h_1 = [-b_1 + (b_1{}^2 - 4b_0 b_2)^{1/2}]/2b_0 \qquad (10.1.4)$$

and

$$h_2 = [-b_1 - (b_1{}^2 - 4b_0 b_2)^{1/2}]/2b_0 . \qquad (10.1.5)$$

Since there are two roots, three distinct cases are possible depending on the relative values of the coefficients b_0, b_1, and b_2. As mentioned in Section 6.8, these are:

1. For $b_1{}^2 > 4b_0 b_2$, two real unequal roots;
2. For $b_1{}^2 = 4b_0 b_2$, a pair of real equal roots;
3. For $b_1{}^2 < 4b_0 b_2$, a pair of conjugate complex (or imaginary if $b_1 = 0$) roots.

For cases 1 and 3, solution by either variation of parameters or the z transform yields

$$x(i) = \frac{1}{h_1 - h_2} \left[(-h_2 h_1{}^i + h_1 h_2{}^i) x(0) + (h_1{}^i - h_2{}^i) x(1) \right.$$

$$\left. + \frac{1}{b_0} \sum_{n=0}^{i-1} f(n)(h_1^{i-1-n} - h_2^{i-1-n}) \right] \qquad (10.1.6)$$

or, alternatively,

$$x(i) = \frac{1}{h_1 - h_2} \left[(-h_2 h_1{}^i + h_1 h_2{}^i) x(0) + (h_1{}^i - h_2{}^i) x(1) \right.$$

$$\left. + \frac{1}{b_0} \sum_{n=0}^{i-1} f(i - 1 - n)(h_1{}^n - h_2{}^n) \right]. \qquad (10.1.7)$$

Note that for $i = 0$ and $i = 1$, these expressions reduce to identities in $x(0)$ and $x(1)$, respectively. For case 2, in which we let

$$h = h_1 = h_2 = -b_1/2b_0, \qquad (10.1.8)$$

the solution is

$$x(i) = -(i - 1)h^i x(0) + ih^{i-1} x(1) + \frac{1}{b_0} \sum_{n=0}^{i-2} f(n)(i - 1 - n)h^{i-2-n} \qquad (10.1.9)$$

or

$$x(i) = -(i - 1)h^i x(0) + ih^{i-1}x(1) + \frac{1}{b_0}\sum_{n=0}^{i-2} f(i - 2 - n)(n + 1)h^n. \quad (10.1.10)$$

These also reduce to identities in $x(0)$ and $x(1)$ for $i = 0$ and $i = 1$. Equations (10.1.6), (10.1.7), (10.1.9), and (10.1.10) can of course also be expressed fully in terms of the coefficients b_0, b_1, and b_2 by appropriate substitution of (10.1.4), (10.1.5), and (10.1.8).

For a second-order system to be stable, the magnitudes of both z_1 and z_2 must be less than unity. The relationships which must exist among the three coefficients for this to occur are easily obtained by applying the Jury test to (10.1.3). This yields the three conditions:

$$b_0 + b_1 + b_2 > 0, \quad (10.1.11)$$

$$b_0 - b_1 + b_2 > 0, \quad (10.1.12)$$

and

$$|b_2| < |b_0|. \quad (10.1.13)$$

The implications of these conditions and the operating characteristics of second-order systems can be more meaningfully described in terms of a specific category of controller. For this purpose, we will work primarily with the general second-order, rapid-response process controller introduced in Section 2.3.

10.2   The Generalized Second-Order, Rapid-Response Process Controller

Basic Controller Characteristics

The decision rule for this controller is, from (2.3.31),

$$c(i) = K_0 \rho(i) + K_1 \rho(i - 1) \quad (10.2.1)$$

and the resulting system difference equation, from (2.3.32), is

$$m(i + 1) - (1 - K_0)m(i) + K_1 m(i - 1)$$
$$= (K_0 + K_1)D - K_0 \varepsilon(i) - K_1 \varepsilon(i - 1) + p(i). \quad (10.2.2)$$

For ease in presentation, we let $D = 0$ as we did in Chapters VII and VIII. We also increase the discrete-time argument i by one to get the equation in standard form for solution. Incorporating these changes in (10.2.2) yields

$$m(i + 2) - (1 - K_0)m(i + 1) + K_1 m(i) = -K_0 \varepsilon(i + 1) - K_1 \varepsilon(i) + p(i + 1).$$
$$(10.2.3)$$

Comparison with (10.1.2) shows that for (10.2.3): $b_0 = 1$, $b_1 = -(1 - K_0)$, $b_2 = K_1$,

$$f(i) = -K_0 \varepsilon(i + 1) - K_1 \varepsilon(i) + p(i + 1), \qquad (10.2.4)$$

$$h_1 = [1 - K_0 + ((1 - K_0)^2 - 4K_1)^{1/2}]/2, \qquad (10.2.5)$$

and

$$h_2 = [1 - K_0 - ((1 - K_0)^2 - 4K_1)^{1/2}]/2. \qquad (10.2.6)$$

The Jury test stability criteria, (10.1.11) through (10.1.13), now become:

$$K_0 + K_1 > 0, \qquad (10.2.7)$$

$$K_0 - K_1 < 2, \qquad (10.2.8)$$

$$|K_1| < 1, \qquad (10.2.9)$$

the last of which can be expanded to yield

$$-1 < K_1 < 1. \qquad (10.2.10)$$

The combinations of K_0 and K_1 which yield stable system performance are indicated by the shaded region of Fig. 10.2.1.

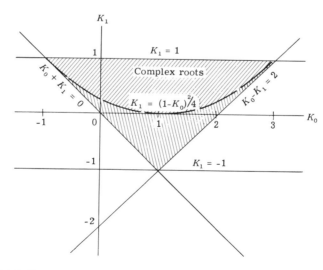

FIG. 10.2.1. Types of roots and region of stability of general second-order, rapid-response process controller as a function of control-system constants K_0 and K_1.

The conditions for each of the three cases expressed in terms of K_0 and K_1 are

1. For $(1 - K_0)^2 > 4K_1$, two real unequal roots;
2. For $(1 - K_0)^2 = 4K_1$, a pair of real equal roots;
3. For $(1 - K_0)^2 < 4K_1$, a pair of conjugate complex (or imaginary if $K_0 = 1$) roots.

The equation for case 2 is a parabola in the K_0, K_1 plane. The portion of this curve which lies in the stable range is plotted in Fig. 10.2.1. The area below this curve corresponds to combinations of K_0 and K_1 which result in real unequal roots (case 1), and the area above to combinations resulting in complex roots (case 3).

The solution of the difference equation (10.1.8) for cases 1 and 3 is found by substitution of (10.2.4), (10.2.5), and (10.2.6) into (10.1.6) or (10.1.7) for $f(i)$, h_1, and h_2, respectively, and $m(i)$ for $x(i)$. Instead of substituting for h_1 and h_2, it is more convenient for now to leave these symbols in the solution because of their rather involved form. Therefore, from (10.1.6) we get

$$m(i) = \frac{1}{h_1 - h_2} \left\{ (-h_2 h_1{}^i + h_1 h_2{}^i) m(0) + (h_1{}^i - h_2{}^i) m(1) \right.$$
$$\left. + \sum_{n=0}^{i-1} [-K_0 \varepsilon(n+1) - K_1 \varepsilon(n) + p(n+1)](h_1^{i-1-n} - h_2^{i-1-n}) \right\}. \quad (10.2.11)$$

A comparable result would be obtained from (10.1.7).

Although (10.2.11) fully describes $m(i)$ for both cases 1 and 3, it might prove instructive to express these results for case 3 in terms of sinusoidal functions. Therefore, as discussed in both Chapters IV and VI, for $(1 - K_0)^2 < 4K_1$ we write

$$h_1 = (1 - K_0)/2 + (j/2)(4K_1 - (1 - K_0)^2)^{1/2} \quad (10.2.12)$$

and

$$h_2 = (1 - K_0)/2 - (j/2)(4K_1 - (1 - K_0)^2)^{1/2}. \quad (10.2.13)$$

These may, therefore, be expressed as

$$h_1 = Re^{j\theta} = R(\cos \theta + j \sin \theta) \quad (10.2.14)$$

and

$$h_2 = Re^{-j\theta} = R(\cos \theta - j \sin \theta), \quad (10.2.15)$$

where the magnitude R is given by

$$R = \tfrac{1}{2}\{(1 - K_0)^2 + [4K_1 - (1 - K_0)^2]\}^{1/2} = (K_1)^{1/2}, \quad (10.2.16)$$

and the phase angle θ is

$$\theta = \tan^{-1}\{[4K_1 - (1 - K_0)^2]^{1/2}/(1 - K_0)\}. \quad (10.2.17)$$

Substitution in (10.2.11) yields, after some manipulation,

$$m(i) = \frac{R^i}{\sin \theta} \left\{ -\sin \theta(i-1)m(0) + \frac{1}{R} \sin \theta i\, m(1) \right.$$

$$\left. + \frac{1}{R^2} \sum_{n=0}^{i-1} [-K_0 \varepsilon(n+1) - K_1 \varepsilon(n) + p(n+1)]R^{-n} \sin \theta(i-1-n) \right\}.$$

$$(10.2.18)$$

In general, we will avoid use of this form even for complex roots unless it is absolutely necessary to do otherwise. Usually, (10.2.11) will be used to represent both cases 1 and 3.

For case 2, the double root can be expressed in either of the equivalent forms

$$h = (1 - K_0)/2 = (K_1)^{1/2}. \qquad (10.2.19)$$

We will, however, retain the symbol h for ease in presentation. Appropriate substitution in (10.1.9) yields

$$m(i) = -(i-1)h^i m(0) + ih^{i-1}m(1)$$

$$+ \sum_{n=0}^{i-2} [-K_0 \varepsilon(n+1) - K_1 \varepsilon(n) + p(n+1)](i-1-n)h^{i-2-n}. \qquad (10.2.20)$$

An alternative expression could of course be obtained from (10.1.10).

It is essential for the calculation of variances and helpful in determining expected values if (10.2.11) and (10.2.20) are revised by combining measurement-error terms with like arguments. For (10.2.11), i.e., for $h_1 \neq h_2$, this results in

$$m(i) = \frac{1}{h_1 - h_2} \{ (-h_2 h_1{}^i + h_1 h_2{}^i)m(0) + (h_1{}^i - h_2{}^i)m(1)$$

$$- K_1(h_1^{i-1} - h_2^{i-1})\, \varepsilon(0)$$

$$- \sum_{n=1}^{i-1} [K_0(h_1^{i-n} - h_2^{i-n}) + K_1(h_1^{i-1-n} - h_2^{i-1-n})]\varepsilon(n)$$

$$+ \sum_{n=1}^{i-1} (h_1^{i-n} - h_2^{i-n})p(n)\}$$

$$= \frac{1}{h_1 - h_2} \{ (-h_2 h_1{}^i + h_1 h_2{}^i)m(0)$$

$$+ (h_1{}^i - h_2{}^i)m(1) - K_1(h_1^{i-1} - h_2^{i-1})\varepsilon(0)$$

$$- \sum_{n=1}^{i-1} [(K_0 h_1 + K_1)h_1^{i-1-n} - (K_0 h_2 + K_1)h_2^{i-1-n}]\varepsilon(n)$$

$$+ \sum_{n=1}^{i-1} (h_1^{i-n} - h_2^{i-n})p(n)\}, \qquad (10.2.21)$$

and for (10.2.20), corresponding to $h = h_1 = h_2$,

$$m(i) = -(i - 1)h^i m(0) + ih^{i-1}m(1) - K_1(i - 1)h^{i-2}\varepsilon(0)$$

$$- \sum_{n=1}^{i-1} [K_0(i - n)h^{i-1-n} + K_1(i - 1 - n)h^{i-2-n}]\varepsilon(n)$$

$$+ \sum_{n=1}^{i-1} (i - n)h^{i-1-n}p(n)$$

$$= -(i - 1)h^i m(0) + ih^{i-1}m(1) - K_1(i - 1)h^{i-2}\varepsilon(0)$$

$$- \sum_{n=1}^{i-1} [K_0 h(i - n) + K_1(i - 1 - n)]h^{i-2-n}\varepsilon(n)$$

$$+ \sum_{n=1}^{i-1} (i - n)h^{i-1-n}p(n). \tag{10.2.22}$$

Mean and Variance of Process Output

The mean of $m(i)$ is simply the expected value of (10.2.21) or (10.2.22) as appropriate. Such expressions are identical in form to the corresponding equation for $m(i)$ with the expected value of each variable replacing the variable itself. There is thus little to be gained in writing them out explicitly. The steady-state mean is found from the limit of $E[m(i)]$ as i increases without bound. For K_0, K_1 combinations in the stable range, the initial condition terms involving $m(0)$, $m(1)$, and $\varepsilon(0)$ vanish. The steady-state values of the remaining terms in $E[\varepsilon(i)]$ and the terms in $E[p(i)]$, all of which are convolved with the roots of the characteristic equation, depend on the exact patterns of the $E[\varepsilon(i)]$ and $E[p(i)]$. Therefore, the best we can do in providing a general expression for the steady-state mean M is to write

$$M = \lim_{i \to \infty} \frac{1}{h_1 - h_2} \left\{ \sum_{n=1}^{i-1} [(K_0 h_1 + K_1)h_1^{i-1-n} - (K_0 h_2 + K_1)h_2^{i-1-n}]E[\varepsilon(n)] \right.$$

$$+ \left. \sum_{n=1}^{i-1} (h_1^{i-n} - h_2^{i-n})E[p(n)] \right\} \tag{10.2.23}$$

for $h_1 \neq h_2$, and

$$M = \lim_{i \to \infty} \left\{ -\sum_{n=1}^{i-1} [K_0 h(i - n) + K_1(i - 1 - n)]h^{i-2-n}E[\varepsilon(n)] \right.$$

$$+ \left. \sum_{n=1}^{i-1} (i - n)h^{i-1-n}E[p(n)] \right\} \tag{10.2.24}$$

for $h = h_1 = h_2$.

The variance of $m(i)$ is of course defined as the second moment of $m(i)$ around its mean. As explained in Section 7.6, determinations of variances can usually be facilitated by taking advantage of the fact that the variance is equivalent to the second moment of the variable around zero minus the square of the expected value of the variable. For the output variables of linear systems, considerable help can be obtained by taking advantage of super-position and separating each forcing function variable into two components, one a deterministic component equal to its expected value and the other an unbiased random variable. The expected response of the system to the unbiased random components is then zero, meaning the output variance is just the second moment around zero of the system response to these unbiased random components.

Specifically, suppose each system perturbation $p(i)$ is characterized by a mean $E[p(i)]$ and variance $V[p(i)]$ and each measurement error by a mean $E[\varepsilon(i)]$ and variance $V[\varepsilon(i)]$. If we now define

$$\bar{p}(i) = p(i) - E[p(i)] \tag{10.2.25}$$

and

$$\bar{\varepsilon}(i) = \varepsilon(i) - E[\varepsilon(i)], \tag{10.2.26}$$

then $\bar{p}(i)$ and $\bar{\varepsilon}(i)$ have means of zero and variances $V[p(i)]$ and $V[\varepsilon(i)]$, respectively. The variance of the system output, $m(i)$, can now be found by substituting $\bar{p}(i)$ for $p(i)$ and $\bar{\varepsilon}(i)$ for $\varepsilon(i)$ in (10.2.21) or (10.2.22), as appropriate, squaring the result, and taking the expected value of the expression obtained. For $h_1 \neq h_2$, the system response to $\bar{p}(i)$ and $\bar{\varepsilon}(i)$, referred to here as $\bar{m}(i)$, is given by

$$\bar{m}(i) = \frac{1}{h_1 - h_2} \left\{ -\sum_{n=1}^{i-1} [(K_0 h_1 + K_1)h_1^{i-1-n} - (K_0 h_2 + K_1)h_2^{i-1-n}]\bar{\varepsilon}(n) \right.$$

$$\left. + \sum_{n=1}^{i-1} (h_1^{i-n} - h_2^{i-n})\bar{p}(n) \right\}. \tag{10.2.27}$$

The expected value of $\bar{m}(i)$ is of course zero. Squaring results in

$$\bar{m}(i)^2 = \frac{1}{(h_1 - h_2)^2} \left\{ \left[-\sum_{n=1}^{i-1} [(K_0 h_1 + K_1)h_1^{i-1-n} - (K_0 h_2 + K_1)h_2^{i-1-n}]\bar{\varepsilon}(n) \right]^2 \right.$$

$$- 2 \left[\sum_{n=1}^{i-1} [(K_0 h_1 + K_1)h_1^{i-1-n} - (K_0 h_2 + K_1)h_2^{i-1-n}]\bar{\varepsilon}(n) \right]$$

$$\left. \cdot \left[\sum_{n=1}^{i-1} (h_1^{i-n} - h_2^{i-n})\bar{p}(n) \right] + \left[\sum_{n=1}^{i-1} (h_1^{i-n} - h_2^{i-n})\bar{p}(n) \right]^2 \right\}. \tag{10.2.28}$$

We now examine each of the three basic terms of $\bar{m}(i)^2$ individually.

The first term of $\bar{m}(i)^2$ involves measurement errors only. We therefore designate it as $\bar{m}_\varepsilon(i)^2$ and note that it is of the form $[-\sum_{n=1}^{i-1} a_\varepsilon(i, n)\bar{\varepsilon}(n)]^2$, where

$$a_\varepsilon(i, n) = \frac{1}{(h_1 - h_2)} [(K_0 h_1 + K_1)h_1^{i-1-n} - (K_0 h_2 + K_1)h_2^{i-1-n}]. \quad (10.2.29)$$

Expansion results in $(i - 1)^2$ terms, $i - 1$ of which involve squares of the $\bar{\varepsilon}(n)$ and the rest cross products of pairs of $\bar{\varepsilon}(n)$. There will be two such terms for each cross product as discussed in Section 7.6. Thus,

$$\bar{m}_\varepsilon(i)^2 = \sum_{n=1}^{i-1} a_\varepsilon(i, n)^2 \bar{\varepsilon}(n)^2 + 2 \sum_{n=2}^{i-1} \sum_{q=1}^{n-1} a_\varepsilon(i, n)a_\varepsilon(i, q)\bar{\varepsilon}(n)\bar{\varepsilon}(q). \quad (10.2.30)$$

The expected value of $\bar{m}_\varepsilon(i)^2$ is, therefore,

$$E[\bar{m}_\varepsilon(i)^2] = \sum_{n=1}^{i-1} a_\varepsilon(i, n)^2 V[\varepsilon(n)] + 2 \sum_{n=2}^{i-1} \sum_{q=1}^{n-1} a_\varepsilon(i, n)a_\varepsilon(i, q)\sigma_\varepsilon(n, q), \quad (10.2.31)$$

where $\sigma_\varepsilon(n, q)$ is the autocovariance between $\varepsilon(n)$ and $\varepsilon(q)$. Obviously, $\sigma_\varepsilon(n, n) = V[\varepsilon(n)]$. We note in passing that if the $\varepsilon(i)$ are stationary,

$$V[\varepsilon(i)] = \sigma_\varepsilon^2 \quad \text{for all } i, \quad (10.2.32)$$

and, letting

$$\tau = n - q, \quad (10.2.33)$$

then

$$\sigma_\varepsilon(n, q) = \sigma_\varepsilon(\tau) \quad \text{for all } n, \quad q = n - \tau. \quad (10.2.34)$$

This results in

$$E[\bar{m}_\varepsilon(i)^2] = \sigma_\varepsilon^2 \sum_{n=1}^{i-1} a_\varepsilon(i, n)^2 + 2 \sum_{n=2}^{i-1} \sum_{\tau=1}^{n-1} a_\varepsilon(i, n)a_\varepsilon(i, n - \tau)\sigma_\varepsilon(\tau)$$

$$= \sigma_\varepsilon^2 \sum_{n=1}^{i-1} a_\varepsilon(i, n)^2 + 2 \sum_{\tau=1}^{i-2} \sigma_\varepsilon(\tau) \sum_{n=\tau+1}^{i-1} a_\varepsilon(i, n)a_\varepsilon(i, n - \tau). \quad (10.2.35)$$

The third term of $\bar{m}(i)^2$ involves perturbation terms only. We designate it by $\bar{m}_p(i)^2$ and note that it is of the form $[\sum_{n=1}^{i-1} a_p(i, n)\bar{p}(n)]^2$, where

$$a_p(i, n) = [1/(h_1 - h_2)](h_1^{i-n} - h_2^{i-n}). \quad (10.2.36)$$

Expansion of $\bar{m}_p(i)^2$ by the same procedures just followed for $\bar{m}_\varepsilon(i)^2$ yields

$$\bar{m}_p(i)^2 = \sum_{n=1}^{i-1} a_p(i, n)^2 \bar{p}(n)^2 + 2 \sum_{n=2}^{i-1} \sum_{q=1}^{n-1} a_p(i, n)a_p(i, q)\bar{p}(n)\bar{p}(q). \quad (10.2.37)$$

The expected value of $\bar{m}_p(i)^2$ is, therefore,

$$E[\bar{m}(i)^2] = \sum_{n=1}^{i-1} a_p(i, n)^2 V[p(n)] + 2 \sum_{n=2}^{i-1} \sum_{q=1}^{n-1} a_p(i, n)a_p(i, q)\sigma_p(n, q), \quad (10.2.38)$$

where $\sigma_p(n, q)$ is the autocovariance between $p(n)$ and $p(q)$. Of course, $\sigma_p(n, n) = V[p(n)]$. If the $p(i)$ are stationary,

$$V[p(i)] = \sigma_p{}^2 \qquad \text{for all } i \tag{10.2.39}$$

and

$$\sigma_p(n, q) = \sigma_p(\tau) \qquad \text{for all } n, \quad q = n - \tau, \tag{10.2.40}$$

where τ is defined in (10.2.33). $E[\bar{m}_p(i)^2]$ can then be expressed as

$$E[\bar{m}_p(i)^2] = \sigma_p{}^2 \sum_{n=1}^{i-1} a_p(i, n)^2 + 2 \sum_{n=2}^{i-1} \sum_{\tau=1}^{n-1} a_p(i, n) a_p(i, n - \tau) \sigma_p(\tau)$$

$$= \sigma_p{}^2 \sum_{n=1}^{i-1} a_p(i, n)^2 + 2 \sum_{\tau=1}^{i-2} \sigma_p(\tau) \sum_{n=\tau+1}^{i-1} a_p(i, n) a_p(i, n - \tau). \tag{10.2.41}$$

The middle term of $\bar{m}(i)^2$ involves both measuring errors and perturbations. We designate it by $\bar{m}_{\varepsilon p}(i)^2$ and note that it is of the form

$$-2\left[\sum_{n=1}^{i-1} a_\varepsilon(i, n)\bar{\varepsilon}(n)\right]\left[\sum_{q=1}^{i-1} a_p(i, q)\bar{p}(q)\right].$$

Expansion results in $(i - 1)^2$ terms involving all of the combinations of cross products of $\bar{\varepsilon}(n)$ and $\bar{p}(q)$, specifically,

$$\bar{m}_{\varepsilon p}(i)^2 = -2 \sum_{n=1}^{i-1} \sum_{q=1}^{i-1} a_\varepsilon(i, n) a_p(i, q)\bar{\varepsilon}(n)\bar{p}(q). \tag{10.2.42}$$

The expected value of $\bar{m}_{\varepsilon p}(i)^2$ is, therefore,

$$E[\bar{m}_{\varepsilon p}(i)^2] = -2 \sum_{n=1}^{i-1} \sum_{q=1}^{i-1} a_\varepsilon(i, n) a_p(i, q)\sigma_{\varepsilon p}(n, q), \tag{10.2.43}$$

where $\sigma_{\varepsilon p}(n, q)$ is the cross covariance between $\varepsilon(n)$ and $p(q)$. Should both $\varepsilon(n)$ and $p(q)$ come from stationary processes,

$$\sigma_{\varepsilon p}(n, q) = \sigma_{\varepsilon p}(\tau) \qquad \text{for all } n, \quad q = n - \tau, \tag{10.2.44}$$

and

$$E[\bar{m}_{\varepsilon p}(i)^2] = -2 \sum_{n=1}^{i-1} \sum_{\tau=-(i-1-n)}^{n-1} a_\varepsilon(i, n) a_p(i, n - \tau)\sigma_{\varepsilon p}(\tau)$$

$$= -2\left[\sum_{\tau=-(i-2)}^{-1} \sigma_{\varepsilon p}(\tau) \sum_{n=1}^{i-1+\tau} a_\varepsilon(i, n) a_p(i, n - \tau)\right.$$

$$\left. + \sum_{\tau=0}^{i-2} \sigma_{\varepsilon p}(\tau) \sum_{n=\tau+1}^{i-1} a_\varepsilon(i, n) a_p(i, n - \tau)\right]. \tag{10.2.45}$$

Combining (10.2.31), (10.2.43), and (10.2.38) gives the following expression for $E[\overline{m}(i)^2]$, which is the variance of $m(i)$ for $h_1 \neq h_2$:

$$V[m(i)] = E[\overline{m}(i)^2]$$

$$= \sum_{n=1}^{i-1} a_\varepsilon(i, n)^2 V[\varepsilon(n)] + 2 \sum_{n=2}^{i-1} \sum_{q=1}^{n-1} a_\varepsilon(i, n) a_\varepsilon(i, q) \sigma_\varepsilon(n, q)$$

$$- 2 \sum_{n=1}^{i-1} \sum_{q=1}^{i-1} a_\varepsilon(i, n) a_p(i, q) \sigma_{\varepsilon p}(n, q)$$

$$+ \sum_{n=1}^{i-1} a_p(i, n)^2 V[p(n)] + 2 \sum_{n=2}^{i-1} \sum_{q=1}^{n-1} a_p(i, n) a_p(i, q) \sigma_p(n, q). \quad (10.2.46)$$

This equation represents the most general form possible for $V[m(i)]$ in that it holds for nonstationary measurement errors and perturbations and includes the effects of autocorrelations among both the measurement errors and perturbations and cross correlations between them. Should the measurement error come from a stationary process, (10.2.35) could be used instead of (10.2.31) in (10.2.46) to represent $E[\overline{m}_\varepsilon(i)^2]$. Should the random component of the perturbations be stationary, (10.2.41) could be used instead of (10.2.38) to represent $E[\overline{m}_p(i)^2]$. Should both the measurement errors and the perturbations be stationary, $V[m(i)]$ could be represented by combining (10.2.35), (10.2.45), and (10.2.41) which results in (again for $h_1 \neq h_2$)

$$V[m(i)] = \sigma_\varepsilon^2 \sum_{n=1}^{i-1} a_\varepsilon(i, n)^2 + 2 \sum_{\tau=1}^{i-2} \sigma_\varepsilon(\tau) \sum_{n=\tau+1}^{i-1} a_\varepsilon(i, n) a_\varepsilon(i, n - \tau)$$

$$- 2 \left[\sum_{\tau=-(i-2)}^{-1} \sigma_{\varepsilon p}(\tau) \sum_{n=1}^{i-1+\tau} a_\varepsilon(i, n) a_p(i, n - \tau) \right.$$

$$\left. + \sum_{\tau=0}^{i-2} \sigma_{\varepsilon p}(\tau) \sum_{n=\tau+1}^{i-1} a_\varepsilon(i, n) a_p(i, n - \tau) \right]$$

$$+ \sigma_p^2 \sum_{n=1}^{i-1} a_p(i, n)^2 + 2 \sum_{\tau=1}^{i-2} \sigma_p(\tau) \sum_{n=\tau+1}^{i-1} a_p(i, n) a_p(i, n - \tau). \quad (10.2.47)$$

The corresponding expressions for the situation in which $h = h_1 = h_2$ are derived in similar fashion. For the general case,

$$V[m(i)] = \sum_{n=1}^{i-1} b_\varepsilon(i, n)^2 V[\varepsilon(n)] + 2 \sum_{n=2}^{i-1} \sum_{q=1}^{n-1} b_\varepsilon(i, n) b_\varepsilon(i, q) \sigma_\varepsilon(n, q)$$

$$- 2 \sum_{n=1}^{i-1} \sum_{q=1}^{i-1} b_\varepsilon(i, n) b_p(i, q) \sigma_{\varepsilon p}(n, q)$$

$$+ \sum_{n=1}^{i-1} b_p(i, n)^2 V[p(n)] + 2 \sum_{n=2}^{i-1} \sum_{q=1}^{n-1} b_p(i, n) b_p(i, q) \sigma_p(n, q), \quad (10.2.48)$$

where

$$b_\varepsilon(i, n) = [K_0 h(i - n) + K_1(i - 1 - n)] h^{i-2-n} \quad (10.2.49)$$

and

$$b_p(i, n) = (i - n)h^{i-1-n}. \tag{10.2.50}$$

Note that (10.2.48) is identical in form to (10.2.46) with $b_\varepsilon(i, n)$ and $b_p(i, n)$ in place of $a_\varepsilon(i, n)$ and $a_p(i, n)$, respectively. Therefore, the expressions which result when the measurement errors and perturbations are stationary will also be identical in form to the comparable expressions developed for $h_1 \neq h_2$, again with $b_\varepsilon(i, n)$ and $b_p(i, n)$ replacing $a_\varepsilon(i, n)$ and $a_p(i, n)$.

Considerable simplification of the expressions for $V[m(i)]$ results if any of the intercorrelations can be assumed not to exist or to be relatively insignificant. Specifically, if the $\varepsilon(i)$ and $p(i)$ are mutually independent, $\sigma_{\varepsilon p}(n, q) = 0$ for all n, q and the entire set of terms involving cross products can be omitted. If the measuring errors are mutually independent, $\sigma_\varepsilon(n, q) = 0$ for all $n \neq q$, and, if the perturbations are mutually independent, $\sigma_p(n, q) = 0$ for all $n \neq q$, resulting in further elimination of terms.

The expressions for steady-state variance V for K_0, K_1 combinations in the stable range are found by taking the limit of $V[m(i)]$ as i increases without bound. For the general situations expressed by (10.2.46) and (10.2.48), specific knowledge of the patterns of $V[\varepsilon(i)]$, $V[p(i)]$, $\sigma_\varepsilon(i, q)$, $\sigma_p(i, q)$, and $\sigma_{\varepsilon p}(i, q)$ with i would be needed in order to find these limits. When the random components of both the measurement errors and the perturbations are stationary, however, V can be found by taking the limits of the various summations in (10.2.47) and its counterpart for $h = h_1 = h_2$. Needless to say, this process is both time-consuming and tedious, and the many possible forms in which the results can be expressed lead one to wonder whether he has found the "best" one. "Best" could conceivably differ depending on the objectives of the model development effort, e.g., gaining insight concerning overall system operation, comparison with other orders or types of controllers, or just getting a simple expression for publication purposes.

The following expressions for steady-state variance given stationary measurement error and random perturbation have been derived with one eye towards simplicity and the other towards limiting the amount of derivation effort expended. Perhaps more basic or useful forms exist. Frankly, the author is happy just to have some expression to present here. In the formula for the case of $h_1 \neq h_2$, use has been made of certain relationships among the roots such as

$$h_1 + h_2 = 1 - K_0, \tag{10.2.51}$$

$$h_1 - h_2 = [(1 - K_0)^2 - 4K_1]^{1/2}, \tag{10.2.52}$$

$$h_1 h_2 = K_1, \tag{10.2.53}$$

$$h_1{}^2 + h_2{}^2 = (1 - K_0)^2 - 2K_1, \tag{10.2.54}$$

and

$$h_1^2 - h_2^2 = (1 - K_0)[(1 - K_0)^2 - 4K_1]^{1/2}. \qquad (10.2.55)$$

Whenever a root must be raised to a power, however, the symbol h_1 or h_2 is retained. This results in

$$V = \frac{1}{(2K_0 - K_0^2 + 2K_1 + K_1^2)(1 - K_1)}$$

$$\cdot \left\{ \sigma_\varepsilon^2[(K_0^2 + K_1^2)(1 + K_1) + 2K_0 K_1(1 - K_0)] + \frac{2}{[(1 - K_0)^2 - 4K_1]^{1/2}} \right.$$

$$\cdot \sum_{\tau=1}^{i-2} \sigma_\varepsilon(\tau)[(K_0^2 + K_1^2)\{(h_1^{\tau+1} - h_2^{\tau+1}) - K_1^2(h_1^{\tau-1} - h_2^{\tau-1})\}$$

$$+ K_0 K_1\{(h_1^{\tau+2} - h_2^{\tau+2}) + (1 - K_1^2)(h_1^\tau - h_2^\tau) - K_1^2(h_1^{\tau-2} - h_2^{\tau-2})\}]$$

$$- \frac{2}{[(1 - K_0)^2 - 4K_1]^{1/2}} \left[\sum_{\tau=-(i-2)}^{-1} \sigma_{\varepsilon p}(\tau)\{K_0[(h_1^{-\tau+1} - h_2^{-\tau+1}) \right.$$

$$- K_1 (h_1^{-\tau-1} - h_2^{-\tau-1})] + K_1[(h_1^{-\tau} - h_2^{-\tau}) - K_1^2(h_1^{-\tau-2} - h_2^{-\tau-2})]\}$$

$$+ \sum_{\tau=0}^{i-2} \sigma_{\varepsilon p}(\tau)\{K_0[(h_1^{\tau+1} - h_2^{\tau+1}) - K_1^2(h_1^{\tau-1} - h_2^{\tau-1})]$$

$$\left. + K_1[(h_1^{\tau+2} - h_2^{\tau+2}) - K_1^2(h_1^\tau - h_2^\tau)]\} \right] + \sigma_p^2(1 + K_1)$$

$$\left. + \frac{2}{[(1 - K_0)^2 - 4K_1]^{1/2}} \sum_{\tau=1}^{i-2} \sigma_p(\tau)[(h_1^{\tau+1} - h_2^{\tau+1}) - K_1^2(h_1^{\tau-1} - h_2^{\tau-1})] \right\}.$$

$$(10.2.56)$$

For $h = h_1 = h_2$, the formula has been left in terms of h. Thus,

$$V = \frac{1}{(1 - h^2)^3} \left\{ \sigma_\varepsilon^2[(K_0^2 + K_1^2)(1 + h^2) + 4K_0 K_1 h] \right.$$

$$+ 2\sum_{\tau=1}^{i-2} \sigma_\varepsilon(\tau)h^{\tau-1}[(K_0^2 + K_1^2)h\{(1 + h^2) + \tau(1 - h^2)\}$$

$$+ K_0 K_1\{4h^2 + \tau(1 - h^4)\}]$$

$$- 2\left[\sum_{\tau=-(i-2)}^{-1} \sigma_{\varepsilon p}(\tau)h^{-\tau-1}\{K_0 h[(1 + h^2) - \tau(1 - h^2)] \right.$$

$$+ K_1[2h^2 - \tau(1 - h^2)]\}$$

$$\left. + \sum_{\tau=0}^{i-2} \sigma_{\varepsilon p}(\tau)h^{\tau-1}\{K_0 h[(1 + h^2) + \tau(1 - h^2)] + K_1[2h^2 + \tau h^2(1 - h^2)]\} \right]$$

$$\left. + \sigma_p^2(1 + h^2) + 2\sum_{\tau=1}^{i-2} \sigma_p(\tau)h[(1 + h^2) + \tau(1 - h^2)] \right\}, \qquad (10.2.57)$$

which could of course be expressed completely in terms of either K_0 or K_1 by use of (10.2.19). The interested reader is free to pursue these derivations.

Note that for $K_1 = 0$, $h_1 = 1 - K_0$, $h_2 = 0$, and the stable range of K_0 is $0 \le K_0 \le 2$, which are precisely the results obtained for the simple proportional controller. Thus, the simple proportional controller is a special case of the generalized second-order controller and its properties can be obtained by setting K_1 and h_2 to zero and h_1 to $1 - K_0$ in the appropriate equations for $h_1 \ne h_2$ in this section.

It is also of interest to investigate the second-order controller with $K_0 = 0$. This is in effect a proportional controller with a one-period delay in feeding back each measurement. This and other special types of second-order process controllers will be discussed in later sections of this chapter. Meanwhile, let us examine the performance of the generalized second-order process controller for a few common types of perturbations.

10.3 Responses of the Generalized Second-Order Process Controller

The responses of the generalized second-order process controller to impulse, step, sinusoidal, and random perturbations will now be studied and, where appropriate, comparisons made to the first-order system.

Impulse Response

In Section 7.3 we discussed the response of the simple first-order controller to an impulse perturbation occurring between instants $i = 0$ and $i = 1$, i.e., for $p(0) = p$ and $p(i) = 0$ for all $i \ne 0$. We seek to perform the same analysis here for the second-order controller. However, (10.2.21) and (10.2.22), the equations for $m(i)$ for $h_1 \ne h_2$ and $h = h_1 = h_2$, respectively, contain no explicit term for $p(0)$. This is due to incrementing the time argument i by one unit to get the original difference equation into standard form for solution, the effect of which was to put the perturbation forcing function term in the form $p(i + 1)$. Since our solutions hold only for $i \ge 0$, $p(0)$ has been excluded. However, the same incrementing process created a term $m(i + 2)$, the z transform of which involves both $m(0)$ and $m(1)$ as initial conditions. If we consider $m(0)$ as an initial deviation caused by setup error, then through action of the controller,

$$m(1) = m(0) - K_0[m(0) + \varepsilon(0)] + p(0). \tag{10.3.1}$$

That is, $m(0)$, $m(1)$, $\varepsilon(0)$, and $p(0)$ are functionally related, so that only three of the four can be imposed as independent initial conditions. The mathematical procedures leading to the formulas for $m(i)$ have taken this into account by eliminating $p(0)$ as an initial condition term. Actually, it would be convenient for purposes of transient analysis to substitute for $m(1)$ in (10.2.21) and (10.2.22) since $m(1)$ is the result of the combined physical occurrences of $m(0)$, $\varepsilon(0)$, and $p(0)$. Substituting followed by recombining terms yields

$$
m(i) = \frac{1}{(h_1 - h_2)} \left\{ (h_1^{i+1} - h_2^{i+1})m(0) \right.
$$

$$
- \sum_{n=0}^{i-1} [(K_0 h_1 + K_1)h_1^{i-1-n} - (K_0 h_2 + K_1)h_2^{i-1-n}]\varepsilon(n)
$$

$$
\left. + \sum_{n=0}^{i-1} (h_1^{i-n} - h_2^{i-n})p(n) \right\} \tag{10.3.2}
$$

for $h_1 \neq h_2$ and

$$
m(i) = (i+1)h^i m(0) - \sum_{n=0}^{i-1} [K_0 h(i-n) + K_1(i-1-n)]h^{i-2-n}\varepsilon(n)
$$

$$
+ \sum_{n=0}^{i-1} (i-n)h^{i-1-n}p(n) \tag{10.3.3}
$$

for $h = h_1 = h_2$.

Note how much simpler these expressions are compared to their counterparts involving $m(1)$ explicitly since here the terms in $\varepsilon(0)$ and $p(0)$ are of the same form as the general terms in $\varepsilon(i)$ and $p(i)$ and are thus included in the appropriate summations. It must be emphasized, however, that (10.2.21) and (10.2.22) are perfectly general, whereas (10.3.2) and (10.3.3) assume start-up conditions as expressed by (10.3.1). It is of course possible that other procedures could be followed which would lead to a different relationship among $m(0)$, $m(1)$, $\varepsilon(0)$, and $p(0)$. For example, where measurements are made strictly on process output, there would be no measurement and hence no control action until a product emerged at $i = 1$. In this case, there is no $\varepsilon(0)$ and

$$
m(1) = m(0) + p(0). \tag{10.3.4}
$$

Substitution of (10.3.4) for $m(1)$ in $m(i)$ would produce a much different result from that above. We will, however, use (10.3.2) and (10.3.3) as the basis for all transient analyses in this chapter.

To facilitate discussion of the impulse response of the second-order controller, we assume setup and measurement errors are negligible, i.e., $m(0)$ and $\varepsilon(i)$ for all i are taken as zero. Thus, for $p(i)$ an impulse of magnitude p at $i = 0$,

$$
m(i) = \frac{1}{(h_1 - h_2)} (h_1{}^i - h_2{}^i)p \tag{10.3.5}
$$

for $h_1 \neq h_2$ and

$$
m(i) = ih^{i-1}p = i\left(\frac{1 - K_0}{2}\right)^{i-1} p = iK_1^{(i-1)/2}p \tag{10.3.6}
$$

for $h = h_1 = h_2$. For case 3, in which h_1 and h_2 form a complex conjugate pair, (10.3.5) may be written

$$m(i) = K_1^{(i-1)/2}(\sin \theta i/\sin \theta)p, \tag{10.3.7}$$

where θ is given by (10.2.17).

In the absence of manageable closed-form expressions for $m(i)$ in terms of K_0 and K_1, transient analysis of second- and higher-order controllers must generally be done numerically. We have therefore listed in Tables 10.3.1 and 10.3.2 the values of $m(i)$, $i = 1, \ldots, 8$, for a unit impulse ($p = 1$) for several K_0, K_1 combinations in the stable region. The impulse response of the first-order controller for selected values of proportionality constant K are also presented for comparison purposes.

From Table 10.3.1 we see that for case 1 operation, regardless of the sign of K_1, $m(i)$ is positive for all i when $K_0 < 1$ but alternates in sign for $K_0 > 1$. It may be recalled from Chapter VII, specifically Table 7.2.1, that the same is

Table 10.3.1

Response to a Unit-Impulse Perturbation of the Generalized Second-Order Process Controller, Case 1 Operation, and of the Simple Proportional Controller

	Second-order controller				First-order controller		
	$K_1 = -0.16$		$K_1 = 0.16$				
i	$K_0 = 0.4$	$K_0 = 1.6$	$K_0 = 0$	$K_0 = 2$	$K = 0.4$	$K = 0.244$	$K = 0.14$
1	1.000	1.000	1.000	1.000	1.000	1.000	1.000
2	0.600	−0.600	1.000	−1.000	0.600	0.756	0.860
3	0.520	0.520	0.840	0.840	0.360	0.571	0.740
4	0.408	−0.408	0.680	−0.680	0.216	0.432	0.636
5	0.328	0.328	0.546	0.546	0.130	0.327	0.547
6	0.262	−0.262	0.436	−0.436	0.078	0.247	0.470
7	0.210	0.210	0.350	0.350	0.047	0.187	0.405
8	0.168	−0.168	0.280	−0.280	0.028	0.141	0.348

true for the first-order controller. The magnitude of $m(i)$ depends on $|1 - K_0|$ and K_1. Comparison of the second-order controller having $K_1 = -0.16$ and $K_0 = 0.4$ with the first-order controller having $K = 0.4$ shows that even though the two start out the same by reducing $m(2)$ to 0.6, with further increases in i the first-order system reduces $m(i)$ toward zero much faster than the second-order controller. The reverse is true when the second-order system with $K_1 = 0.16$ and $K_0 = 0$ is compared to a (hypothetical) first-order system with $K = 0$ for which $m(i) = 1$ for all i. For $i > 2$, the second-order system results in reduction of $m(i)$.

The first-order system with $K = 0.244$ was selected to produce essentially the same output at $i = 5$ as the second-order system with $K_1 = -0.16$ and $K_0 = 0.4$. Note that the second-order system produces much faster response than this first-order system up to $i = 5$ but a slower response thereafter. Similarly, the first-order system with $K = 0.14$ was selected to produce essentially the same output at $i = 5$ as the second-order system with $K_1 = 0.16$ and $K_0 = 0$. Here, however, the second-order system has a slower response than the first-order system up to $i = 5$ but a faster one thereafter. Thus, negative values of K_1 tend to quicken the immediate response at the expense of slowing it for higher values of i, while for positive K_1 the reverse is true. An additional item worthy of note is that as K_0 deviates farther from unity, the general speed of response decreases as it does with a first-order system. Had K_0 been negative or greater than 2, the magnitude of $m(2)$ would have been greater than one.

Although no combinations corresponding to case 2 are shown in Table 10.3.1, the same patterns of results are obtained.

Table 10.3.2 shows the impulse responses of several second-order systems for case 3 operation. Here $m(i)$ is calculated from (10.3.7). Because of the $\sin \theta i$ factor, which of course can vary from plus to minus one, the patterns

Table 10.3.2

Unit-Impulse Response of Generalized Second-Order Process Controller, Case 3 Operation

i	$K_1 = 0.18$		$K_1 = 0.25$		$K_1 = 0.65$	
	$K_0 = 0.4,$ $\theta = 45°$	$K_0 = 1.6,$ $\theta = 135°$	$K_0 = 0.4,$ $\theta = 53.13°$	$K_0 = 1.6,$ $\theta = 126.87°$	$K_0 = -0.4,$ $\theta = 29.73°$	$K_0 = 2.4,$ $\theta = 150.27°$
1	1.000	1.000	1.000	1.000	1.000	1.000
2	0.600	−0.600	0.600	−0.600	1.398	−1.398
3	0.180	0.180	0.112	0.112	1.310	1.310
4	0.000	0.000	−0.084	0.084	0.924	−0.924
5	−0.032	−0.032	−0.078	−0.078	0.442	0.442
6	−0.020	0.020	−0.026	0.026	0.018	−0.018
7	−0.006	−0.006	0.004	0.004	−0.261	−0.261
8	0.000	0.000	0.009	−0.009	−0.377	0.377

of signs of $m(i)$ with i are no longer regular. However, even though the pattern of signs depends on the specific value of θ, it may be noted that for a given K_1 the magnitude of each $m(i)$ depends on $|1 - K_0|$, and the signs of the $m(i)$ are the same for odd values of i for both $1 - K_0 < 0$ and $1 - K_0 > 0$, but are opposite for i even.

We again see that the farther K_0 deviates from one the slower the general response and that when K_0 is negative or greater than two, $m(i)$ initially increases in magnitude as i increases. Since $K_1^{(i-1)/2}$ is the amplitude of the sinusoid at instant i, it forms an upper bound on the magnitude of $m(i)$. The actual value of $m(i)$ depends heavily of course on θ since $K_1^{(i-)/2}$ is multiplied by $\sin \theta i/\sin \theta$. Since θ is an arc tangent function of K_0 and K_1, this makes analytic determination of any sort of optimal values of these control parameters extremely unlikely.

From what has been shown here it is obvious that if the only concern were the speed of response to impulse-type disturbances, the best controller would be a first-order controller with $K = 1$. It is because of other performance criteria and other forms of perturbation that it is often desirable to design more sophisticated controllers.

Response to a Step Perturbation

Again, to facilitate discussion we will work with $m(i)$ as given by (10.3.2) and (10.3.3). Setting $m(0)$ to zero and assuming negligible measuring error, we obtain for a step perturbation of magnitude p,

$$m(i) = \frac{p}{(K_0 + K_1)(h_1 - h_2)} [(h_1 - h_2) - (1 - h_2)h_1^{i+1} + (1 - h_1)h_2^{i+1}]$$

(10.3.8)

for case 1 operation,

$$m(i) = \frac{p}{(1 - h)^2} \{1 - [1 + (1 - h)i]h^i\}$$

(10.3.9)

for case 2 operation, and

$$m(i) = \frac{p}{(K_0 + K_1) \sin \theta} [\sin \theta + K_1^{(i+1)/2} \sin \theta i - K_1^{i/2} \sin \theta(i + 1)]$$

(10.3.10)

for case 3 operation. Each equation is in a form for reasonably efficient calculation of $m(i)$ for any desired values of K_0, K_1, and i. In (10.3.8) and (10.3.10) use has been made of the relationship $(1 - h_1)(1 - h_2) = K_0 + K_1$. It is also true that $(1 - h)^2 = K_0 + K_1$, which could be substituted in (10.3.9) if desired.

Major interest in system responses to step perturbations is usually focused on steady-state performance. In all three cases it is apparent that for a stable system

$$m = \lim_{i \to \infty} m(i) = \frac{p}{(K_0 + K_1)} \quad \text{(2nd-order controller)}, \quad (10.3.11)$$

which has strong resemblance to the steady-state output of the first-order controller subjected to a step perturbation

$$m = \lim_{i \to \infty} m(i) = p/K \qquad \text{(1st-order controller)}. \qquad (10.3.12)$$

Whereas the best m that can be obtained by a first-order controller is limited by stability considerations to a value slightly greater than $p/2$, the second-order controller can operate stably with K_0 close to three and K_1 close to one thus reducing m to just over $p/4$. Thus the second-order controller offers inherently better performance with respect to the steady-state response to a step perturbation than the first-order controller. In Section 10.5 we will see how the proportional-plus-summation controller, which we have shown to be a special type of second-order controller, can reduce this steady-state response to zero.

A brief look at the transient response to a step perturbation is now in order. Table 10.3.3 lists the values of $m(1)$ through $m(8)$ of the second-order system for a unit step perturbation ($p = 1$) for a few combinations of K_0 and K_1. For comparison purposes, similar values of $m(i)$ for the first-order controller are presented in Table 10.3.4 for selected values of K.

From the examples in Table 10.3.3 we see that for $K_0 < 1$, $m(i)$ either increases (for $m > 1$) or decreases (for $m < 1$) smoothly from its original value of $m(1) = 1$ toward m. This is readily apparent in examples 1 and 3 but is masked by the superimposed sinusoid present in the examples of case 3 operation, 5 and 6. For $K_0 > 1$, the approach to m is oscillatory, alternately jumping above and below m as i increases. The same patterns exist for the first-order system in Table 10.3.4. Although the rapidity with which a system

Table 10.3.3

Unit Step Function Response of Generalized Second-Order Process Controller

	Example number	1	2	3	4	5	6
	K_0	0.40	1.60	0.60	1.40	0.40	−0.40
	K_1	−0.16	−0.16	0.04	0.04	0.25	0.65
	Case number	1	1	2	2	3	3
i	m	4.167	0.694	1.5625	0.694	1.538	4.00
1		1.000	1.000	1.000	1.000	1.000	1.000
2		1.600	0.400	1.400	0.600	1.600	2.400
3		2.121	0.920	1.520	0.720	1.710	3.702
4		2.529	0.512	1.552	0.688	1.626	4.629
5		2.854	0.840	1.560	0.696	1.548	5.073
6		3.117	0.578	1.5619	0.694	1.523	5.089
7		3.329	0.788	1.5624	0.695	1.526	4.831
8		3.496	0.620	1.5625	0.694	1.535	4.460

Table 10.3.4

Unit Step Function Response of First-Order Process Controller

i \ K / m	0.24 / 4.167	0.40 / 2.50	0.64 / 1.5625	1.44 / 0.694	1.60 / 0.625
1	1.000	1.000	1.000	1.000	1.000
2	1.760	1.600	1.360	0.560	0.400
3	2.338	1.960	1.490	0.754	0.760
4	2.777	2.176	1.536	0.668	0.544
5	3.110	2.306	1.553	0.706	0.674
6	3.364	2.383	1.559	0.689	0.596
7	3.556	2.430	1.5613	0.697	0.642
8	3.703	2.458	1.5621	0.693	0.615

reaches steady state is probably not of great importance with step perturbations, the fluctuations encountered in this process could be. Comparison of examples 2 and 4 from Table 10.3.3 and the example for $K = 1.44$ in Table 10.3.4, all of which correspond to an m of 0.694, shows example 2 with $|1 - K_0| = 0.60$ fluctuates the most while example 4 with $|1 - K_0| = 0.40$ fluctuates the least. The first-order system with $|1 - K| = 0.44$ plots a middle ground. Comparison of example 3 with the first-order system with $K = 0.64$, both of which have an $m = 1.5625$, also shows the effects of $|1 - K_0|$ and $|1 - K|$, respectively, on system response. Example 3, with $|1 - K_0| = 0.6$, jumps to $m(2) = 1.4$ and leads the first-order system slightly for all i, although the rates of convergence to m are very close. Example 5, however, with an m almost the same as that of example 3, even though it also converges at about the same rate, varies much more widely in the interim because of the superimposed sinusoid with $R = K_1^{1/2} = 0.5$. Similar observations may be made by comparing examples 1 and 6 from Table 10.3.3 and $K = 0.24$ in Table 10.3.4. The examples for $K = 0.4$ and $K = 1.6$ in Table 10.3.4 are included for completeness in case comparisons with second-order controllers with $K_0 = 0.4$ or 1.6 are desired.

Because of the many combinations of K_0 and K_1 available and the subtle differences in response which can occur, it might be helpful for a system designer to compile a set of responses in terms of K_1 for specific numerical values of K_0 or vice versa. For example, for $K_0 = 1$ it can be shown that for $K_1 < 0$ (case 1 operation),

$$m(i) = \begin{cases} \dfrac{p}{1 + K_1}\left[1 - (\sqrt{-K_1})^{i+1}\right] & \text{for } i \text{ odd,} \\[4mm] \dfrac{p}{1 + K_1}\left[1 - (\sqrt{-K_1})^{i}\right] & \text{for } i \text{ even,} \end{cases} \tag{10.3.13}$$

and for $K_1 > 0$ (case 3 operation),

$$m(i) = \frac{p}{1 + K_1}\left\{1 + (\sqrt{K_1})^{i+1} \sin \frac{\pi i}{2} - (\sqrt{K_1})^i \sin \frac{\pi(i+1)}{2}\right\}. \quad (10.3.14)$$

For case 2 operation, $K_1 = 0$ and $m(i) = p$ for all i. All three cases, however, give the same results, which are listed in Table 10.3.5 as a function of K_1. In this particular case the simple expression

$$m(i) = m(2n - 1) = m(2n) = \frac{p}{1 + K_1}[1 - (-K_1)^n] \quad (10.3.15)$$

provides a ready means of evaluating $m(i)$ for any i; however, in general a series of tables of the general form of Table 10.3.5 would be needed because of the complexity of the expressions involved when K_0 is other than one.

Table 10.3.5

Step Function Response for Generalized
Second-Order Controller with $K_0 = 1$

i	$m(i)$
1, 2	1
3, 4	$(1 - K_1^2)/(1 + K_1) = 1 - K_1$
5, 6	$(1 + K_1^3)/(1 + K_1)$
7, 8	$(1 - K_1^4)/(1 + K_1)$

Frequency Response

In general, the transient response to a sinusoidal perturbation is of little interest to the system designer both because of the bounded nature of the sinusoid and the complexity of the expressions for $m(i)$ in these circumstances. We therefore limit our attention to the steady-state response.

In this derivation we omit, without loss of generality, the phase angle θ of the perturbation. Following the procedures used for the first-order controller in Section 7.5, we express the z transform of $m(i)$ as

$$Z[m(i)] = \frac{z^2 p \sin w}{(z - e^{jw})(z - e^{-jw})(z - h_1)(z - h_2)} \quad (10.3.16)$$

for $h_1 \neq h_2$ and

$$Z[m(i)] = \frac{z^2 p \sin w}{(z - e^{jw})(z - e^{-jw})(z - h)^2} \quad (10.3.17)$$

for $h = h_1 = h_2$, where p is the amplitude of the perturbation sinusoid. Partial fraction expansion followed by inverse transformation back to the i domain yields

$$m(i) = \frac{pe^{jw}}{2j[e^{j2w} - (h_1 + h_2)e^{jw} + h_1 h_2]} e^{jwi}$$

$$- \frac{pe^{-jw}}{2j[e^{-j2w} - (h_1 + h_2)e^{-jw} + h_1 h_2]} e^{-jwi} + c_1 h_1{}^i + c_2 h_2{}^i \quad (10.3.18)$$

for $h_1 \neq h_2$ and

$$m(i) = \frac{pe^{jw}}{2j[e^{jw} - 2he^{jw} + h^2]} e^{jwi} - \frac{pe^{-jw}}{2j[e^{-j2w} - 2he^{-jw}]} e^{-jwi} + c_3 ih^{i-1} + c_4 h^i$$

$$(10.3.19)$$

for $h = h_1 = h_2$. Note that no effort has been made to evaluate the coefficients of the terms involving h_1, h_2, and h since these terms vanish as i increases. Noting that $h_1 + h_2 = 2h = 1 - K_0$ and that $h_1 h_2 = h^2 = K_1$, we can express the steady-state output for both $h_1 \neq h_2$ and $h_1 = h_2$ as

$$m_i = \frac{pe^{jw}}{2j[e^{j2w} - (1 - K_0)e^{jw} + K_1]} e^{jwi}$$

$$- \frac{pe^{-jw}}{2j[e^{-j2w} - (1 - K_0)e^{-jw} + K_1]} e^{-jwi}. \quad (10.3.20)$$

Substitution of $\cos 2w + j \sin 2w$ and $\cos w + j \sin w$ for e^{j2w} and e^{jw}, respectively, in $p/[e^{j2w} - (1 - K_0)e^{jw} + K_1]$ permits writing it in polar form as $|m| e^{j\phi}$, where

$$|m| = p\{[\cos 2w - (1 - K_0)\cos w + K_1]^2 + [\sin 2w - (1 - K_0)\sin w]^2\}^{-1/2}$$

$$= p[4K_1 \cos^2 w - 2(1 - K_0)(1 + K_1)\cos w + (1 - K_0)^2 + (1 - K_1)^2]^{-1/2}$$

$$(10.3.21)$$

and

$$\phi = \tan^{-1} \frac{-[\sin 2w - (1 - K_0)\sin w]}{\cos 2w - (1 - K_0)\cos w + K_1}. \quad (10.3.22)$$

Similarly, $p/[e^{-j2w} - (1 - K_0)e^{-jw} + K_1]$ can be expressed as $|m| e^{-j\phi}$. Substitution into (10.3.20) yields

$$m_i = \frac{|m|}{2j}\{e^{j[w(i+1)+\phi]} - e^{-j[w(i+1)+\phi]}\} = |m| \sin[w(i+1) + \phi]. \quad (10.3.23)$$

For the general case of a perturbation sinusoid with phase angle θ, m_i must be expressed as

$$m_i = |m| \sin[w(i+1) + \theta + \phi].$$

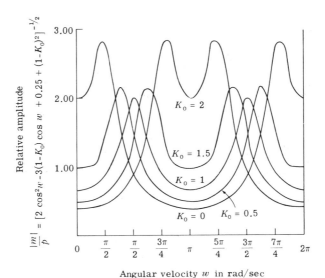

FIG. 10.3.1. Amplitude frequency response of generalized second-order process controller for $K_1 = 0.5$ and selected values of K_0.

The items of interest in systems analysis or design are of course the frequency-dependent relative amplitude

$$|m|/p = [4K_1 \cos^2 w - 2(1 - K_0)(1 + K_1) \cos w \\ + (1 - K_0)^2 + (1 - K_1)^2]^{-1/2} \qquad (10.3.24)$$

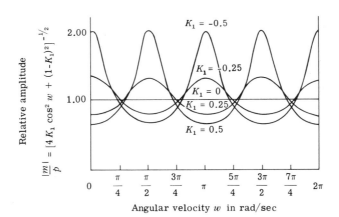

FIG. 10.3.2. Amplitude frequency response of generalized second-order process controller for $K_0 = 1.0$ and selected values of K_1.

and the phase shift angle ϕ. Figure 10.3.1 shows the amplitude frequency response based on (10.3.24) for $K_1 = 0.5$ and selected values of K_0. Figure 10.3.2 is the amplitude frequency response for $K_0 = 1$ and selected values of K_1. Together these figures provide a sampling over the stable region of K_0, K_1 combinations. The response curves are all symmetrical around $w = \pi$ rad. For $K_1 > 0$ and $|(1 - K_0)(1 + K_1)/4K_1| < 1$, each curve obtains a maximum value of $\{(1 - K_0)^2 + (1 - K_1)^2 - [(1 - K_0)^2(1 + K_1)^2/4K_1]\}^{-1/2}$ at the two frequencies for which $w = \cos^{-1}[(1 - K_0)(1 + K_1)/4K_1]$. Note that as K_0 deviates farther from one, the location of this maximum deviates farther from $\pi/2$ and the maximum value of $|m|/p$ increases. Each curve obtains local minima of $1/(K_0 + K_1)$ at $w = 0$ and of $1/(2 - K_0 + K_1)$ at $w = \pi$. Thus the global minimum is at $w = 0$ for $K_0 > 1$ and at π for $K_0 < 1$. For $K_1 < 0$, on the other hand, $1/(K_0 + K_1)$ and $1/(2 - K_0 + K_1)$ become local maxima and the two points at which the curve equals

$$\{(1 - K_0)^2 + (1 - K_1)^2 - [(1 - K_0)^2(1 - K_1)^2/4K_1]\}^{-1/2}$$

become the minima.

Response to Random Perturbations and Measurement Error

All the equations needed for complete transient and steady-state analysis of the response of the generalized second-order process controller to random perturbations and measurement errors were developed in Section 10.2. It was found that very little could be said quantitatively about the mean and variance of process output unless the random variables involved were stationary or at least adhered to some fairly simple known function of i. We will limit attention here to stationary random variables. We will further limit attention to unbiased random variables since, as previously discussed, biased variables can be considered as the sum of an unbiased random variable and a deterministic one. In the case of a stationary variable, the deterministic component is equivalent to a step function, the response to which has already been discussed. Any constant bias in measurement error can be handled by similar techniques. Finally, we will limit attention to steady-state conditions.

Mutually Independent, Identically Distributed Measurement Error

Consider first the situation in which the measurement errors are mutually independent as well as being identically distributed and unbiased and neglect any randomness in the perturbations $p(i)$. This set of conditions would be of interest where very little is known about the patterns of the perturbation to which our process might be subjected but must be ready to react when perturbations do occur. This is the situation dealt with in Sections 8.6 and 8.7 relative to the first-order process controller. In this situation, the steady-state mean M will be zero since the $\varepsilon(i)$ are assumed to be unbiased and the steady-state variance is

$$V = \frac{[(K_0^2 + K_1^2)(1 + K_1) + 2K_0 K_1(1 - K_0)]\sigma_\varepsilon^2}{(2K_0 - K_0^2 + 2K_1 + K_1^2)(1 - K_1)}. \qquad (10.3.25)$$

Equation (10.3.25) holds for both $h_1 \neq h_2$ and $h_1 = h_2$ and is obtained from either (10.2.56) or (10.2.57) by setting all variance and covariance terms to zero except those involving σ_ε^2. Note that for $K_1 = 0$, V reduces to

$$V = K_0^2 \sigma_\varepsilon^2/(2K_0 - K_0^2) = K_0 \sigma_\varepsilon^2/(2 - K_0), \qquad (10.3.26)$$

which is the same as (8.2.7) for the steady-state variance of the first-order controller.

One way to get a feel for how V varies with K_0 and K_1 would be simply to calculate values of V for various combinations of K_0 and K_1 which sufficiently cover the stable region. An alternative which we have adopted here is to construct equivariance contours for selected values of V. Actually, it is more convenient to work with the relative steady-state variance $\hat{V} = V/\sigma_\varepsilon^2$. Thus

$$\hat{V} = \frac{(K_0^2 + K_1^2)(1 + K_1) + 2K_0 K_1(1 - K_0)}{(2K_0 - K_0^2 + 2K_1 + K_1^2)(1 - K_1)}, \qquad (10.3.27)$$

which can be readily solved for K_0 in terms of V and K_1. This yields

$$K_0 = \frac{2\hat{V} - (\hat{V} + 1)K_1(1 + K_1)}{(\hat{V} + 1)(1 - K_1)}. \qquad (10.3.28)$$

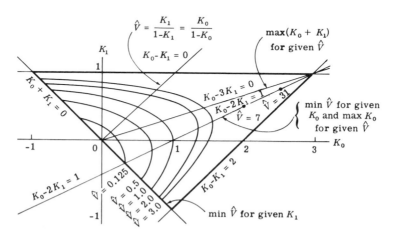

FIG. 10.3.3. Relative steady-state variance of generalized second-order process controller as a function of K_0 and K_1.

To compute the contour for a given $\hat{V}$ one first substitutes the $\hat{V}$ of interest in (10.3.28) and then calculates K_0 for sufficient values of K_1 to allow determination of the contour. Several such contours are plotted in Fig. 10.3.3. Only those portions of the curves lying in the stable region are shown since only in this region is $\hat{V}$ meaningful.

Note that each contour begins and ends on the boundary of the stable region given by $K_0 + K_1 = 0$, i.e., $K_0 = -K_1$. Since (10.3.27) is indeterminant at $K_0 = -K_1$, the limit of $\hat{V}$ as $K_0 \to K_1$ must be found using L'Hospital's rule. Thus

$$\hat{V}\Big|_{K_0 = -K_1} = \lim_{K_0 \to -K_1} \frac{(K_0{}^2 + K_1{}^2)(1 + K_1) + 2K_0 K_1(1 - K_0)}{(2K_0 - K_0{}^2 + 2K_1 + K_1{}^2)(1 - K_1)}$$

$$= \lim_{K_0 \to -K_1} \frac{2(1 - K_1)K_0 + 2K_1}{2(1 - K_1{}^2)} = \frac{K_1{}^2}{1 - K_1{}^2}, \qquad (10.3.29)$$

from which the intercepts of each contour with the boundary of the stable region are found to be

$$K_1 = \pm[\hat{V}/(\hat{V} + 1)]^{1/2}. \qquad (10.3.30)$$

Note that for $K_1 = 0$, $\hat{V} = 0$, a seemingly highly desirable result. However, under these conditions K_0 also equals zero, which means there is no control whatsoever, and we are completely at the mercy of whatever external perturbations may occur. As $\hat{V}$ increases the contours lie farther from the origin and approach the boundaries $K_0 - K_1 = 2$ and $K_1 = 1$ of the stable region as $\hat{V}$ continues to increase without bound. Therefore, we again see, as we did with the first-order controller, that increases in the control variables cause rapid increases in $\hat{V}$.

It is interesting to note that for any specific value of K_1, $\hat{V}$ is minimized for $K_0 = -K_1$. Thus the minimum value of $\hat{V}$ for a given K_1 can be obtained from (10.3.29). This can be verified by examination of $\partial \hat{V}/\partial K_0 = 0$. Since $K_0 = -K_1$ is the lower limit for K_0 in the stable region, we see that for any given K_1, the higher K_0 is made, the higher the resulting value of $\hat{V}$. Similarly, from $\partial \hat{V}/\partial K_1 = 0$, one finds that $\hat{V}$ is minimized for a given value of K_0 at

$$K_1 = -(1 - K_0)/2 \qquad (10.3.31)$$

which is shown in Fig. 10.3.3 by the line $K_0 - 2K_1 = 1$. The farther K_1 deviates from this line, the larger $\hat{V}$ becomes. It may also be shown that this minimum $\hat{V}$ for a given K_0 is

$$\hat{V}\Big|_{K_1 = -(1 - K_0)/2} = \frac{-3K_0{}^3 + 19K_0{}^2 - 9K_0 + 1}{3K_0{}^3 - 19K_0{}^2 + 33K_0 - 9}. \qquad (10.3.32)$$

There may be occasions when either K_0 or K_1 is specified from other considerations and, if so, the above relationships can be used to obtain the smallest steady-state variance possible under the circumstances.

In passing we note two additional items of interest. First, at $K_0 = K_1$, shown by the line $K_0 - K_1 = 0$ in Fig. 10.3.3,

$$\hat{V}\Big|_{K_0 = K_1} = K_1/(1 - K_1). \tag{10.3.33}$$

Second, the partial derivative of (10.3.28) with respect to K_1 indicates that the maximum value of K_0 for any given $\hat{V}$ occurs at

$$K_1 = 1 - [2/(\hat{V} + 1)]^{1/2} \tag{10.3.34}$$

and equals

$$K_0 = 3 - 2[2/(\hat{V} + 1)]^{1/2} = 1 + 2K_1, \tag{10.3.35}$$

which is the same as $K_0 - 2K_1 = 1$, the locus of values of K_1 for which $\hat{V}$ is minimized for a given K_0. These two relationships could be helpful in faring in additional contours if such should be needed.

Knowledge of the relationship between $\hat{V}$ and K_0 and K_1 can be extremely valuable in controller design. Consider the situation of a generalized second-order process controller with an unbiased, mutually independent, identically distributed measurement error with variance σ_ε^2 subject to a step perturbation of magnitude p. Suppose we wish to minimize the expected steady-state process output M while limiting the steady-state variance V to some specified value $V_{\max}$. In our earlier discussion of the steady-state output for a step perturbation, we found that $m = p/(K_0 + K_1)$. Since $\varepsilon(i)$ is unbiased,

$$M = m + 0 = p/(K_0 + K_1). \tag{10.3.36}$$

Thus, for a given p, M is minimized when $K_0 + K_1$ is maximized. This maximization is constrained by both stability and variance considerations. Recall, however, that (10.3.28) relates K_0 to $\hat{V} = V/\sigma_\varepsilon^2$ and K_1. Therefore,

$$K_0 + K_1 = \frac{2\hat{V}_{\max} - (\hat{V}_{\max} + 1)K_1(1 + K_1)}{(\hat{V}_{\max} + 1)(1 - K_1)} + K_1 = \frac{2[\hat{V}_{\max} - (\hat{V}_{\max} + 1)K_1^2]}{(\hat{V}_{\max} + 1)(1 - K_1)}. \tag{10.3.37}$$

Setting the first derivative of (10.3.37) to zero and solving for K_1 yields

$$K_1 = 1 - (\hat{V} + 1)^{-1/2}. \tag{10.3.38}$$

A second solution $K_1 = 1 + (\hat{V} + 1)^{-1/2} > 1$ for all $\hat{V}$, so does not correspond to a stable system. Substitution of (10.3.38) in (10.3.28) gives $K_0 = 3K_1$ (shown by the line $K_0 - 3K_1 = 0$ in Fig. 10.3.3), hence

$$(K_0 + K_1)_{\max}\Big|_{\hat{V}} = 4K_1 = 4[1 - (\hat{V} + 1)^{-1/2}]. \tag{10.3.39}$$

Therefore, for the situation cited,

$$K_1 = 1 - \left(\frac{V_{max}}{\sigma_\varepsilon^2} + 1\right)^{-1/2} = 1 - \left(\frac{\sigma_\varepsilon^2}{\sigma_\varepsilon^2 + V_{max}}\right)^{1/2}, \qquad (10.3.40)$$

$K_0 = 3K_1$, $(K_0 + K_1)_{max} = 4K_1$, and

$$M_{min} = p \bigg/ 4 \left[1 - \left(\frac{\sigma_\varepsilon^2}{\sigma_\varepsilon^2 + V_{max}}\right)^{1/2}\right]. \qquad (10.3.41)$$

Mutually Independent, Unbiased, Identically Distributed Perturbation and Measurement Error

The situation considered here is the same as in the previous subsection with the addition of a mutually independent, unbiased, stationary random perturbation with variance σ_p^2. If we also assume the perturbations to be statistically independent of the measurement error, the steady-state variance of process output is, from (10.2.56),

$$V = \frac{[(K_0^2 + K_1^2)(1 + K_1) + 2K_0 K_1(1 - K_0)]\sigma_\varepsilon^2 + (1 + K_1)\sigma_p^2}{(2K_0 - K_0^2 + 2K_1 + K_1^2)(1 - K_1)}. \qquad (10.3.42)$$

Detailed analysis of the relationships among V, K_0, K_1, σ_ε^2, and σ_p^2 is obviously a time-consuming business. However, for a given ratio

$$\bar{r} = \sigma_p^2/\sigma_\varepsilon^2, \qquad (10.3.43)$$

one can substitute $\bar{r}\sigma_\varepsilon^2$ for σ_p^2 and then divide through by σ_ε^2 to obtain the relative variance defined by

$$\hat{V} = \frac{V}{\sigma_\varepsilon^2} = \frac{(K_0^2 + K_1^2 + \bar{r})(1 + K_1) + 2K_0 K_1(1 - K_0)}{(2K_0 - K_0^2 + 2K_1 + K_1^2)(1 - K_1)}. \qquad (10.3.44)$$

Solving for K_0 yields the unpleasant result

$$K_0 =$$

$$\frac{\hat{V} - (\hat{V} + 1)K_1 \pm [(\hat{V} + 1)^2 K_1^4 + (\hat{V} + 1)(\bar{r} - 2\hat{V})K_1^2 + \hat{V}^2 - (\hat{V} + 1)\bar{r}]^{1/2}}{(1 - K_1)(\hat{V} + 1)},$$

$$(10.3.45)$$

meaning that except for fortuitous combinations of values of V, σ_ε^2, and σ_p^2 contour determination will definitely require computer assistance.

Finally, it is easily shown that if $\varepsilon(i)$ has a constant bias ε, a term equal to $-\varepsilon$ is introduced into the steady-state mean M. Note that this effect is independent of the control parameters K_0 and K_1.

Other Possibilities

Many possibilities exist for the types of random variables which could enter a control system. The effects of an intercorrelated random perturbation were illustrated relative to the first-order controller in Chapter VII. The same analysis approach developed there would apply here. It would also apply, with appropriate modification, to intercorrelated measurement errors.

Although significant general cross correlations between measurement error and perturbations is extremely unlikely, a special case in which measurement error is correlated with the concurrent value of the perturbation can arise. This would result when both the operation of the process and the measurement of its output are sensitive to the same exogenous factor. For example, ambient temperature or humidity could affect both the output of a chemical reaction and the instrumentation used to measure it. This would be especially true where instrument sensors are in close proximity to or even immersed in the material to be measured. Mechanical vibration could also affect both a cylinder wall and an on-line optical device used to determine its diameter, straightness, and roundness. Since we have defined $p(i)$ as the net effect of perturbations between time i and $i + 1$, it seems logical to think in terms of a relationship between $p(i - 1)$ and $\varepsilon(i)$, the measurement made at time i, although other possibilities certainly exist. In this case the cross correlation between $\varepsilon(i)$ and $p(i)$ is zero except for $\tau = 1$, so the variance expression will contain a term in $\sigma_{\varepsilon p}(1)$.

Another unlikely though possible source of random variation could be the signal representing the desired level of operation D. It is not that the actual desired level would be randomly varying, but where information concerning this level must be continually or periodically transmitted to the control system by an electrical signal or mechanical positioning device, noise can often be superimposed in the transmission process.

We have looked extensively at the generalized second-order process controller. Let us now consider a few specific configurations of this type of controller, specifically the proportional-plus-difference controller, the proportional-plus-summation controller, and the proportional controller with a one-period feedback delay.

10.4   Proportional-Plus-Difference Control

The system difference equation for the proportional-plus-difference rapid-response process controller was derived in Section 4.2. After incrementing the time argument to put the equation in standard form for solution, we obtained (see Eq. (4.2.55))

$$m(i + 2) - (1 - K - L)m(i + 1) - Lm(i)$$
$$= KD - (K + L)\varepsilon(i + 1) + L\varepsilon(i) + p(i + 1). \tag{10.4.1}$$

The solution of (10.4.1) is readily obtained from the solution of the difference equation for the generalized second-order process controller by substituting $(K + L)$ for K_0 and $-L$ for K_1. Thus, the roots of the characteristic equation are

$$h_1 = \frac{1 - K - L + [(1 - K - L)^2 + 4L]^{1/2}}{2} \tag{10.4.2}$$

and

$$h_2 = \frac{1 - K - L - [(1 - K - L)^2 + 4L]^{1/2}}{2}, \tag{10.4.3}$$

which are the defining equations for h_1 and h_2 throughout this section. From (10.2.21), for $h_1 \neq h_2$,

$$m(i) = \frac{1}{(h_1 - h_2)} \left\{ (-h_2 h_1{}^i + h_1 h_2{}^i) m(0) + (h_1{}^i - h_2{}^i) m(1) + L(h_1^{i-1} - h_2^{i-1}) \varepsilon(0) \right.$$

$$- \sum_{n=1}^{i-1} [(K + L)(h_1^{i-n} - h_2^{i-n}) - L(h_1^{i-1-n} - h_2^{i-1-n})] \varepsilon(n)$$

$$\left. + \sum_{n=1}^{i-1} (h_1^{i-n} - h_2^{i-n}) p(n) \right\}, \tag{10.4.4}$$

and from (10.2.22), for $h = h_1 = h_2$,

$$m(i) = -(i - 1) h^i m(0) + i h^{i-1} m(1) + L(i - 1) h^{i-2} \varepsilon(0)$$

$$- \sum_{n=1}^{i-1} [(K + L) h(i - n) - L(i - 1 - n)] h^{i-2-n} \varepsilon(n)$$

$$+ \sum_{n=1}^{i-1} (i - n) h^{i-1-n} p(n). \tag{10.4.5}$$

Comparable expressions based on (10.3.2) and (10.3.3) can be similarly derived.

From (10.2.7) through (10.2.9), the conditions for stability are:

$$K > 0, \tag{10.4.6}$$

$$K + 2L < 2, \tag{10.4.7}$$

and

$$-1 < L < 1. \tag{10.4.8}$$

This region of stability is indicated by the shaded area in Fig. 10.4.1. It may be observed that in Fig. 10.4.1 the boundary of the stable region $L = -1$ corresponds to the boundary $K_1 = 1$ in Fig. 10.3.3, $K + 2L = 2$ corresponds to $K_0 - K_1 = 2$, and the L axis corresponds to $K_0 + K_1 = 0$ since

$$K_0 + K_1 = K + L - L = K. \tag{10.4.9}$$

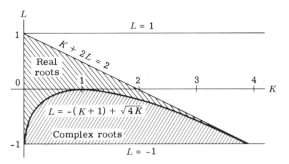

FIG. 10.4.1. Types of roots and region of stability of proportional-plus-difference rapid-response process controller as a function of proportionality constant K and difference constant L.

From (10.4.2) and (10.4.3) it is apparent that the case of the double real root $h = h_1 = h_2$ occurs for $(1 - K - L)^2 + 4L = 0$, which is possible only for $L \leq 0$. In the stable region, this is more conveniently expressed as

$$L = -(K + 1) + (4K)^{1/2} \tag{10.4.10}$$

or as

$$K = 1 - L \pm (-4L)^{1/2}, \qquad L \leq 0. \tag{10.4.11}$$

The double root h is thus given by either

$$h = 1 - (K)^{1/2} \tag{10.4.12}$$

or

$$h = \pm(-4L)^{1/2}, \qquad L \leq 0. \tag{10.4.13}$$

Note that when (10.4.13) is used, $h = +(-4L)^{1/2}$ corresponds to $K = 1 - L - (-4L)^{1/2}$ and $h = -(4L)^{1/2}$ corresponds to $K = 1 - L + (-4L)^{1/2}$.

For $L < -(K + 1) + (4K)^{1/2}$ or $1 - L - (-4L)^{1/2} < K < 1 - L + (-4L)^{1/2}$, h_1 and h_2 are complex. For $L > -(K + 1) + (4K)^{1/2}$ or K outside the range $1 - L \pm (-4L)^{1/2}$, two real unequal roots result. These regions are indicated in Fig. 10.4.1.

The performance of the proportional-plus-difference controller for any set of assumed operating conditions can be found by proper substitution of K and L in the appropriate expressions describing the operation of the generalized second-order controller. Nevertheless, it should prove informative to consider a few aspects of proportional-plus-difference control explicitly.

Steady-State Effects of Measurement Error

The steady-state variance of process output caused by the feedback of a stationary, mutually independent measurement error was given for the generalized second-order controller in (10.3.25). For the proportional-plus-difference controller this becomes

$$V = \frac{[K(1 + L) + 2L^2]\sigma_\varepsilon^2}{[2(1 - L) - K](1 + L)}. \tag{10.4.14}$$

Contours of the relative variance $\hat{V} = V/\sigma_\varepsilon^2$ based on (10.4.14) are shown in Fig. 10.4.2. These are most easily computed from

$$K = \frac{2[\hat{V} - L^2(\hat{V} + 1)]}{(\hat{V} + 1)(1 + L)}. \tag{10.4.15}$$

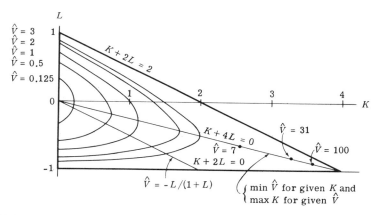

FIG. 10.4.2. Relative steady-state variance of proportional-plus-difference process controller as a function of K and L.

As expected from comparison with Fig. 10.3.3, the contours originate and terminate on the L axis at values of L given by (10.3.30). As $\hat{V}$ increases, they lie farther from the origin and approach the other two boundaries of the stable region as $\hat{V}$ increases without bound. The maximum value of K on each contour, found from differentiating (10.4.15) with respect to L, occurs at

$$L = -[1 - (\hat{V} + 1)^{-1/2}] \tag{10.4.16}$$

and equals

$$K = 4[1 - (\hat{V} + 1)^{-1/2}] = -4L. \tag{10.4.17}$$

This is also the locus of points corresponding to the minimum value of $\hat{V}$ for a given K, found by differentiating (10.4.14) with respect to L. Not unexpectedly, $\hat{V}$ is minimized for a given value of L at $K = 0$.

Transient Analysis

The effect of the proportional and difference parameters on the transient response following an impulse is perhaps best discussed in terms of combinations of K and L lying on the same $\hat{V}$ contour. For this analysis we will use as our point of departure the revised equations for $m(i)$, (10.3.2) and (10.3.3), and assume a unit impulse at time $i = 0$. Table 10.4.1 lists $m(1)$ through $m(8)$ for several K, L pairs on the $\hat{V} = 0.5$ and $\hat{V} = 1$ contours. Note that systems for which $L = 0$ are actually simple proportional controllers. Although the individual patterns are hard to compare, especially with the preponderance of combinations which result in case 3 operation where both magnitude, $(-L)^{(i-1)/2}$, and angle, $i\theta$, affect the results, a few facts seem clear. First, where speed of response following an impulse perturbation is the sole criterion of system performance, the first-order system with $K = 1$ is the best choice possible. If the speed of response is potentially limited by a maximum allowable $\hat{V} < 1$ such that $K = 1$ is not allowed, the first-order system with $K = 2\hat{V}/(\hat{V} + 1)$ is probably as good as any. Finally, if for any reason some difference control is required, positive values of L lead to quicker initial response and in general comparable responses for higher values of i than do negative values of L.

The points of intersection between the $\hat{V}$ contours and the L axis result in marginal stability since for $K = 0$, $h_1 = 1$ and $h_2 = -L$. The transient response is thus

$$m(i) = [1 - (-L)^i]/(1 + L) \tag{10.4.18}$$

which gives $m(2) = 1 - L$, $m(3) = (1 + L^3)/(1 + L)$, etc. For $L < 0$, there is an asymptotic buildup toward the steady-state value $1/(1 + L)$. For $L > 0$, the output first dips below $1/(1 + L)$ then alternates above and below $1/(1 + L)$. Since $K = 0$ corresponds to a pure differencing controller, it is apparent that such a controller should generally be avoided from the standpoint of impulse response.

In spite of the somewhat poor showing made by proportional-plus-difference control relative to simple proportional control in its speed of response to an impulse perturbation, it is still a useful mode of control in several types of situations. One of these is where smoothness or some other measure of the response pattern is of more importance than sheer speed, perhaps to avoid overextending control drives or linkages. Another might

Table 10.4.1

Unit-Impulse Response of Proportional-Plus-Difference Process Controller

ν	0.5						1.0				
K	0.333	0.667	0.732	0.667	0.333	0.111	0.333	1.000	1.172	1.000	0.333
L	−0.500	−0.333	−0.183	0.000	0.333	0.500	−0.667	−0.500	−0.293	0.000	0.500
i											
1	1.000	1.000	1.000	1.000	1.000	1.000	1.000	1.000	1.000	1.000	1.000
2	1.167	0.643	0.451	0.333	0.333	0.389	1.333	0.500	0.121	0.000	0.167
3	0.862	0.107	0.020	0.111	0.444	0.652	1.110	−0.250	−0.278	0.000	0.527
4	0.422	−0.138	−0.073	0.037	0.259	0.449	0.591	−0.376	−0.068	0.000	0.171
5	0.062	−0.122	−0.037	0.012	0.235	0.500	0.047	−0.062	0.073	0.000	0.292
6	−0.139	−0.036	−0.003	0.004	0.165	0.419	−0.332	0.157	0.029	0.000	0.134
7	−0.193	0.015	0.005	0.001	0.133	0.413	−0.473	0.109	−0.018	0.000	0.169
8	−0.155	0.022	0.003	0.000	0.099	0.380	−0.411	−0.023	−0.011	0.000	0.093
1/K	3.000	1.500	1.366	1.500	3.000	9.000	3.000	1.000	0.853	1.000	3.000

311

be where speed of response must be weighed against $\hat{V}$ and perhaps frequency response or step function response. Specifically, suppose $\hat{V}$ were limited to 0.5 and we wished to minimize the steady-state output for a step perturbation and still react reasonably promptly to an impulse. Since steady-state response to a step perturbation of magnitude p is $p/(K_0 + K_1)$, which from (10.4.9) is p/K, we might very well use the system with the maximum K commensurate with $\hat{V}$ not exceeding 0.5. From Table 10.4.1, this is $K = 0.732$, $L = -0.183$, the impulse response for which seems very close to that of $K = 0.667$, $L = 0$, at least for $i > 2$.

In the not too unheard of case where some knowledge is available concerning the patterns of perturbations which are likely to occur, including highly intercorrelated random effects, considerable benefits can be obtained by effective matching of the controller impulse response to the patterns anticipated. Consider the somewhat hypothetical example in which

$$p(i) = (-\tfrac{1}{2})^{i-1}. \tag{10.4.19}$$

This is a convergent geometric sequence of alternating sign, which could serve as an oversimplified model of a surge in line voltage, liquid in a Fourdrinier headbox, or perhaps patterns of customer orders for a new prestige product with reliability problems. The resulting process output is given by

$$m(i) = [1/(h_1 - h_2)] \sum_{n=0}^{i-1} (h_1^{i-n} - h_2^{i-n})(-\tfrac{1}{2})^{n-1}. \tag{10.4.20}$$

Table 10.4.2 lists $m(1)$ through $m(8)$ for six of the K, L combinations whose impulse responses are included in Table 10.4.1, three from both the $\hat{V} = 0.5$ and $\hat{V} = 1$ contours.

Table 10.4.2

Transient Response to $p(i) = (-\tfrac{1}{2})^{i-1}$ of Proportional-Plus-Difference Process Controller

$\hat{V}$		0.5			1.0	
K	0.667	0.732	0.667	1.000	1.172	1.000
i $\quad$ L	-0.333	-0.183	0.000	-0.500	-0.293	0.000
1	1.000	1.000	1.000	1.000	1.000	1.000
2	0.143	-0.049	-0.167	0.000	-0.379	-0.500
3	0.035	0.044	0.194	-0.250	-0.089	0.250
4	-0.156	-0.095	0.061	0.251	-0.024	-0.125
5	-0.051	0.012	0.042	0.063	0.085	0.063
6	-0.010	-0.008	-0.017	0.125	-0.013	-0.031
7	0.020	0.010	0.010	0.045	0.011	0.016
8	0.012	-0.003	-0.006	-0.047	-0.005	-0.008

Selection among the candidates shown requires specification of a performance criterion. If we simply wanted to get $|m(i)|$, the magnitude of $m(i)$ to stay below some threshold, say 0.010, as quickly as possible without regard for what happens prior to this time, either the $K = 0.732$, $L = -0.183$ or $K = 1$, $L = 0$ controllers would be selected since both pass this threshold at $i = 8$. In the more likely case, however, where an individual value of $m(i)$ outside of some set of tolerance limits, m_L and m_U, results in rejection of the product, we would pick the K, L values to minimize the total number of occurrences of $m(i) < m_L$ and $m(i) > m_U$. For tolerances of ± 0.100, $K = 0.732$, $L = -0.183$ is a clear winner. For tolerances of ± 0.016, both $K = 0.667$, $L = -0.333$ and $K = 0.732$, $L = -0.183$ result in $m(i)$ within tolerances for $i \geq 2$. In general, one would probably select $K = 0.732$, $L = -0.183$ since the actual values of $m(i)$ are smaller. This provides a greater margin in case other types of perturbations are also present. However, if for some reason changes in sign of $m(i)$ are undesirable, possibly because of increased wear on the adjustment mechanism or nonlinear factors such as gear backlash or hysteresis, the controller with $K = 0.667$, $L = -0.333$ might be chosen. Note the fortunate result that generally the K, L combinations from the $\hat{V} = 0.5$ contour result in smaller values of $m(i)$ than those from the $\hat{V} = 1$ contour. It also seems that with respect to most reasonable performance criteria, for the $p(i)$ pattern assumed here, the best K, L combination for a given $\hat{V}$ is the one involving the maximum value of K on the contour, which also results in minimum steady-state output following a step perturbation.

It is conceivable that a more thorough search of K, L combinations on a greater variety of $\hat{V}$ contours, possibly concentrating on combinations for which $K = -4L$ (the maximum K on a given contour), would yield even better responses than those shown above. Obviously a numerical search is required to find controllers with good transient responses to any perturbation pattern or combination of patterns.

Amplitude Frequency Response

Figures 10.3.1 and 10.3.2 show the increasing sensitivity of the amplitude of the steady-state output of the generalized second-order controller to the angular velocity of a sinusoidal perturbation as K_0 deviates farther from 1 and K_1 deviates farther from 0. It has been pointed out that where the angular velocity of the perturbation can be anticipated, advantage can be taken of the controller sensitivity by choosing control parameters for which the relative amplitude of the output is small for the w involved. The same is obviously true for the proportional-plus-difference controller. In situations where the angular frequency of potential sinusoidal perturbations is not known, however, one must consider many or even all possibilities. Thus, the maximum value of the relative amplitude is often of interest.

Table 10.4.3 lists the maximum value of $|m|/p$ and the value of w at which it occurs for each of the K, L combinations from Table 10.4.1. Also shown to provide a measure of consistency are the corresponding values of $(|m|/p)_{min}$

Table 10.4.3

Maximum and Minimum Relative Amplitudes for Proportional-Plus-Difference Controller

K	L	$\hat{V}$	$|m|_{max}$	w at $|m|_{max}$	$|m|_{min}$	w at $|m|_{min}$
0.333	−0.500	0.5	3.534	0.161π, 1.839π	0.375	π
0.667	−0.333	0.5	1.835	0.268π, 1.732π	0.500	π
0.732	−0.183	0.5	1.441	0.240π, 1.760π	0.612	π
0.667	0.000	0.5	1.500	0, 2π	0.750	π
0.333	0.333	0.5	3.000	0, 2π	0.720	0.553π, 1.447π
0.111	0.500	0.5	9.000	0, 2π	0.648	0.531π, 1.469π
0.333	−0.667	1.0	5.263	0.187π, 1.813π	0.333	π
1.000	−0.500	1.0	2.137	0.378π, 1.622π	0.500	π
1.172	−0.293	1.0	1.422	0.458π, 1.542π	0.707	π
1.000	0.000	1.0	1.000	all w	1.000	all w
0.333	0.500	1.0	3.000	0, 2π	0.664	0.513π, 1.487π

and the values of w at which they occur. We note that in general as K deviates from 1 and L from 0 the difference between $(|m|/p)_{max}$ and $(|m|/p)_{min}$ increases. Also, as shown earlier, $|m|/p = 1$ for all w when $K = 1$ and $L = 0$, which would make this the preferred controller if restrictions on $\hat{V}$ and other factors permit. Were $\hat{V}$ limited to 0.5, the smallest $(|m|/p)_{max}$ occurs for $K = 0.732$, $L = −0.183$. As expected, extrema occurring at $w = 0$ or 2π equal $1/K$; those at $w = \pi$ equal $1/(2 − K − 2L)$. Thus $K = 0$ and $K = 2(1 − L)$ are not good design values from a general frequency response standpoint.

Selection of K and L

Selection of K and L for a proportional-plus-difference controller that is in some way optimal is obviously not an easy task, even with clearly defined performance criteria and specifically known or assumed patterns of perturbations. Except in very simple cases, such as minimizing the steady-state mean following a step perturbation given some constraint on $\hat{V}$, some sort of numerical analysis is required. In general, transient responses are more difficult to analyze, and hence to design for, than steady-state performance.

The designer must usually, therefore, consider the general type or configuration needed to work within the imposed environment, try to get reasonable values for the control variables, and build in some capability for additional tuning during laboratory tests and even on the job. Simulations can be very helpful in selection of the so-called reasonable values of control parameters.

10.5 Proportional-Plus-Summation Control

The system difference equation for the proportional-plus-summation rapid-response process controller can be obtained by setting L to zero in (4.2.59), the equation for the proportional-plus-difference-plus-summation controller. This gives

$$
\begin{aligned}
m(i+2) &- (2 - K - N)m(i+1) + (1 - K)m(i) \\
&= Nm(0) + N\varepsilon(0) + ND - (K + N)\varepsilon(i+1) + K\varepsilon(i) + p(i+1) - p(i).
\end{aligned}
$$

$$(10.5.1)$$

Recall that (4.2.59) was obtained by taking the first forward difference of the original system equation which contained summation terms. As a result, (10.5.1) is not a special case of the generalized second-order process controller because of the different forcing function format. This is readily apparent from comparing (10.5.1) with (10.2.3). The roots of the characteristic equation are easily found to be

$$
h_1 = \frac{2 - K - N + [(2 - K - N)^2 - 4(1 - K)]^{1/2}}{2}
$$

$$(10.5.2)$$

and

$$
h_2 = \frac{2 - K - N - [(2 - K - N)^2 - 4(1 - K)]^{1/2}}{2},
$$

$$(10.5.3)$$

and the solution of (10.5.1) for $D = 0$ in terms of these roots can be derived by either the variation of parameters or the z transform as

$$
\begin{aligned}
m(i) = \frac{1}{(h_1 - h_2)} \Big\{ &[(h_1 - 1)h_1{}^i - (h_2 - 1)h_2{}^i]m(0) + K(h_1^{i-1} - h_2^{i-1})\varepsilon(0) \\
&+ \sum_{n=1}^{i-1} [\{-(K + N)h_1 + K\}h_1^{i-1-n} \\
&- \{-(K + N)h_2 + K\}h_2^{i-1-n}]\varepsilon(n) - (h_1^{i-1} - h_2^{i-1})p(0) \\
&+ \sum_{n=1}^{i-1} [(h_1 - 1)h_1^{i-1-n} - (h_2 - 1)h_2^{i-1-n}]p(n) \Big\}
\end{aligned}
$$

$$(10.5.4)$$

for $h_1 \neq h_2$. The solution for $h = h_1 = h_2$ is left as an exercise for the reader.

The Jury stability criteria yield the following conditions for the proportional-plus-summation controller to be stable:

$$N > 0, \tag{10.5.5}$$

$$2K + N < 4, \tag{10.5.6}$$

$$0 < K < 2. \tag{10.5.7}$$

The resulting region of stability is indicated by the shaded area in Fig. 10.5.1.

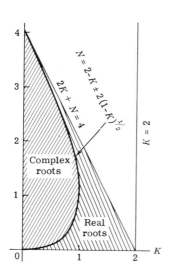

FIG. 10.5.1. Types of roots and region of stability of proportional-plus-summation controller as a function of proportionality constant K and summation constant N.

From (10.5.2) and (10.5.3), the case of the double root $h = h_1 = h_2$ occurs for $(2 - K - N)^2 - 4(1 - K) = 0$, which can be expressed for convenience of computation as

$$N = 2 - K \pm 2(1 - K)^{1/2}. \tag{10.5.8}$$

h is given most conveniently as

$$h = (2 - K - N)/2 = \pm (1 - K)^{1/2}, \tag{10.5.9}$$

the plus sign corresponding to $N = 2 - K - 2(1 - K)^{1/2}$ and the minus sign to $N = 2 - K + 2(1 - K)^{1/2}$. The areas within the stable region corresponding to real and complex roots are indicated in Fig. 10.5.1.

Although the procedures for analysis of system performance are the same as those for any second-order control system, it should still prove interesting to mention briefly a few features of the proportional-plus-summation controller.

Steady-State Effects of Measurement Error

The steady-state variance of process output caused by the feedback of a stationary, mutually independent measurement error is

$$V = \frac{[2K^2 + KN + 2N]\sigma_\varepsilon^2}{(4 - 2K - N)K}. \tag{10.5.10}$$

Remember that this expression cannot be obtained by substitution of functions of K and N for K_0 and K_1 in (10.3.27) because of the different relationship existing between the coefficients of dependent variables and those of the forcing function terms. One could now compute contours of $\hat{V}$ or in some manner explore the relationship among $\hat{V}$, K, and N as has been done for the generalized second-order controller and the proportional-plus-difference controller.

It is interesting to note that for $N = 0$,

$$V = \frac{2K^2\sigma_\varepsilon^2}{(4 - 2K)K} = \frac{K\sigma_\varepsilon^2}{2 - K}, \tag{10.5.11}$$

the expression for steady-state variance of the simple proportional controller. On the other hand, V becomes unbounded for $K = 0$ regardless of the value of N, indicating the inappropriateness of pure summation control.

General Steady-State Response

The proportional-plus-summation controller is extremely valuable because it can achieve a zero steady-state response to a step perturbation. This is readily apparent from (10.5.1) where the only function of the perturbation which appears is its first forward difference which must equal zero for a step. Thus not only is the steady-state response equal to zero, but so is the transient response. The zero steady-state deviation can also be shown by applying the final-value theorem to the z transform of $M(z)$. To facilitate discussion, let D, $m(0)$, and all $\varepsilon(i)$ be zero. Starting with

$$\Delta m(i) + Km(i) + N \sum_{n=0}^{i} m(n) = p(i), \tag{10.5.12}$$

the z transform becomes

$$(z - 1)M(z) + KM(z) + N \frac{z}{z - 1} M(z) = P(z). \tag{10.5.13}$$

The coefficient of $M(z)$ is

$$z - 1 + K + N \frac{z}{z - 1} = \frac{z^2 - (2 - K - N)z + (1 - K)}{z - 1} = \frac{1}{G(z)}. \quad (10.5.14)$$

Thus

$$Z[m(i)] = G(z)P(z) = \frac{(z - 1)P(z)}{z^2 - (2 - K - N)z + (1 + K)}. \quad (10.5.15)$$

Steady-state output for $P(z)$ such that $Z[m(i)]$ meets the conditions specified for the final-value theorem in Section 5.2 can be found from

$$m = \lim_{i \to \infty} m(i) = \lim_{z \to 1} (z - 1)Z[m(i)] = \lim_{z \to 1} \frac{(z - 1)^2 P(z)}{z^2 - (2 - K - N)z + (1 - K)}.$$

$$(10.5.16)$$

For $p(i)$ an impulse of finite magnitude, m is clearly zero. The only way it could be otherwise is for the characteristic equation to have a double root of one, which is impossible for a stable system. For a step of magnitude p, $P(z) = pz/(z - 1)$, and $(z - 1)Z[m(i)]$ becomes $pz(z - 1)/[z^2 - (2 - K - N)z + (1 - K)]$, so the limit as $z \to 1$ is indeed zero. For $p(i) = ip$, a ramp function which could be caused by accelerating orders for a popular product, $P(z) = pz/(z - 1)^2$ and $(z - 1)M(z)$ becomes $pz/[z^2 - (2 - K - N)z + (1 - K)]$. The limit as $z \to 1$ is p/N, a finite limit which decreases as N increases. Note the similarity in the roles of N in the proportional-plus-summation controller and K in the proportional-plus-difference controller. Also recall that the generalized second-order controller cannot follow a ramp perturbation. The deviation increases without bound. The $(z - 1)$ term in the numerator of $G(z)$ is what permits the proportional-plus-summation controller to maintain a zero steady-state deviation for a step and a finite deviation for a ramp. It follows that to obtain a zero steady-state deviation for a ramp and a finite one for a perturbation of the form $i^2 p$, a double summation must be built into the control rule. To generalize, to maintain a zero steady-state deviation for a perturbation of the form $i^k p$ and a finite steady-state deviation for one of the form $i^{k+1} p$, the controller must contain $k + 1$ orders of summation.

Amplitude Frequency Response

For a sinusoidal perturbation $p(i) = p \sin wi$, $P(z) = z \sin w/(z - e^{jw})$ $\times (z - e^{-jw})$. Substituting in (10.5.15), we obtain

$$Z[m(i)] = \frac{(z - 1)zp \sin w}{(z - e^{jw})(z - e^{-jw})(z - h_1)(z - h_2)}, \quad (10.5.17)$$

where $(z - h_1)(z - h_2)$ represents the denominator of $G(z)$ in factored form. Comparison with (10.3.16) shows an identical form except the $(z - 1)$ in (10.5.17) replaces a z in (10.3.16). Thus, one readily obtains by the same procedures

$$m_i = \frac{|m|}{2j} \{(e^{jw} - 1)e^{jwi}e^{j\phi} - (e^{-jw} - 1)e^{-jwi}e^{-j\phi}\}$$

$$= |m|\{\sin[w(i + 1) + \phi] - \sin[wi + \phi]\}, \qquad (10.5.18)$$

where

$$|m| = p\{4h_1h_2 \cos^2 w - 2(h_1 + h_2)(1 + h_1h_2) \cos w$$
$$+ [(h_1 + h_2)^2 + (1 - h_1h_2)^2]\}^{-1/2}$$
$$= p\{4(1 - K) \cos^2 w - 2(2 - K - N)(2 - K) \cos w$$
$$+ [(2 - K - N)^2 + K^2]\}^{-1/2} \qquad (10.5.19)$$

and

$$\phi = \tan^{-1} \frac{-[\sin 2w - (h_1 + h_2) \sin w]}{\cos 2w - (h_1 + h_2) \cos w + h_1h_2}$$

$$= \tan^{-1} \frac{-[\sin 2w - (2 - K - N) \sin w]}{\cos 2w - (2 - K - N) \cos w + (1 - K)}. \qquad (10.5.20)$$

However,

$$\sin[w(i + 1) + \phi] - \sin[wi + \phi] = 2 \cos\left(wi + \frac{w}{2} + \phi\right) \sin \frac{w}{2}$$

$$= 2 \sin \frac{w}{2} \sin\left(wi + \frac{w}{2} + \frac{\pi}{2} + \phi\right), \qquad (10.5.21)$$

so that in essence the magnitude of the output sinusoid is $2|\sin(w/2)| |m|$. Plots of relative magnitude $2|\sin(w/2)| |m|/p$ vs w constitute the amplitude frequency response for this controller. The phase frequency response is complicated somewhat by the fact that $\sin(w/2)$ changes sign every 2π rad, although it remains positive for w between zero and 2π.

10.6 Proportional Control with One-Period Delay

Delays of various sorts are very common in systems of all kinds. The most common sources of delay in process control are in obtaining the output to be observed, in performing measurements, in feeding measured values back to the controller, and in implementing control action. The first of these sources is often due to what is called "transportation lag." A striking example of transportation lag is in the manufacture of paper on a Fourdrinier machine where the time required for a point in the continuous paper sheet to travel from the headbox to the windup reel is in the order of three to five minutes. Since many of the quality characteristics of the paper must be measured near the windup reel while many of the adjustments which must be made to correct deviations from specifications affect the process near the headbox, what is being measured at any instant was caused by factors which had been in effect several minutes before. This can be equivalent to as many as six scanning periods of the density or moisture gage. When items must be removed from a line for laboratory testing, severe delays can result. Oven-dry tests of moisture in paper, which were used before the advent of the moisture gage, sometimes took up to eight hours to perform, a fact which essentially elimi-nated any possibility of on-line control. Fortunately, not all delays are quite so severe. Yet, as we shall see both here and in the next chapter, even relatively short delays can have severely limiting effects on our ability to control a process.

Consider a proportional controller with proportionality constant K with a one-period delay in feeding back each measurement of process output. The measurement upon which control decision $c(i)$ is based is the one involving $m(i-1)$. Thus, the system difference equation is

$$\Delta m(i) = K\rho(i-1) + p(i) = KD - Km(i-1) - K\varepsilon(i-1) + p(i). \quad (10.6.1)$$

With $D = 0$ and time arguments incremented by one, this becomes

$$m(i+2) - m(i+1) + Km(i) = -K\varepsilon(i) + p(i+1), \quad (10.6.2)$$

which is a special case of the generalized second-order process controller with $K_0 = 0$ and $K_1 = K$. From our studies of the generalized controller we realize that we have a relatively difficult control problem on our hands.

First we see from Fig. 10.2.1 that for stable operation, K (see K_1 in the figure) is constrained to the range $0 < K < 1$. The one impulse response shown in Table 10.3.1 with $K_0 = 0$ shows a relatively poor speed of response for $K = K_1 = 0.16$. Other values of K will unfortunately give similar results. Steady-state response to a step perturbation cannot be too good since for $K < 1$, m will be greater than the size of the step. We also see from Fig. 10.3.1 that the amplitude frequency response of a controller with $K_0 = 0$ and $K_1 = 0.5$ is highly frequency sensitive. This could be an advantage if all forcing sinusoids had angular frequencies near π, but a possible disaster if w were near zero or 2π.

The effects of the one-period delay on the steady-state variance given stationary, mutually independent measurement errors can also be severe. For the first-order rapid-response controller with parameter K, $\hat{V}$ has been shown to be $K/(2 - K)$. $\hat{V}$ for the proportional controller with the one-period delay is found by letting $K_0 = 0$ and $K_1 = K$ in (10.3.27) to be

$$\hat{V} = \frac{K(1 + K)}{(2 + K)(1 - K)} \qquad \text{(one-period delay).} \qquad (10.6.3)$$

The values of V for the controller both with and without the delay are plotted against K in Fig. 10.6.1. The effects of the delay increase rapidly as K increases but are significant even for small K. At $K = 0.1$, $\hat{V} = 0.058$ for the delay system and 0.053 for the system without delay, almost identical. However, at $K = 0.3$ the $\hat{V}$ values are 0.242 and 0.176, a ratio of 1.375; and at $K = 0.6$ the values are 0.923 and 0.429, a ratio of over two. The ratio increases without bound as K approaches one.

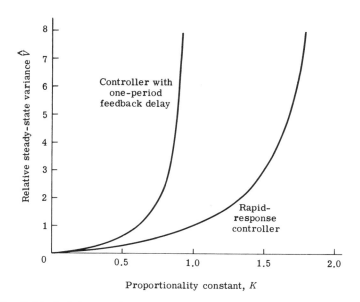

FIG. 10.6.1. Relative steady-state variance vs proportionality constant for rapid-response and one-period delay proportional controllers.

In summary it must be concluded that the one-period delay can cause severe limitations in system performance. In the next chapter we will see how even greater delays affect both proportional and other types of controllers.

Exercises

10.1 For the generalized second-order process controller whose output for $h_1 \neq h_2$ is given by (10.2.21), derive expressions for the expected value $E[m(i)]$ and variance $V[m(i)]$ of $m(i)$ for the following situations:

a. $p(i) = 0$ and $\varepsilon(i)$ identically and independently distributed with mean zero and variance σ_ε^2 for all i.

b. The same as (a) except $E[\varepsilon(i)] = \varepsilon$.

c. $p(i)$ and $\varepsilon(i)$ identically and independently distributed with means of zero and variances σ_p^2 and σ_ε^2, respectively, for all i. $p(i)$ and $\varepsilon(i)$ mutually independent.

d. The same as (c) except $p(i)$ and $\varepsilon(i)$ cross correlated with cross covariance function $\sigma_{\varepsilon p}(\tau)$ equal to c for $\tau = 1$ and 0 for $\tau \neq 1$.

e. $p(i)$ and $\varepsilon(i)$ identically distributed with means of zero, variances σ_p^2 and σ_ε^2, respectively, and autocovariance functions $\sigma_p(\tau) = A^{-|\tau|}\sigma_p^2$ and $\sigma_\varepsilon(\tau) = B^{-|\tau|}\sigma_\varepsilon^2$ for all i, where $|A| > 1$ and $|B| > 1$. $p(i)$ and $\varepsilon(i)$ mutually independent.

f. The same as (e) except $p(i)$ and $\varepsilon(i)$ cross correlated with cross covariance function $\sigma_{\varepsilon p} = C^{-|\tau|}\sigma_{\varepsilon p}$ for all i, where $|C| > 1$ and $\sigma_{\varepsilon p}$ is a constant.

10.2 **(a)** through **(f)**.   The same as the corresponding part of Exercise 10.1 except for case 2 operations, i.e., for $h = h_1 = h_2$

10.3 **(a)** through **(f)**.   The same as the corresponding part of Exercise 10.1, i.e., for $h_1 \neq h_2$, except the expressions to be derived are for the steady-state mean and variance of process output, M and V, respectively. Where appropriate, the expressions for V may be compared with (10.2.56).

10.4 **(a)** through **(f)**.   The same as the corresponding part of Exercise 10.2, i.e., for $h = h_1 = h_2$, except the expressions to be derived are for the steady-state mean and variance of process output, M and V, respectively. Where appropriate, the expressions for V may be compared with (10.2.57).

10.5

a. For the situation in Exercise 10.3(e), over what combinations of K_0 and K_1 will the steady-state variance V be greater than it would have been had $p(i)$ and $\varepsilon(i)$ been independently distributed?

b. The same as part (a) except for the situation in 10.4(e).

c. The same as part (a) except for the situation in 10.3(f).

d. The same as part (a) except for the situation in 10.4(f).

10.6 For the generalized second-order process controller with $D = m(0) = \varepsilon(i) = 0$, find the number of adjustments n required to reduce an impulse perturbation $p(i)$ to a fraction no greater than 10% of its original value for each of the following values of K_0 and K_1:

a.	$K_0 = 0.30, K_1 = 0.10$	**b.**	$K_0 = 1.70, K_1 = 0.10$
c.	$K_0 = 0.40, K_1 = 0.09$	**d.**	$K_0 = 1.60, K_1 = 0.09$
e.	$K_0 = 0.5, K_1 = 0.07$	**f.**	$K_0 = 1.5, K_1 = 0.07$

10.7 **(a)** through **(f)**. The same as 10.6(a) through 10.6(f), respectively, except considering the impulse in D instead of $p(i)$.

10.8 **(a)** through **(f)**. For the generalized second-order process controller, plot the output $m(i)$ for $i = 1$ through 8 given $m(0) = D = \varepsilon(i) = 0$ and a unit step perturbation occurring at $i = 0$ for the values of K_0 and K_1 in the corresponding part of Exercise 10.6. Also show the steady-state value of $m(i)$.

10.9 Derive an expression for the output $m(i)$ of a generalized second-order process controller for which $m(0) = D = \varepsilon(i) = 0$, $p(i)$ is a unit impulse at $i = 0$, and $K_0 = 1$, as a function of K_1 and i.

10.10 The cost of operation of a generalized second-order process controller depends primarily upon the deviation of the desired and actual output and the cost of making an adjustment. Thus, the operating cost for period i can be modeled as $C(i, K_0, K_1) = C_D[D - m(i)] + C_C[a(i)]$, where $C(i, K_0, K_1)$ is the operating cost for period i given control parameters K_0 and K_1, $C_D[\cdot]$ is the functional relationship between the deviation of process output $m(i)$ from D and the resulting cost, and $C_C[\cdot]$ is the functional relationship between the adjustment made and the resulting cost. For $m(0) = D = \varepsilon(i) = 0$ and $p(i)$ a step of size p at $i = 0$, find the values of the parameters K_0 and K_1 which will minimize the steady-state cost per period for the following cost functions:

a.	$C(i, K_0, K_1) = C_1[m(i)]^2 + C_2[a(i)]^2,$	C_1, C_2 constants
b.	$C(i, K_0, K_1) = C_1[m(i)]^2 + C_2\lvert a(i)\rvert,$	C_1, C_2 constants
c.	$C(i, K_0, K_1) = C_1\lvert m(i)\rvert + C_2\lvert a(i)\rvert,$	C_1, C_2 constants

10.11 **(a)** through **(c)**. The same as 10.10 except instead of minimizing the steady-state cost per period we seek to minimize the total cost for the first three periods.

10.12 Derive an expression for the output $m(i)$ of a generalized second-order process controller for which $m(0) = D = \varepsilon(i) = 0$, $p(i)$ is a unit step at $i = 0$, and $K_0 = 1$, as a function of K_1 and i.

10.13 Show that the amplitude frequency response of a generalized second-order process controller:

a. Obtains a maximum value of

$$\{(1 - K_0)^2 + (1 - K_1)^2 - [(1 - K_0)^2(1 + K_1)^2/4K_1]\}^{-1/2}$$

at the two frequencies for which $w = \cos^{-1}[(1 - K_0)(1 + K_1)/4K_1]$ for $K_1 > 0$ and $|(1 - K_0)(1 + K_1)/4K_1| < 1$. (Where do the maxima occur when $|(1 - K_0)(1 + K_1)/4K_1| > 1$?)

b. Obtains local minima of $1/(K_0 + K_1)$ at $w = 0$ and of $1/(2 - K_0 + K_1)$ at $w = \pi$ when $K_1 > 0$.

c. Obtains local maxima of $1/(K_0 + K_1)$ at $w = 0$ and of $1/(2 - K_0 + K_1)$ at $w = \pi$ when $K_1 < 0$.

d. Obtains a minimum value of

$$\{(1 - K_0)^2 + (1 - K_1)^2 - [(1 - K_0)^2(1 + K_1)^2/4K_1]\}^{-1/2}$$

at the two frequencies for which $w = \cos^{-1}[(1 - K_0)(1 + K_1)/4K_1]$ for $K_1 < 0$ and $|(1 - K_0)(1 + K_1)/4K_1| < 1$. (Where do the minima occur when $|(1 - K_0)(1 + K_1)/4K_1| > 1$?)

10.14 Extend the range of Fig. 10.3.1, the amplitude frequency response for a generalized second-order process controller with $K_1 = 0.5$, by deriving the response curves for $K_0 = -0.3$ and $K_0 = 2.3$. Show specifically the maxima and minima of each curve and the value of each curve at $w = 0$ and $w = \pi$.

10.15 A generalized second-order process controller is subject to a sinusoidal perturbation with amplitude p and angular frequency w radians per time period. A cost $C_m[m]$ is incurred when the magnitude of the steady-state sinusoid is m and a cost $C_C[K_0, K_1]$ results when the control parameters have values of K_0 and K_1. Determine the necessary and sufficient conditions for the cost $C(K_0, K_1) = C_m[m] + C_C[K_0, K_1]$ to be minimized for the following cost functions. Where possible, analytically determine an expression for the optimum values of K_0 and K_1 which minimize $C(K_0, K_1)$. Where analytic solution for K_0 and K_1 is not possible, indicate a procedure to follow to determine them for given values of w and p.

a. $C(K_0, K_1) = C_1 m^2 + C_2 K_0^2 + C_3 K_1^2$, C_1, C_2, C_3 constants
b. $C(K_0, K_1) = C_1 m^2 + C_2(K_0 + K_1)^2$, C_1, C_2 constants
c. $C(K_0, K_1) = C_1 m^2 + C_2(|K_0| + |K_1|)^2$, C_1, C_2 constants

10.16 (a) through (c). Find the optimal values of K_0 and K_1 for the corresponding part of 10.15 for $w = \pi/4$ rad and $p = 1$.

10.17 Plot the phase frequency response for the generalized second-order process controller for $K_1 = 0.5$ and the following values of K_0:

a. −0.3   **b.** 0   **c.** 0.5   **d.** 1.0
e. 1.5   **f.** 2.0   **g.** 2.3

10.18 Plot the phase frequency response for the generalized second-order process controller for $K_0 = 1.0$ and the following values of K_1:

a. −0.5   **b.** −0.25   **c.** 0   **d.** 0.25   **e.** 0.50

10.19 For the generalized second-order process controller with $m(0) = D = \varepsilon(i) = 0$, $p(i) = p\sin(2\pi/3 + \theta)$, $K_0 = 0.30$, and $K_1 = 0.10$, find the value of θ which minimizes the following steady-state cost criteria (n is a large positive integer):

a. $C(m_i) = \displaystyle\sum_{i=3n}^{3n+2} |m_i|$ **b.** $C(m_i) = \displaystyle\sum_{i=3n}^{3n+2} m_i^2$ **c.** $C(m_i) = \displaystyle\sum_{i=3n}^{3n+2} |m_i|^3$

10.20 (a) through (c). The same as 10.19 except for $K_0 = 0.40$ and $K_1 = 0.09$.

10.21 For a generalized second-order process controller with $m(0) = D = p(i) = 0$ and $\varepsilon(i)$ a stationary, unbiased, independently distributed random variable with variance $\sigma_\varepsilon^2 = 4 \times 10^{-6}$ in.^{2}, find the set of values of K_0 and K_1 within the stable range which result in a steady-state variance of the process output equal to

a. 1×10^{-6} in.^{2} **b.** 20×10^{-6} in.^{2}

10.22 For a generalized second-order process controller with $m(0) = D = \varepsilon(i) = 0$ and $p(i)$ a stationary, unbiased, independently distributed random variable with variance 4×10^{-6} in.^{2}, find the set of values of K_0 and K_1 within the stable range which result in a steady-state variance of process output equal to

a. 2×10^{-6} in.^{2} **b.** 4×10^{-6} in.^{2} **c.** 8×10^{-6} in.^{2}

10.23 For a generalized second-order process controller with $m(0) = D = 0$ and $p(i)$ and $\varepsilon(i)$ stationary, unbiased, independently distributed random variables with variances $\sigma_p^2 = 4 \times 10^{-6}$ in.^{2} and $\sigma_\varepsilon^2 = 2 \times 10^{-6}$ in.^{2}, respectively, $p(i)$ and $\varepsilon(i)$ mutually independent, find the set of values of K_0 and K_1 within the stable range which result in a steady-state variance of process output equal to

a. 1×10^{-6} in.^{2} **b.** 2×10^{-6} in.^{2} **c.** 4×10^{-6} in.^{2}
d. 8×10^{-6} in.^{2}

10.24 For a generalized second-order process controller with $m(0) = D = p(i) = 0$ and $\varepsilon(i)$ a stationary, unbiased, random variable with variance $\sigma_\varepsilon^2 = 4 \times 10^{-6}$ in.^{2} and autocovariance function $\sigma_\varepsilon(\tau) = 2^{-|\tau|}\sigma_\varepsilon^2$, find the set of values of K_0 and K_1 within the stable range which result in a steady-state variance of process output equal to

a. 2×10^{-6} in.^{2} **b.** 4×10^{-6} in.^{2} **c.** 8×10^{-6} in.^{2}
d. 20×10^{-6} in.^{2}

10.25 **(a)** through **(m)**.   The same as the corresponding part of Exercise 8.6 except based on the generalized second-order process controller with $m(0) = D = p(i) = 0$.

10.26 **(a)** through **(m)**.   The same as the corresponding part of Exercise 8.7 except based on the generalized second-order process controller with $m(0) = D = p(i) = 0$.

10.27
    a.   For the situation in Exercise 10.3(e), find the values of K_0 and K_1 for which the steady-state variance V is minimized.
    b.   The same as part (a) except for the situation in 10.4(e)
    c.   The same as part (a) except for the situation in 10.3(f)
    d.   The same as part (a) except for the situation in 10.4(f)

10.28 The ideal operating temperature for a chemical reactor is 500°F. Company policy requires that actual temperature must fall between specified tolerance limits at least 95% of the time. Temperatures are read at evenly spaced time intervals and adjustments made by a second-order process controller with control parameters K_0 and K_1. If the sensing device has an unbiased, stationary, independently and normally distributed instrumentation error with variance 225 deg^2, and the effect of an impulse perturbation is measured by $\sum_{i=1}^{5} \{E[m(i)]\}^2$ (assuming the impulse at $i = 0$), what values should be selected for K_0 and K_1 to minimize the effects of impulse perturbations while adhering to company policy given the following tolerances:

    a.   460°F to 540°F   **b.**   470°F to 530°F   **c.**   480°F to 520°F

10.29 **(a)** through **(c)**.   The same as 10.28 except the effect of the impulse perturbation is measured by $\sum_{i=1}^{5} |E[m(i)]|$, assuming the impulse occurred at $i = 0$.

10.30 A generalized second-order controller with parameters K_0 and K_1 is used to control the length of spacer pins at 3.000 cm. The standard deviation of these lengths is constant at 0.005 cm. Each adjustment is based on the average of a sample of five pins. Assuming instrumentation error is negligible, find the values of K_0 and K_1 which will limit the fraction of defective pins when $m(i) = D = 3.000$ cm to no more than 2%, yet will minimize

$\sum_{i=1}^{5} \{E[m(i)]\}^2$ should a shift in process mean occur at $i = 0$ given the following tolerances on individual pins:

a. 2.980 to 3.020 cm   **b.** 2.988 to 3.012 cm   **c.** 2.989 to 3.011 cm

10.31 **(a)** through **(c).** The same as 10.30 except we wish to minimize $\sum_{i=1}^{5} |E[m(i)]|$ for an impulse at $i = 0$.

10.32 **(a)** through **(c).** Assume the chemical reactor in Exercise 10.28 is also subject to a random, unbiased, stationary, independently and normally distributed perturbation $p(i)$ with variance $\sigma_p^2 = 100$ deg², while all other conditions remain as in 10.28. $p(i)$ and $\varepsilon(i)$ are mutually independent. Solve the corresponding part of 10.28 given this modified set of conditions.

10.33 **(a)** through **(c).** The same as 10.32 except using the measure of the effect of the impulse given in 10.29.

10.34 **(a)** through **(c).** Assume that the spacer-pin production process in Exercise 10.30 is also subject to a random, unbiased, stationary, independently and normally distributed perturbation $p(i)$ with standard deviation $\sigma_p = 0.001$ cm, while all other conditions remain as in 10.30. $p(i)$ and $\varepsilon(i)$ are mutually independent. Solve the corresponding part of 10.30 for this modified set of conditions.

10.35 **(a)** through **(c).** The same as 10.34 except using the measure of the effect of the impulse given in 10.31.

10.36 A generalized second-order process controller with parameters K_0 and K_1 is to be used to control a process subject to a step perturbation of magnitude p. An unbiased, stationary, independently distributed measuring error $\varepsilon(i)$ with variance σ_ε^2 is also present with each measurement of the process output. For the following relationships between σ_ε and p, find the values of K_0 and K_1 which minimize the second moment Q of the steady-state distribution of the process output around the desired level of output and the value of Q_{min} as a function of p:

a. $\sigma_\varepsilon = \sqrt{2}\,p$ **b.** $\sigma_\varepsilon = p$ **c.** $\sigma_\varepsilon = (\sqrt{2}/2)p$

10.37 **(a)** through **(c).** For the system described in Exercise 10.36, find the values of K_0 and K_1 which maximize the probability, Γ, of the steady-state process mean being between the limits m_U and m_L and find the value of Γ_{max} given $m_U = 2$, $m_L = -2$, and $p = 2$. Assume $\varepsilon(i)$ is normally distributed.

10.38 **(a)** through **(c).** For the system described in Exercise 10.36, assume that $\sigma_\varepsilon = \sigma_y/2$, where σ_y is the standard deviation of the distribution of individual items around the mean (the items are identically, independently, and normally distributed) and the sample size is 4. Find the values of K_0 and K_1

which maximize the steady-state probability Γ_T of individual items lying between the limits y_U and y_L and find the value of $\Gamma_{T\,max}$ given $y_U = 2$, $y_L = -2$, and $p = 2$.

10.39 **(a)** through **(c)**.   The same as 10.15 except $p(i) = p_1\mu(i) + p_2 \sin wi$.

10.40 **(a)** through **(c)**.   The same as 10.16 except $p(i) = p_1\mu(i) + p_2 \sin wi$.

10.41 **(a)** through **(c)**.   The same as 10.19 except

$$p(i) = 2p\mu(i) + p \sin(2\pi i/3 + \theta).$$

10.42 **(a)** through **(c)**.   The same as 10.20 except

$$p(i) = 2p\mu(i) + p \sin(2\pi i/3 + \theta).$$

10.43 **(a)** through **(c)**.   The same as 10.28 except instead of an impulse, the effect of a perturbation of form $p(i) = (-\frac{1}{2})^{i-1}$ is to be minimized.

10.44 **(a)** through **(c)**.   The same as 10.29 except instead of an impulse, the effect of a perturbation of form $p(i) = (-\frac{1}{2})^{i-1}$ is to be minimized.

10.45 **(a)** through **(c)**.   For the chemical reactor and tolerances in Exercise 10.28, find values for K_0 and K_1 which minimize the relative amplitude $|m|/p$ of the steady-state output for a sinusoidal perturbation of angular frequency $\pi/6$ radians per period.

10.46 **(a)** through **(c)**.   The same as 10.45 except we wish to find the values of K_0 and K_1 which minimize the sum of the relative amplitudes $|m|/p$ of the steady-state outputs for sinusoidal perturbations of $\pi/6$ and $2\pi/3$ radians per period.

10.47 **(a)** through **(c)**.   The same as 10.45 except we wish to find the values of K_0 and K_1 which minimize the maximum relative amplitude $|m|/p$ of the steady-state outputs for all sinusoidal perturbations between $\pi/4$ and $3\pi/4$ radians per period. For what value(s) of w will this maximum relative amplitude occur?

10.48 through **10.94**   Repeat Exercises 10.1 through 10.47 for the proportional-plus-difference controller with parameters K and L. Where numerical values for K_0 and K_1 are given, calculate K and L from $K = K_0 + K_1$ and $L = -K_1$.

10.95   Solve Eq. (10.5.1), the system difference equation for the proportional-plus-summation controller for $h = h_1 = h_2$, i.e., for case 2 operation, by the following methods. Express all boundary conditions as initial conditions, i.e., $m(0)$, $m(1)$, etc., as appropriate.

    a.   Variation of parameters   **b.**   The z transform

10.96 For a proportional-plus-summation controller with $D = m(0) = \varepsilon(i) = 0$ and $p(i)$ a unit impulse at $i = 0$, find the values of process output $m(i)$ for $i = 1$ through 8 for:

    a. $K = 1.5, N = 0.5$ **b.** $K = 1.0, N = 1.0$ **c.** $K = 0.5, N = 1.5$

10.97 (**a**) through (**c**). The same as Exercise 10.96 except for $p(i)$ a unit step at $i = 0$.

10.98 (**a**) through (**c**). The same as Exercise 10.96 except for $p(i) = (-\tfrac{1}{2})^{i-1}$.

10.99 (**a**) through (**c**). The same as Exercise 10.96 except for a ramp perturbation, $p(i) = ip$.

10.100 through **10.104** The same as Exercises 10.1 through 10.5 except for the proportional-plus-summation controller with parameters K and N.

10.105 Plot the $\hat{V} = V/\sigma_\varepsilon^2$ contours in the stable range of K and N for the proportional-plus-summation controller with $D = m(0) = p(i) = 0$ and an unbiased, stationary, independently distributed measurement error $\varepsilon(i)$ with variance σ_ε^2 for the following values of $\hat{V}$:

    a. 0.125   **b.** 0.5   **c.** 1.0   **d.** 2.0   **e.** 3.0   **f.** 7.0

10.106 Derive the conditions for which the amplitude frequency response of the proportional-plus-summation controller obtain their maxima and minima, and the values of these maxima and minima as functions of K, N, and w.

10.107 through **10.139** Repeat 10.15 through 10.47 for the proportional-plus-summation controller with parameters K and N. Where numerical values for K_0 and K_1 are given, calculate K and N from $K = 1 - K_1$ and $N = K_0 + K_1$.

10.140 through **10.142** The same as 10.36 through 10.38 except instead of the step perturbation of magnitude p there is a ramp perturbation $p(i) = ip$.

10.143 A rapid-response process is subject to a step perturbation of magnitude p units per minute. The nature of the process and measuring device are such that any information obtained during any given measurement cannot be acted upon (fed back for comparison with $D = 0$) for two minutes. Two control systems are proposed. First is a simple proportional controller with a sampling period of two minutes. Second is a proportional controller with a sampling interval of one minute and a one-period feedback delay. If the measurement error $\varepsilon(i)$ in either case is unbiased, stationary, and independently distributed with variance $\sigma_\varepsilon^2 = rp^2$ and the maximum allowable steady-state variance of process output $V_{max} = sp^2$, for what combinations of values of r and s will each proposed controller produce the smaller steady-state mean of process output M?

Chapter XI | *n*th-Order and Complex Systems

Although the methodology developed thus far is perfectly general and applies to any discrete, linear, time-invariant system, the specific cases that have been studied have been for the most part single feedback loop, first- and second-order systems. Since real world systems can conform to an unlimited number of configurations of almost any order, the analyst-designer must be able to apply his tools and techniques to general categories of discrete, linear, time-invariant systems. In this, our concluding chapter, we extend some of the material covered earlier to general types of systems.

Of prime importance is the determination of the appropriate transfer function to relate any input, no matter where or how it enters the system, to system output or to the performance of any particular component of interest within the system. For example, we might be interested in the effect of customer orders on the inventory level in a production inventory control system. We also, however, might be interested in the effect of these orders on the production facility, or of changes in desired inventory level on the actual level or on production. Though all these situations involve the same system, a different difference equation must be derived to relate the desired dependent variable to the appropriate forcing function, resulting in a separate, unique transfer function for each case of interest.

Difference equations and hence transfer functions can be derived directly from the relationships expressed by the block diagram of the system. For complex systems, however, the process of transfer function development is often enhanced by representing the system by what is called a "signal flow graph" or, more generally, a "flow graph." We, therefore, first introduce the properties of flow graphs and then illustrate their use in the derivation of transfer functions of linear, time-invariant systems. The chapter closes with presentations of a few generalizations of results obtained in earlier chapters for specific systems.

11.1 Signal Flow Graphs

The concept of the signal flow graph as a convenient means of representing and analyzing the topological properties of systems was introduced by Mason (1953) and extended in a follow-on paper (1956). Although the context in which the development was presented was that of electrical and electronic networks modeled as continuous-time, linear, time-invariant systems, the approach is equally applicable to discrete-time systems. Flow graphs were thus used by Howard as the basic means of system representation and model manipulation in his development in 1963 of the theory of discrete, linear systems.†

The original concepts have been greatly extended and formalized, both procedurally and analytically. Our development here, however, will be relatively introductory in nature. We will present just enough of the methodology to supply sufficient insight to the analyst or designer to allow him to develop transfer functions for systems he might reasonably expect to encounter in practice. Those aspiring to greater efficiency in this task are referred to the literature (Scarf *et al.*, 1963; Howard, 1971a; Mason, 1953, 1956).

Basic Structure and Definitions

A flow graph is a convenient, compact means of representing linear, time-invariant systems. It is a direct alternate to the block diagram in that all relationships among system components and between the system and its environment are expressed in similar ways in both approaches.

In Fig. 1.2.1 we presented a general system in block diagram form showing an input vector $r = \{r(0), r(1), \ldots, r(i), \ldots\}$, an output vector $c = \{c(0), c(1), \ldots, c(i), \ldots\}$, and a system operator S representing the transformation performed by the system on r to produce c. To facilitate our current discussion this form is repeated here in Fig. 11.1.1a. In Section 6.7 it was shown that for linear, time-invariant systems the operation of the system on the input r is

---

† Howard introduced his discrete, linear systems theory in Scarf *et al.* (1963). Howard authored Chapter 6 entitled "System Analysis of Linear Models." A greatly expanded version of this material is in Chapter 2, "Systems Analysis of Linear Processes" of Howard's own book (1971a).

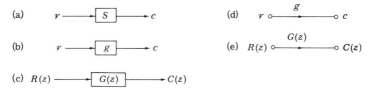

FIG. 11.1.1. System representation. (a) Discrete system with system operator S, $c = S(r)$. (b) Discrete, linear, time-invariant system with unit-impulse response $g(i)$, $c = r * g$. (c) z transform representation of system with transfer function $G(z) = Z[g(i)]$, $C(z) = R(z)G(z)$. (d) Flow graph representation of system with unit-impulse response $g(i)$, $c = r * g$. (e) Flow graph of z transform representation of system with transfer function $G(z) = Z[g(i)]$, $C(z) = R(z)G(z)$.

fully described by the unit-impulse response $g(i)$ and the output is $c(i) = \sum_{n=0}^{i} r(n)g(i - n)$, i.e., the convolution of the vector r and the vector $g = \{g(0), g(1), \ldots, g(i), \ldots\}$. This case is shown in Fig. 11.1.1b. The convolution of two vectors, or "signals" in Mason's terms, is often expressed by writing an asterisk between the symbols for the vectors whose components are convolved. Thus, for a discrete, linear, time-invariant system, we may conveniently write

$$c = r * g, \tag{11.1.1}$$

while for the general system the best we can do is

$$c = S(r), \tag{11.1.2}$$

i.e., to say that c is a function S of the input vector r. Since many analysts prefer to work whenever possible in the z domain, we have included as Fig. 11.1.1c the transformed version of the system in Fig. 11.1.1b. Here $R(z)$, $G(z)$, and $C(z)$ replace r, g, and c, respectively, and of course

$$C(z) = R(z)G(z). \tag{11.1.3}$$

Figures 11.1.1d and 11.1.1e are the flow graph versions of 11.1.1b and 11.1.1c, respectively. The major differences are the replacement of the box describing the system operation in the block diagram form by a simple arrow showing the direction from input to output in the flow graph form and the introduction of the small circles at the points of system input and output. These changes may seem trivial, especially since the unit-impulse response or transfer function must still be listed alongside the arrow where the block used to be, but experience has shown considerable advantage in working with diagrams in this form.

Formally, each circle is called a *node*, and is labeled by the symbol for the vector describing the system variable at that point, e.g., the node labeled r

in Fig. 11.1.1d represents the input to the system and indeed is often referred to as an input node. This would be $R(z)$ in Fig. 11.1.1e. Similarly, the node labeled c or $C(z)$, as appropriate, can be thought of as the output node for the system. The line joining the two nodes is called an *arc*, and the allowable direction of flow from input to output is indicated by the direction of the arrowhead. The unit-impulse response or transfer function of the system represented is of course given by the arc label g or $G(z)$.

The flow graph has several advantages over the block diagram. First, there is a consistency in the representation of all kinds of components, so that once one determines the transfer function of each type of functional unit the arc for that component is complete. Further, the nodes form natural junction and branching points for joining components together in larger subsystems and systems wherein the arc for each component becomes a branch of the larger network. Finally, this sort of configuration provides the structure to guide the reduction of the complex network formed to an equivalent single arc, the transfer function of which is the transfer function for the overall system, i.e., the function we were trying to determine in the first place. Throughout this process, the simple format of the arcs and their labels, free of boxes, blocks, summers, and the like, considerably reduces the amount of labor involved.

Having looked at the basic building blocks of the flow graph, let us now determine how one derives the single-arc equivalents of some basic patterns of networks involving two or more arcs. Later we will discuss how one constructs the flow graph of an actual system using as examples some of the systems discussed in earlier chapters.

Flow Graph Reduction

Several basic patterns of interconnected arcs are common in systems of all kinds. We will consider reduction of several of these to single-arc equivalents. First, however, note the following relationships which will be used extensively later.

When two or more graphs have a common output node as shown in Fig. 11.1.2a, the total output is the sum of the outputs from each of the branches. This situation could occur in an assembly operation where mating parts come together or where similar types of products are routed through a common inspection station. When two or more branches have a common input node as shown in Fig. 11.1.2b, the output of each branch is the convolution of the common input r and the unit-impulse response of that branch. This could correspond to a production inventory control system where customer orders both deplete inventory and are fed forward to be used for immediate production decisions.

Note that we could have used z transform representations in Fig. 11.1.2. In what follows we will deal with signals in the i domain and the unit-impulse responses of the arcs involved except where it is required to use the z transform to obtain the answer being sought. Let us now examine some basic patterns of arcs.

FIG. 11.1.2. Branches with connected inputs or outputs. (a) Branches with connected outputs, $c = r_1 * g_1 + r_2 * g_2$. (b) Branches with connected inputs, $c_1 = r * g_1$, $c_2 = r * g_2$.

Parallel Connection

When two or more arcs have a common input and a common output as shown in Fig. 11.1.3a they are said to be connected in parallel. As just discussed, the output of each arc is the convolution of the common input and

FIG. 11.1.3. Reduction of parallel connection. (a) Parallel connection of n arcs. (b) Single-arc equivalent.

the unit-impulse function of the arc, while the total output is the sum of the outputs of each individual arc. Thus,

$$c = r * g_1 + r * g_2 + \cdots + r * g_n$$
$$= r * (g_1 + g_2 + \cdots + g_n) = r * \left(\sum_{k=1}^{n} g_k \right), \tag{11.1.4}$$

which indicates that a set of parallel arcs can be replaced by a single arc whose transfer function is the sum of the transfer functions of the individual arcs. This equivalent is shown in Fig. 11.1.3b.

Series Connection

When the output of one arc is the input to another as shown in Fig. 11.1.4a, the arcs are said to be connected in series. The outputs of the two arcs are respectively, $c_1 = r * g_1$ and $c = c_1 * g_2$, which, by straightforward substitution for c_1, yields

$$c = c_1 * g_2 = (r * g_1) * g_2 = r * (g_1 * g_2). \tag{11.1.5}$$

Therefore, the series connection can be replaced by a single arc whose unit-impulse function is the convolution of the impulse functions of the series-connected arcs. This equivalent is shown in Fig. 11.1.4b. This result can of

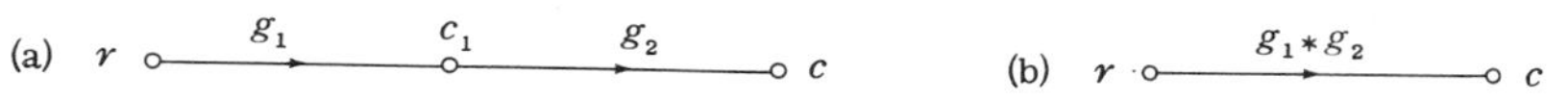

(a) r o———g_1———c_1 o———g_2———o c (b) r o———$g_1 * g_2$———o c

FIG. 11.1.4. Reduction of series connection. (a) Series connection of arcs. (b) Single-arc equivalent.

course be extended to cover any number of series-connected arcs. Note that the order of the arcs in the series connection does not affect the results since $g_1 * g_2 = g_2 * g_1$.

Interlocking Connection

When parallel and series connections of arcs overlap, the result is called an interlocking connection. Although an infinite number of interlocking configurations is possible, one of the simpler possibilities is shown in Fig. 11.1.5a. A production line making several similar products which share some common production processes but in different combinations could lead to such a model.

As shown, no series or parallel branch reductions can be performed directly. However, the basic arc input–output relationships provide a basis for proceeding. Specifically,

$$c_1 = r * g_1, \tag{11.1.6}$$

$$c_2 = r * g_2 + c_1 * g_3, \tag{11.1.7}$$

$$c = c_1 * g_4 + c_2 * g_5. \tag{11.1.8}$$

Substitution of (11.1.6) into (11.1.7) gives

$$c_2 = r * g_2 + r * g_1 * g_3 = r * (g_2 + g_1 * g_3) \tag{11.1.9}$$

and substitution of (11.1.6) and (11.1.9) into (11.1.8) gives

$$c = r * g_1 * g_4 + r * (g_2 + g_1 * g_3) * g_5 = r * [g_1 * g_4 + (g_2 + g_1 * g_3) * g_5]. \tag{11.1.10}$$

The expression in brackets in (11.1.10) is of course the unit-impulse response of the single-arc equivalent of the original interlocking network. Thus we have performed the reduction we set out to do. Of more importance, however, is to identify from what was done a procedure that can be relied upon in similar situations so that a complete brute-force algebraic analysis of each new case can be avoided. We note that the form of the impulse response indicates

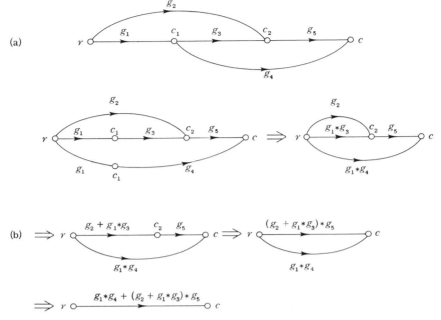

FIG. 11.1.5. Reduction of interlocking connection. (a) Interlocking connection of arcs.
(b) Derivation of single-arc equivalent.

equivalence to two parallel arcs from r to c, one involving a series connection
of g_1 and g_4 and the other a series–parallel connection involving g_1, g_2, g_3,
and g_5. Thus the original network could have been drawn in the form of two
parallel subnetworks as indicated. This has been done in the first diagram of
Fig. 11.1.5b, which results in effectively duplicating the node c_1 and the arc g_1.
Once the network has been thus modified, further reduction to a single arc
using step-by-step series and parallel reduction is straightforward.

The concept to be realized from this exercise is that it is always permissible
(and sometimes necessary) when one or more arcs are members of more than
one path through all or part of the network to duplicate the arc and one or
both of its nodes as needed to make all such paths explicit. The resulting configuration will always be equivalent to the original one and the steps involved
in completing the reduction to a single arc will often be more readily identified.
Often numerous choices of expansions of this sort are possible, all of which
will eventually yield the same single-arc equivalent, but some will do so much
more readily than others. Thus, experience and luck sometimes do play roles
in the process. Note that either of the initial expansions shown in Fig. 11.1.6

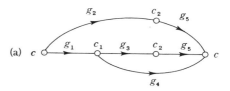

FIG. 11.1.6. Alternate initial expansions for reduction of network in Fig. 11.1.5a.

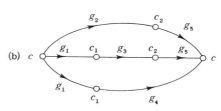

could have been used to derive the single-arc equivalent to the network in Fig. 11.1.5a. The second of these is a complete expansion into a parallel connection of all paths leading from r to c, a form much preferred by some analysts. It is left to the reader to verify the equivalence of the results with those of Fig. 11.1.5b. For those who prefer the transform format, the transform of (11.1.10) is of course

$$C(z) = R(z)\{G_1(z)G_4(z) + [G_2(z) + G_1(z)G_3(z)]G_5(z)\}. \qquad (11.1.11)$$

Simple Feedback Loop

An arc which has the same node as both its input and output constitutes what is called a simple feedback loop. In discussing such an arc in isolation from any larger network, it is helpful to consider the node of the arc as being fed from a source node r through an auxiliary arc with a transfer function of unity, i.e., $g = \delta$, and providing an output signal to a node c through another auxiliary arc which also has a transfer function of unity. These auxiliaries, called identity arcs, are not needed when the feedback loop is imbedded in a larger network. A simple loop and its auxiliary arcs are shown in Fig. 11.1.7a. The basic arc equations yield

$$e = r * \delta + e * h = r + e * h \qquad (11.1.12)$$

FIG. 11.1.7. Reduction of simple feedback loop. (a) Simple feedback loop. (b) Transform version of single-arc equivalent.

and

$$c = e * \delta = e, \qquad (11.1.13)$$

which give

$$c = r + c * h. \qquad (11.1.14)$$

Solution for c as an explicit function of r and h is not possible in general terms. However, if we take z transforms to obtain

$$C(z) = R(z) + C(z)H(z), \qquad (11.1.15)$$

we can solve for $C(z)$ as

$$C(z) = R(z)/[1 - H(z)]. \qquad (11.1.16)$$

Thus, the transfer function of the simple feedback loop is

$$G_f(z) = 1/[1 - H(z)], \qquad (11.1.17)$$

where $H(z)$ is the transfer function of the arc forming the loop. The unit-impulse response of the loop, if it were desired, could be found by inverse transformation of $G_f(z)$.

Compound Feedback Loop

When two nodes are connected by a pair of arcs such that one node is the input for one arc and the output for the other and for the second node the reverse is true, the pair constitutes what is called a compound feedback loop. As in the case of the simple feedback loop, discussion of the compound loop in isolation from any larger network is facilitated by use of auxiliary input and output identity arcs as shown in Fig. 11.1.8a.

As was done with the interlocking connection, we make use of the fact that paths sharing the same arcs may be expressed individually by duplicating the representation of the shared arcs and whatever nodes are involved. We therefore duplicate the arc from e to b and node b as shown in Fig. 11.1.8b. We thus identify a simple feedback loop with impulse response $g * h$ and an output arc with response g, pictured in 11.1.8c. At this point we must invoke the z transform. From our previous discussion of the simple feedback loop, the transfer function of this loop is $1/[1 - G(z)H(z)]$. Since the loop is in series with the arc with transfer function $G(z)$, as shown in 11.1.8d, the transfer function of the single-arc equivalent of the compound feedback loop is

$$G_{cf}(z) = G(z)/[1 - G(z)H(z)]. \qquad (11.1.18)$$

The unit-impulse response could of course be found by inverse transformation of $G_{cf}(z)$.

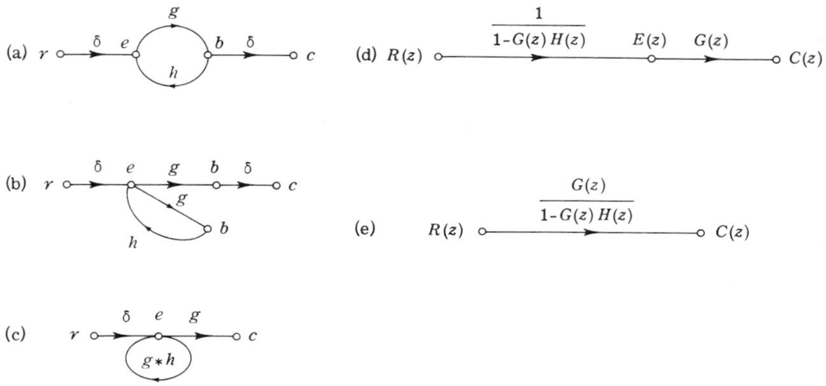

FIG. 11.1.8. Reduction of compound feedback loop. (a) Compound feedback loop. (b) Expanded compound feedback loop. (c) Simple-loop equivalent. (d) Transform version of reduction of simple feedback loop. (e) Transform version of single-arc equivalent.

This concludes our examination of the basic approaches to the reduction of complex flow graphs. Used astutely in proper combination, they should permit the reduction of any flow graph to an equivalent single arc, admitting that the more complicated the original network the more ingenuity that may be required to get the job done. Invariably where feedback is present transform versions must be used. We now turn our attention to the representation of systems in flow graph form.

Flow Graph Representation of Systems

The representation of systems by flow graphs is essentially straightforward, especially if we work in the z domain with components whose transfer functions are known. The i domain equivalents of some circuit elements, however, may require some development, although their use can usually be avoided by working completely with z transforms.

As an example, consider the summation of the increments $\Delta m(i)$ to produce the output $m(i)$ which occurs in discrete-control systems such as those shown in Figs. 2.3.1 and 2.4.1 for the process control and the production inventory control systems, respectively. This component represents the relationship

$$m(i) = m(0) + \sum_{n=0}^{i-1} \Delta m(n). \tag{11.1.19}$$

It can be represented by a flow graph arc with input vector Δm and output vector m, or in the z domain by $Z[\Delta m(i)]$ and $Z[m(i)]$, respectively. The transfer function is of course that of a summation $z/(z-1)$ followed by a one-period delay z^{-1}, since the upper limit of the summation is $i-1$. The resulting flow graph and its single-arc equivalent are shown in Figs. 11.1.9a and

(a) $Z[\Delta m(i)]$ 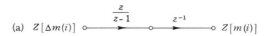 $Z[m(i)]$

(b) $Z[\Delta m(i)]$ 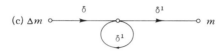 $Z[m(i)]$

(c) Δm 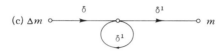 m

FIG. 11.1.9. Flow graph representation of Eq. (11.1.19). (a) Summation followed by one-period delay. (b) Single-arc equivalent. (c) i domain version.

11.1.9b, respectively.† To obtain the i domain equivalent, we must take the inverse transformations of both the summation and delay operations. The one-period delay is most easily represented in the i domain by the symbol δ^1, where in general the vector

$$\delta^k = \begin{cases} 1 & \text{for } i=k, \\ 0 & \text{elsewhere,} \end{cases} \tag{11.1.20}$$

i.e., a unit-impulse delayed by k time periods. Note that $\delta^0 = 1$ for $i = 0$; i.e., $\delta^0 = \delta$, the identity function, and therefore δ^k reproduces the input at the output node with a k-period delay. The inverse transform of the summation transfer function $z/(z-1)$ is most easily determined if we divide numerator and denominator by z to obtain $1/(1-z^{-1})$, which is in the form of a simple feedback loop with $H(z) = z^{-1}$ or $h = \delta^1$. This configuration is shown in Fig. 11.1.9c. We will consider the representations of other functional operations in the context of specific categories of systems.

---

† Note the consistency with the transfer function needed to transform an input m to an output Δm. Since $Z[\Delta m(i)] = (z-1)Z[m(i)]$, the transfer function is $z-1$, the reciprocal of the one above.

The Generalized Discrete Process Controller

The discrete generalized process controller introduced in Fig. 2.3.1 is used to illustrate the derivation of a system transfer function by means of flow graphs. Since we are already familiar with its performance characteristics, we have a means of checking results and comparing procedures. The system block diagram is repeated here as Fig. 11.1.10a. The flow graph representation is in Fig. 11.1.10b. Note that each exogenous signal is entered through an

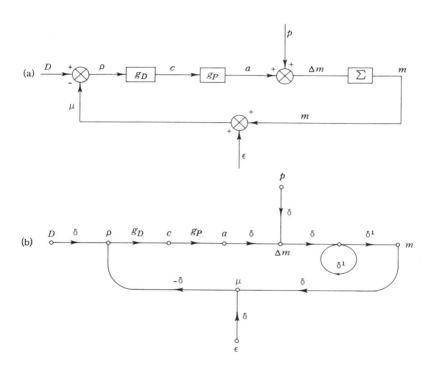

FIG. 11.1.10. Representations of discrete generalized process controller. (a) Block diagram. (b) Flow graph.

identity arc. Identity arcs have also been used elsewhere for convenience in physically separating circuit points having identical signals such as nodes *m* and *μ* and nodes *a* and Δ*m*. Many experienced analysts would draw this network without these arcs, connecting the arc g_P directly to node Δ*m* and the measurement-error signal *ε* to *m*. Thus signals *a* and *μ* would not appear which, even though the eventual outcome is unaffected, could cause confusion when checking the network for correspondence with the real world or a preliminary block diagram. Note also the representations of the summation block by the feedback loop followed by a one-period delay and the subtraction of μ from D to form ρ by the " $-\delta$ " impulse function on the arc connecting nodes μ and ρ.

Since, strictly speaking, a transfer function is defined between a pair of points in a system, we must decide which pair or pairs are of interest. In general, especially with control systems, we are primarily concerned about the effects of the various exogenous variables on the output. Thus we might need to develop a "perturbation transfer function" relating p and m, a "measurement-error transfer function" relating ε and m, or a "desired-level transfer function" (perhaps more aptly called a "reference-level transfer function") relating D and m, although any pair of functionally connected nodes could be picked. We will now derive each of the three transfer functions just mentioned by reduction of the flow graph to the appropriate single-arc equivalent.

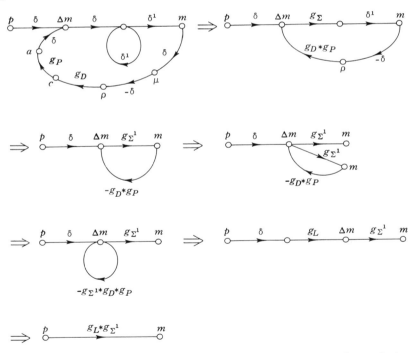

FIG. 11.1.11. Reduction of flow graph of generalized process controller to single arc relating p and m.

Figure 11.1.11 shows the various steps in deriving the perturbation transfer function. We have somewhat arbitrarily chosen to develop the single-arc equivalent completely in the i domain and take z transforms at the end. This necessitates defining g_Σ as the unit-impulse response of the summation and hence g_Σ^1 as this function delayed by one period and letting g_L be the response of the major feedback loop attached to node Δm in the fifth step of the reduction. This procedure, though somewhat cumbersome compared to earlier use of the z transform, gives added exposure to the flow graph procedures for illustrative purposes. Note that in dealing with the input p, we ignore the inputs D and ε. We also arbitrarily follow tradition by writing the flow graph with the input p on the left and the output m on the right. In the second step the simple feedback loop has been replaced by an arc labeled g_Σ and several series connections of arcs reduced. The third step completes the reduction of all series-connected arcs resulting in a compound feedback loop with the delayed summation g_Σ^1 as its forward response and $-g_D * g_P$, the negative of the convolved decision rule and physical properties of the process responses in the feedback branch. This compound loop is reduced to a simple loop in steps 4 and 5 by duplicating arc g_Σ^1 and node m. The simple loop is replaced by an arc labeled g_L in step 6. Final reduction yields the single-arc equivalent with response $g_L * g_\Sigma^1$.

Since both g_L and g_Σ^1 are simply symbols representing the responses of feedback loops, it is necessary to use z transforms to express the response of this single arc in terms of actual system parameters. This of course provides the transfer function which in many situations is the objective sought. Otherwise, inverse transformation by the methods of Chapter VI must follow. The z transform of the single arc in Fig. 11.1.11 is

$$M(z) = P(z)G_L(z)G_\Sigma(z)z^{-1}. \tag{11.1.21}$$

From Fig. 11.1.9, $G_\Sigma(z) \cdot z^{-1}$ is $1/(z-1)$, and from Fig. 11.1.11,

$$G_L(z) = \frac{1}{1 - [-G_\Sigma(z) \cdot z^{-1}G_D(z)G_P(z)]} = \frac{1}{1 + [1/(z-1)]G_D(z)G_P(z)}. \tag{11.1.22}$$

Therefore,

$$M(z) = P(z)\frac{1}{1 + [1/(z-1)]G_D(z)G_P(z)} \cdot \frac{1}{z-1} = P(z)\frac{1}{z-1+G_D(z)G_P(z)}, \tag{11.1.23}$$

which yields the perturbation transfer function

$$\frac{M(z)}{P(z)} = \frac{1}{z-1+G_D(z)G_P(z)}. \tag{11.1.24}$$

A similar derivation of the single-arc equivalent relating ε and m is shown in Fig. 11.1.12. Here we let $p = D = 0$ and draw the network with ε on the

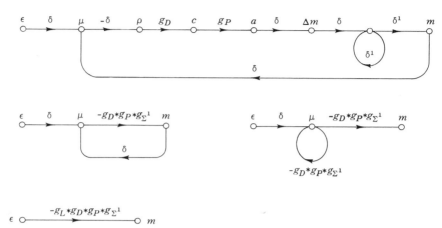

FIG. 11.1.12. Reduction of flow graph of generalized process controller to single arc relating ε and m.

left and m on the right. The same definitions apply as before. Here we see the necessity of the identity arc between m and μ to provide the required separation between ε and m. From the single-arc equivalent derived we get

$$M(z) = -E(z)G_L(z)G_D(z)G_P(z)G_\Sigma(z)z^{-1}, \tag{11.1.25}$$

which after appropriate substitution gives

$$M(z) = E(z)\frac{-G_D(z)G_P(z)}{z - 1 + G_D(z)G_P(z)}. \tag{11.1.26}$$

The measurement-error transfer function is thus

$$\frac{M(z)}{E(z)} = \frac{-G_D(z)G_P(z)}{z - 1 + G_D(z)G_P(z)}. \tag{11.1.27}$$

It is left as an exercise for the reader to show that the desired-level transfer function is

$$\frac{M(z)}{D(z)} = \frac{G_D(z)G_P(z)}{z - 1 + G_D(z)G_P(z)}, \tag{11.1.28}$$

which is precisely the negative of the measurement-error transfer function.

It is not surprising that all three transfer functions have the same denominator, which they of course must since the denominator is the z transform of the characteristic equation. Note that for a rapid-response proportional controller $g_P = \delta$ and $g_D = K_0 \delta$, from which $G_P(z) = 1$, $G_D(z) = K_0$, and $1/[z - 1 + G_D(z)G_P(z)] = 1/[z - (1 - K_0)]$, the now familiar expression we have been loosely calling the system transfer function of the simple proportional controller. Although this terminology is not incorrect, we are now in a position to sharpen our definition of transfer function as it relates to two specific variables. Specifically, the more precise meaning of the term transfer function includes the numerator terms as well as the denominator (the characteristic equation) terms that relate $M(z)$ to the transform of a particular exogenous variable. For the simple proportional controller, the three transfer functions given by (11.1.24), (11.1.27), and (11.1.28) become $1/[z - (1 - K_0)]$, $-K_0/[z - (1 - K_0)]$, and $K_0/[z - (1 - K_0)]$, respectively. Combining gives

$$M(z) = \frac{K_0 D(z) - K_0 E(z) + P(z)}{z - (1 - K_0)}, \qquad (11.1.29)$$

which can be seen by inspection of (8.1.4), the system difference equation of the simple proportional controller, to be the solution for $M(z)$ of the transformation of the difference equation. The results are thus the same, although the use of the term transfer function is somewhat different. Because of these differences in usage, the analyst must be careful to determine precisely what is implied each time he comes across the term.

Flow Graphs of Specific System Components

The derivations just completed were done in terms of a general control decision rule g_D and physical properties of the process g_P. This provided a transfer function in terms of $G_D(z)$ and $G_P(z)$ into which the transforms of any rules and properties of interest could be substituted to obtain transfer functions of specific systems. This procedure is highly recommended whenever it is applicable. There may be occasions, however, when it is advantageous to represent a specific decision rule or physical property in flow graph form. We, therefore, examine a few of the more common types.

We have already mentioned that a delay of k periods is represented by δ^k in the i domain and z^{-k} in the z domain. This relationship is basic to most of what follows. The control rule for the generalized nth-order process controller was given in (2.3.33) as

$$c(i) = \sum_{k=0}^{n-1} K_k \rho(i - k). \qquad (11.1.30)$$

Physically this means that the control signal at time i is the sum of the most recent measured deviation $\rho(i)$, multiplied by K_0, the next most recent measured deviation $\rho(i - 1)$, multiplied by K_1, and so forth until a total of n

terms are added in. Thus $\rho(i - 1)$ has to be delayed one period before being introduced into the sum, $\rho(i - 2)$ delayed two periods, and $\rho(i - k)$ delayed k periods. The flow graph relating ρ to c for the nth-order controller is shown in Fig. 11.1.13. The z transform version would be basically the same with $R(z)$

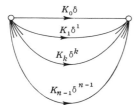

FIG. 11.1.13. Flow graph of control rule for generalized nth-order process controller.

and $C(z)$ replacing ρ and c and with z^{-k} replacing δ^k. Since difference control is a special case of the generalized second-order system, it can be represented by the graph in Fig. 11.1.13 with $K_0 = L$, $K_1 = -L$, and K_2 through K_{n-1} equal to zero. Alternatively, it could be represented by a parallel connection of arcs with responses δ and $-\delta^1$ followed by an arc with response $L\delta$. Proportional-plus-difference control can be represented by the graph in Fig. 11.1.13 with $K_0 = K + L$, $K_1 = -L$, and the remaining $K_k = 0$.

When summation control is used, it is necessary to introduce a feedback loop as previously discussed. For example, the control signal for a proportional-plus-difference-plus-summation controller was given in (2.3.29) as

$$c(i) = K\rho(i) + L\,\Delta\rho(i - 1) + N \sum_{n=0}^{i} \rho(n). \qquad (11.1.31)$$

A flow graph representation is shown in Fig. 11.1.14. Double summation would require the cascading of two summing loops, i.e., connecting the output

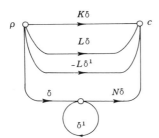

FIG. 11.1.14. Flow graph of control rule for proportional-plus-difference-plus-summation controller.

of one summer to the input of the next with an identity arc or a proportional arc ($g = N\delta$, N a constant) as appropriate. High-order differences could also be represented by cascading pairs of difference arcs, introducing the constant

L at any convenient stage. They can also be represented by flow graphs based on the definitional formula given in (3.1.20). Thus a second difference could be given as a special case of Fig. 11.1.13 with $K_0 = L$, $K_1 = -2L$, $K_2 = L$, and the remaining K_k set to zero.

Having now developed a general and reasonably straightforward set of procedures for deriving transfer functions for discrete, linear, time-invariant systems, we close with three somewhat miscellaneous topics involving extensions of some of the results of earlier chapters to general types of systems. Although analytic determination of the performance of third- and higher-order systems in terms of control-system parameters is rarely possible, there are a few results derived for relatively simple first- and second-order systems that can be generalized to a degree. Those to be discussed here are the steady-state response of the generalized nth-order process controller to a step input, the effects of feedback delay on various configurations of process controllers, and the development of the frequency response of any system with known transfer function.

11.2   Step Function Response of nth-Order Generalized Process Controller

The system difference equation for the nth-order generalized process controller was given in (2.3.34) as

$$m(i + 1) - (1 - K_0)m(i) + \sum_{k=1}^{n-1} K_k\, m(i - k)$$

$$= D\sum_{k=0}^{n-1} K_k - \sum_{k=0}^{n-1} K_k \varepsilon(i - k) + p(i). \tag{11.2.1}$$

To obtain the standard form for solution, we must increment the time argument by $n - 1$. This gives

$$m(i + n) - (1 - K_0)m(i + n - 1) + \sum_{k=1}^{n-1} K_k\, m(i + n - 1 - k)$$

$$= D\sum_{k=0}^{n-1} K_k - \sum_{k=0}^{n-1} K_k \varepsilon(i + n - 1 - k) + p(i + n - 1), \tag{11.2.2}$$

whose z transform is

$$\left[z^n - (1 - K_0)z^{n-1} + \sum_{k=1}^{n-1} K_k z^{n-1-k} \right] M(z)$$

$$= \left(D\sum_{k=0}^{n-1} K_k \right) \frac{z}{z - 1} - \sum_{k=0}^{n-1} K_k z^{n-1-k} E(z) + z^{n-1}P(z) + I(z), \tag{11.2.3}$$

where $M(z)$, $E(z)$, and $P(z)$ are the z transforms of $m(i)$, $\varepsilon(i)$, $p(i)$, and $I(z)$ represents the various initial condition terms and their z multipliers.

One of the few system responses that can be analytically determined for any order of controller is the steady-state response to a step perturbation. To illustrate, let $D = 0$ and assume all measurement errors are negligible. Since our interest is in a steady-state response we also ignore $I(z)$. For a step of magnitude p, $P(z) = pz/(z - 1)$, so that

$$M(z) = \frac{pz^n}{(z - 1)[z^n - (1 - K_0)z^{n-1} + \sum_{k=1}^{n-1} K_k z^{n-1-k}]}. \quad (11.2.4)$$

The steady-state response is, therefore,

$$m = \lim_{i \to \infty} m(i) = \lim_{z \to 1} (z - 1)M(z) = \frac{p}{\sum_{k=0}^{n-1} K_k}, \quad (11.2.5)$$

i.e., the magnitude of the perturbation divided by the sum of the control-system parameters. Note that this is compatible with the results previously obtained for the first- and second-order systems. Although the magnitudes of the K_k are limited by stability and perhaps other factors, one can see the possibility of lowering the steady-state response by adding terms to the controller if for some reason a summation device is not used.

11.3   Effects of Delay on Process Controller Stability

In Section 10.6 we discussed the difficulties imposed by a one-period feedback delay in controlling a process with a proportional controller. A natural question to ask is what might be the effects of delays of two or more periods. Although in-depth answers to this question are difficult to determine since the system equations involved are of third and higher order, a lot can be said about the limits imposed on the proportionality constant K by stability considerations. We first discuss these limitations on K for a proportional controller and then look briefly at the effects of delay on the proportional-plus-difference controller.

Proportional Control

The system difference equation for a proportional controller with a feedback delay of τ periods is

$$m(i + 1) - m(i) + Km(i - \tau) = KD - K\varepsilon(i - \tau) + p(i), \quad (11.3.1)$$

from which the characteristic equation written in terms of the transform variable z is

$$B(z) = z^{\tau+1} - z^{\tau} + K. \tag{11.3.2}$$

The first two steps in the Jury test yield

$$B(z = 1) = 1 - 1 + K = K > 0 \tag{11.3.3}$$

for all τ and

$$B(z = -1) = \begin{cases} 1 + 1 + K = 2 + K > 0 & \text{for } \tau \text{ odd,} \\ -1 - 1 + K = -2 + K < 0 & \text{for } \tau \text{ even.} \end{cases} \tag{11.3.4}$$

Equation (11.3.3) is obviously extremely helpful, whereas (11.3.4) which "limits" K to either $-2 < K$ or $K < 2$ is not. Each succeeding step in the Jury process exists only for increasingly higher values of τ. Thus we need $|b_{\tau+1}| < |b_0|$ only for $\tau \geq 1$, $|c_\tau| > |c_0|$ only for $\tau \geq 2$, $|d_{\tau-1}| > |d_0|$ only for $\tau \geq 3$, etc. This is because the last row in the Jury table, which is row $2n - 3$ or in this case the $(2\tau - 1)$st row, must have exactly three entries. Thus, for $\tau = 1$ there is only one row and only the test involving the b's is involved. For $\tau = 2$ there are three rows and the two tests involving b's and c's are necessary, and so forth. It is also easy to show that, when they exist, all b_0, $b_{\tau+1}$, c_0, c_τ, d_0, $d_{\tau-1}$, etc. are independent of τ. The specific parameter values for the first four tests are given in Table 11.3.1. Note the rapid increase in complexity as

Table 11.3.1

Jury-Test Parameters for Proportional Process Controller with Proportionality Constant K and τ-Period Delay

Minimum τ for existence	Parameters	
1	$b_{\tau+1} = K$	$b_0 = 1$
2	$c_\tau = K^2 - 1$	$c_0 = -K$
3	$d_{\tau-1} = K^4 - 3K^2 + 1$	$d_0 = K$
4	$e_{\tau-2} = K^8 - 6K^6 + 11K^4 - 7K^2 + 1$	$e_0 = -K^3 + K$

one proceeds down the table. Remembering that the actual tests involve the absolute values of the parameters listed makes us realize the difficulty of arriving at a completely general set of limits for K as a function of τ by means of the Jury test.

Some important insights, however, have emerged from some very tedious work involving the entries in Table 11.3.1. Furthermore, these insights have been confirmed by some equally tedious work based on the Routh–Hurwitz stability criteria. These may be summarized as follows. First, all of the tests involving the parameters in the table impose positive upper bounds and negative lower bounds on K. Since $B(z = 1) > 0$ placed a lower bound on K of zero, then indeed zero is the overall lower bound. Second, the upper bound imposed by each test as one moves down through the table is tighter than the one preceding it. This leaves us in the unhappy position of always having to complete the entire Jury table and run the last, most complicated test to determine the actual upper bound for a given τ. On the other hand, it provides an increasingly narrower range over which to run the numerical search for the root. Note that the order of the polynomial whose smallest positive root must be found will always be $2^{\tau-1}$.

The Routh–Hurwitz criteria can also provide some help here although as yet not a complete solution. The bilinear transformation of the characteristic equation in (11.3.2) yields, after multiplying both sides of the equation by $(\lambda - 1)^{\tau+1}$,

$$\hat{B}(\lambda) = (\lambda + 1)^{\tau+1} - (\lambda + 1)^{\tau}(\lambda - 1) + K(\lambda - 1)^{\tau+1} = 0. \quad (11.3.5)$$

Expanding and grouping terms in like powers of λ gives

$$\hat{B}(\lambda) = \sum_{k=0}^{\tau+1} \frac{\tau!}{k!(\tau + 1 - k)!} [2k + (-1)^k(\tau + 1)K]\lambda^{\tau+1-k}. \quad (11.3.6)$$

The entries for the first two rows of the Routh array are, therefore, given by

$$B_k = 2k + (-1)^k(\tau + 1)K. \quad (11.3.7)$$

First we note that B_0 and B_1, the leading entries in these rows, are

$$B_0 = K \quad (11.3.8)$$

and

$$B_1 = 2 - (\tau + 1)K, \quad (11.3.9)$$

both of which must be of the same sign for the system to be stable. For $B_0 = K > 0$, $B_1 > 0$ requires

$$K < 2/(\tau + 1). \quad (11.3.10)$$

For $K < 0$, however, $B_1 < 0$ requires that $K > 2/(\tau + 1) > 0$, which is impossible for negative K. This confirms the nonnegativity of K and also indicates that the leading entries of all rows in the Routh array must be positive. It is interesting to note that (11.3.10) is a sufficient condition for the positivity of all coefficients of $\hat{B}(\lambda)$, thus assuring the satisfaction of the first two Routh–Hurwitz stability requirements. The leading entry of the third row of the Routh array is, if needed,

$$C_0 = \frac{\tau}{3} \cdot \frac{12 - 6\tau K - (\tau^2 + 3\tau + 2)}{2 - (\tau + 1)K}. \quad (11.3.11)$$

Since the denominator is positive by (11.3.10), C_0 will be positive if the numerator is positive. The range of K for which this occurs is found by means of the quadratic formula to be

$$\frac{-3\tau - (21\tau^2 + 36\tau + 24)^{1/2}}{\tau^2 + 3\tau + 2} < K < \frac{-3\tau + (21\tau^2 + 36\tau + 24)^{1/2}}{\tau^2 + 3\tau + 2}. \qquad (11.3.12)$$

The lower bound in (11.3.12) is observed to be negative, so only the upper bound is potentially binding.

Development of general expressions for the leading entries in the fourth and subsequent rows is not readily possible because of the increasing order of the numerators of the terms in each row of the Routh array. From (11.3.7) we see that each B_k is a linear function of K. Thus, so are B_0 and B_1, the leading entries in rows one and two. The numerator of every entry in the third row, i.e., the C's with even subscripts (see page 270), is the difference between the products of two B_k's. Thus all third-row entries will have numerators which are quadratic functions of K. The numerators of every term in the fourth row, i.e., the C's with odd subscripts, involve products of a B_k and a C with an even subscript. Thus the numerators of all fourth-row entries will be cubic functions of K. In general, the order of the numerator polynomial of all entries in any given row is the sum of the orders of the numerator polynomials of the two preceding rows. Letting $O(p)$ represent the order of the numerator polynomial of the elements of row p, this relationship can be represented as

$$O(p) = O(p - 1) + O(p - 2). \qquad (11.3.13)$$

As an interesting aside, (11.3.13) is the recursive relationship for the sequence known as the Fibonacci numbers. If we define $\tilde{F}(n)$ as the nth Fibonacci number and the boundary conditions $\tilde{F}(0) = \tilde{F}(1) = 1$ usually defined for the Fibonacci sequence, $O(p) = \tilde{F}(p - 1)$. Of more importance for our immediate purposes, however, is that (11.3.13), which is a homogeneous second-order linear difference equation, can be solved to yield

$$O(p) = \frac{1}{\sqrt{5}} \left[\left(\frac{1 + \sqrt{5}}{2} \right)^p - \left(\frac{1 - \sqrt{5}}{2} \right)^p \right]. \qquad (11.3.14)$$

The values of $O(p)$ for $p = 1$ through 10 are listed in Table 11.3.2. From this we see that attempts at general expressions beyond the third row are not worthwhile. (One exception to the above relationships involves the element $B_{\tau+1} = 2 + (-1)^{\tau+1}K$, the constant term in $\hat{B}(\lambda)$ which, because of cancellations made possible by the structure of the array, becomes the last element in alternate rows and eventually the final element in the left-hand column.

Table 11.3.2

Orders of Numerator Polynomials of Entries in Various Rows of the Routh Array

Row p	1	2	3	4	5	6	7	8	9	10
Order $O(p)$	1	1	2	3	5	8	13	21	34	55

Without the cancellation, however, all that has been said above still applies here also.)

Some further insights have been provided by Mitchell who worked out the Routh arrays for delays of one through four periods.† For the cases examined he found that, in addition to $K > 0$,

1.   For $\tau \geq 1$, the upper bound on K is prescribed by the smallest positive root of the numerator polynomial of row $\tau + 1$, i.e., the next to last row of the Routh array.

2.   A close approximation to this upper bound is given by row 2 of the array as $2/(\tau + 1)$.

These findings have recently been verified by the author who also extended the numerical results through $\tau = 7$. The values obtained are shown in Table 11.3.3. Calculations for higher values of τ using simple root-finding

Table 11.3.3

Upper Bounds on K for Stable Operation of Proportional Controller as a Function of Feedback Delay

Feedback delay, τ	K_{max} from row $\tau + 1$	Approximation $K = 2/(\tau + 1)$	Difference $[2/(\tau + 1)] - K_{max}$
0	2.0000 (row 2)	2.0000	0
1	1.0000	1.0000	0
2	0.6180	0.6667	0.0487
3	0.4450	0.5000	0.0550
4	0.3473	0.4000	0.0527
5	0.2846	0.3333	0.0487
6	0.2411	0.2857	0.0446
7	0.2086	0.2500	0.0414

† See Mitchell (1964). Referenced results are in Chapter IV of this thesis where, in the context of production inventory control, the symbol τ is a measure of production time. It turns out that τ as used here is equivalent to Mitchell's $\tau - 1$.

programs experienced some difficulty because of the high order of the equations involved.

Although it is doubtful that much need would ever arise in a practical situation for values of K_{max} for larger values of τ, some patterns indicated by the current results may be of interest. If these patterns indeed hold for all τ, then, as with the Jury test, to find the actual upper bound on K one must proceed to row $\tau + 1$, the last computed row of the array (the last row is simply the constant $B_{\tau+1}$ which is never binding for $\tau > 0$) and find the smallest positive root of the largest order numerator polynomial present. The order of this polynomial will be $O(\tau + 1) = \tilde{F}(\tau)$, which from Table 11.3.2 is observed to be equal to or less than $2^{\tau-1}$, the order of the highest Jury table polynomial, for all $\tau > 0$. Furthermore, the difference between the two increases rapidly with increasing τ which, coupled with the fact that the Jury evaluations involve absolute-value statements and the Routh calculation's straightforward polynomials, indicates that determination of upper bounds on K for higher values of τ will undoubtedly be easier using the Routh–Hurwitz format. Of perhaps more importance from the practical point of view is the approximation to K_{max} obtainable from row 2 of the Routh array. From Table 11.3.3 it would appear that the difference between $2/(\tau + 1)$ and K_{max} may have reached a peak at $\tau = 3$, meaning that as τ continues to increase, $2/(\tau + 1)$ could provide a better and better approximation. Verification is left to the interested reader with an above-average amount of spare time.

In conclusion, a safe upper bound for K could be determined from the material in Table 11.3.3 for any $\tau \geq 8$ by subtracting 0.0414 from $2/(\tau + 1)$, assuming of course that the actual difference is less. Regardless, it is apparent that as the length of the delay increases, the more restricted is the designer's selection of K and hence his ability to effect control.

Proportional-Plus-Difference Control

The few results ground out to date pertaining to the effects of delay on the allowable range of K and L for stable operation of a proportional-plus-difference process controller appear similar to those of the proportional controller. The major finding is, naturally, that the allowable range decreases as the length of the delay increases. A few more quantitative results are summarized as follows.

From the Jury criteria we find as we did with the proportional controller that all tests, when they exist, are independent of τ. As before, the larger τ is, the more rows there are in the Jury table and therefore more tests to run. $B(z = 1) > 0$ yields $K > 0$ and $|b_{\tau+1}| < |b_0|$ yields $-1 < L < 1$. Tests based on succeeding rows involve polynomials in K and L whose order is twice that of the one for the previous test. Each row test seems to correspond to a boundary for the stable range which enters the half-plane corresponding to $K > 0$ at $L = 1$, makes a smooth loop, and exits at $L = -1$. This is illustrated in Fig. 11.3.1 for the case of $\tau = 2$. As before, the boundary imposed by each succeeding row is tighter than the preceding one. Thus the boundary imposed by the last row, in this case row $2\tau + 1$, of the Jury table in conjunction with $K > 0$ defines the stable region. The curve to be plotted will thus be a polynomial in K and L of order 2^{τ}. It will also cross the K axis at the value corresponding to the upper bound on K for stable operation of the proportional controller with the same delay.

The Routh–Hurwitz criteria also offer some insight. For $\tau = 2$,

$$\hat{B}(\lambda) = (\lambda + 1)^{\tau + 2} - (\lambda + 1)^{\tau + 1}(\lambda - 1)$$

$$+ (K + L)(\lambda + 1)(\lambda - 1)^{\tau + 1} - L(\lambda - 1)^{\tau + 2}$$

$$= \sum_{k=0}^{\tau + 2} \frac{(\tau + 1)!}{k!(\tau + 2 - k)!} \{2k + (-1)^{k}[(\tau + 2 - 2k)K - 2kL]\}\lambda^{\tau + 2 - k}.$$

$$(11.3.15)$$

The nonnegativity of K is confirmed by the requirement for all coefficients of λ to be of the same sign. It again appears that the numerator polynomial of the next to last row of the Routh array, row $\tau + 2$ in this case, is the binding constraint on K and L. It will be, however, of order $\tilde{F}(\tau + 1)$ instead of 2^{τ} with the Jury table expression, so should be considerably easier to work with even though the resulting curve should be the same.

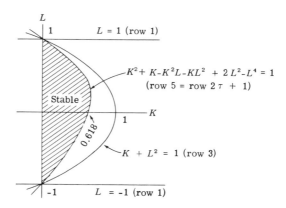

FIG. 11.3.1. Limits on range of K and L imposed by each test from the Jury table for stable operation of a proportional-plus-difference controller with a two-period feedback delay.

A somewhat conservative approximation to this curve could be found by connecting the point on the K axis for K_{max} for the proportional controller with the same delay to the points ± 1 on the L axis by line segments. As τ gets larger, this approximation should improve. Where K_{max} is not known, its approximation as discussed in the previous subsection could be used as the point on the K axis.

For the interested reader, Mitchell (1964) has worked out several cases of proportional-plus-difference and proportional-plus-summation controllers with feedback delay using the Routh–Hurwitz criteria. It might be helpful to note that this method yields the formula $K + K^2 + KL - L + L^2 + L^3 = 1$ for the binding curve in Fig. 11.3.1 which is much easier to use to plot points for an accurate reproduction.

11.4   Determination of Frequency Response from the System Transfer Function

In Chapters VII and X we derived expressions for the amplitude and phase frequency response of several fairly simple systems. In this section we generalize this procedure to general discrete, linear, time-invariant systems.

Derivation Based on Transfer Function

Our derivation assumes knowledge of the transfer function $G(z)$ of the system of interest. It is important that in this context $G(z)$ be defined specifically in terms of a given pair of points in the system so that the complete relationship between the input point and the response point is represented in $G(z)$. A review of the cases considered in earlier chapters will reveal that we have indeed done this without being overly explicit about it.

We assume with no serious loss of generality that

$$G(z) = A(z)/B(z), \qquad (11.4.1)$$

where $A(z)$ and $B(z)$ are polynomials in z, and that $B(z)$ is in the standard form discussed in Chapter VI, specifically, that the coefficient of b_0 is unity. Assume an input signal

$$r(i) = \sin wi, \qquad (11.4.2)$$

i.e., a unit sine wave of angular frequency w. For linear systems, the effect of an amplitude p is obviously to multiply the output caused by the unit sine wave by p. Furthermore, for time-invariant systems, the effect of a phase angle θ in $r(i)$ simply shifts the response to (11.4.2) by θ. Thus p and θ can be accounted for after finding the response to (11.4.2). Strictly speaking, however, frequency response is defined as the steady-state response to a unit sine wave with zero phase angle, which is what we wish to derive here.

The z transform of $r(i)$ is

$$R(z) = \frac{z \sin w}{z^2 - 2z \cos w + 1} = \frac{1}{2j} \left[\frac{z}{z - e^{jw}} - \frac{z}{z - e^{-jw}} \right]. \qquad (11.4.3)$$

Thus, the z transform of the output is

$$C(z) = R(z)G(z) = \frac{z(\sin w)A(z)}{[z^2 - 2z \cos w + 1]B(z)}. \qquad (11.4.4)$$

Dividing by z followed by partial fraction expansion yields

$$\bar{C}(z) = \frac{C(z)}{z} = \frac{(\sin w)A(z)}{[z^2 - 2z \cos w + 1]B(z)}$$

$$= \frac{K_+}{z - e^{jw}} + \frac{K_-}{z - e^{-jw}} + \sum_{r=1}^{n} \frac{K_r}{(z - z_r)^q}, \qquad (11.4.5)$$

where the summation includes the n terms representing the roots z_r of the characteristic equation $B(z) = 0$. The q exponent simply represents the fact that multiple roots require expansion terms dealing with higher powers of $z - z_r$. The exact form of these terms is not of consequence here anyway, since in the steady state they all vanish. Now

$$K_+ = \lim_{z \to e^{jw}} \left\{ (z - e^{jw}) \frac{\sin w}{(z - e^{jw})(z - e^{-jw})} G(z) \right\}$$

$$= \lim_{z \to e^{jw}} \left\{ \frac{\sin w}{z - e^{-jw}} G(z) \right\} = \frac{\sin w}{e^{jw} - e^{-jw}} G(e^{jw})$$

$$= \frac{\sin w}{2j \sin w} G(e^{jw}) = \frac{1}{2j} G(e^{jw}), \qquad (11.4.6)$$

where $G(e^{jw})$ represents the transfer function $G(z)$ with e^{jw} substituted for z wherever it appears. In similar fashion,

$$K_- = \frac{1}{-2j} G(e^{-jw}). \qquad (11.4.7)$$

Therefore,

$$C(z) = \frac{G(e^{jw})}{2j} \cdot \frac{z}{z - e^{jw}} - \frac{G(e^{-jw})}{2j} \cdot \frac{z}{z - e^{-jw}} + \sum_{r=1}^{n} \frac{K_r}{(z - z_r)^q} \qquad (11.4.8)$$

so that

$$c(i) = \frac{G(e^{jw})}{2j} (e^{jw})^i - \frac{G(e^{-jw})}{2j} (e^{-jw})^i + \begin{bmatrix} n \text{ terms of the form} \\ K_r i^q z_r{}^i \end{bmatrix}. \qquad (11.4.9)$$

For a steady state to be defined, the system must be stable, meaning $|z_r| < 1$ for all r. Therefore, for a stable system, the steady-state response is

$$c_i = \lim_{i \to \infty} c(i) = \frac{G(e^{jw})}{2j} e^{jwi} - \frac{G(e^{-jw})}{2j} e^{-jwi}. \tag{11.4.10}$$

It is easily shown and has been previously demonstrated that $G(e^{jw})$ and $G(e^{-jw})$ are complex conjugates. Therefore, $G(e^{jw})$ can be written as

$$G(e^{jw}) = \text{Re } G(e^{jw}) + j \text{ Im } G(e^{jw}) \tag{11.4.11}$$

and $G(e^{-jw})$ expressed in terms of the real and imaginary parts of $G(e^{jw})$ as

$$G(e^{-jw}) = \text{Re } G(e^{jw}) - j \text{ Im } G(e^{jw}). \tag{11.4.12}$$

By definition, the magnitude of $G(e^{jw})$ is

$$|G(e^{jw})| = \{[\text{Re } G(e^{jw})]^2 + [\text{Im } G(e^{jw})]^2\}^{1/2} \tag{11.4.13}$$

and the phase shift is

$$\phi = \tan^{-1} \frac{\text{Im}[G(e^{jw})]}{\text{Re}[G(e^{jw})]}. \tag{11.4.14}$$

Thus, we may express

$$G(e^{jw}) = |G(e^{jw})| e^{j\phi} \tag{11.4.15}$$

and

$$G(e^{-jw}) = |G(e^{jw})| e^{-j\phi}, \tag{11.4.16}$$

from which

$$c_i = \frac{|G(e^{jw})|}{2j} e^{j\phi} e^{jwi} - \frac{|G(e^{jw})|}{2j} e^{-j\phi} e^{-jwi}$$

$$= |G(e^{jw})| \left[\frac{e^{j(wi+\phi)} - e^{-j(wi+\phi)}}{2j} \right] = |G(e^{jw})| \sin(wi + \phi). \tag{11.4.17}$$

As noted previously, the output of any linear system to a unit sinusoid is a sine of the same frequency with magnitude $|G(e^{jw})|$ and phase shift ϕ. The main point of our discussion here, however, is that this magnitude and phase shift are directly obtainable from the transfer function of the system involved by substitution in $G(z)$ of

$$z = e^{jw} = \cos w + j \sin w \tag{11.4.18}$$

followed by rationalization to obtain $\text{Re } G(e^{jw})$ and $\text{Im } G(e^{jw})$. $|G(e^{jw})|$ and ϕ are then obtained from (11.4.13) and (11.4.14), respectively. Of great importance is the fact that it is not necessary, or even desirable, to find the roots of $B(z) = 0$. Note in passing the following relationship which some authors prefer to make use of in this context. Multiplication of (11.4.11) by (11.4.12) yields

$$G(e^{jw})G(e^{-jw}) = [\text{Re } G(e^{jw})]^2 + [\text{Im } G(e^{jw})]^2 \qquad (11.4.19)$$

which provides the alternative expression for the magnitude,

$$|G(e^{jw})| = [G(e^{jw})G(e^{-jw})]^{1/2}. \qquad (11.4.20)$$

Before illustrating the calculation of the frequency response for a given $G(z)$, it will be helpful to develop some computational procedures for this purpose.

Procedures for Calculation of Frequency Response

For any given $G(z)$, one could of course make a frontal attack by substitution of $z = e^{jw} = \cos w + j \sin w$ and rationalization of the result to obtain the desired magnitude and phase angle as a function of w. Considerable time and effort can be saved, however, if the necessary manipulation is done before the fact in general terms as follows.

For any $G(z)$, $G(e^{jw})$ can be expressed as

$$G(e^{jw}) = \frac{n_R + jn_I}{d_R + jd_I}, \qquad (11.4.21)$$

where n_R is the real part of the numerator, n_I the imaginary part of the numerator, d_R the real part of the denominator, and d_I the imaginary part of the denominator. Rationalizing produces

$$G(e^{jw}) = \frac{n_R + jn_I}{d_R + jd_I} \cdot \frac{d_R - jd_I}{d_R - jd_I}$$

$$= \frac{[n_R d_R + n_I d_I] + j[-n_R d_I + n_I d_R]}{d_R^2 + d_I^2}. \qquad (11.4.22)$$

The magnitude can therefore be expressed as

$$|G(e^{jw})| = \{[\text{Re } G(e^{jw})]^2 + [\text{Im } G(e^{jw})]^2\}^{1/2}$$

$$= \left[\frac{n_R^2 d_R^2 + 2n_R d_R n_I d_I + n_I^2 d_I^2}{(d_R^2 + d_I^2)^2} + \frac{n_R^2 d_I^2 - 2n_R d_I n_I d_R + n_I^2 d_R^2}{(d_R^2 + d_I^2)^2}\right]^{1/2}$$

$$= \left[\frac{n_R^2 d_R^2 + n_I^2 d_I^2 + n_R^2 d_I^2 + n_I^2 d_R^2}{(d_R^2 + d_I^2)^2}\right]^{1/2} = \left[\frac{n_R^2 + n_I^2}{d_R^2 + d_I^2}\right]^{1/2}, \qquad (11.4.23)$$

and the phase angle as

$$\phi = \tan^{-1}\frac{\text{Im } G(e^{jw})}{\text{Re } G(e^{jw})} = \tan^{-1}\frac{-n_R d_I + n_I d_R}{n_R d_R + n_I d_I}. \qquad (11.4.24)$$

Use of (11.4.23) and (11.4.24) will facilitate considerably the determination of $|G(e^{jw})|$ and ϕ for a given $G(z)$. Nevertheless, the following example will demonstrate that even with these aids, a certain amount of tedium is associated with determining the frequency response of all but the simplest systems.

Example   *Simple Second-Order System*   Consider the simple system represented by

$$G(z) = \frac{z + a}{z^2 + bz + c}, \tag{11.4.25}$$

where a, b, and c are constants such that the system is stable. Therefore,

$$G(e^{jw}) = \frac{\cos w + j \sin w + a}{\cos 2w + j \sin 2w + b \cos w + jb \sin w + c}$$

$$= \frac{[\cos w + a] + j[\sin w]}{[\cos 2w + b \cos w + c] + j[\sin 2w + b \sin w]}, \tag{11.4.26}$$

from which

$$n_R = \cos w + a, \qquad d_R = \cos 2w + b \cos w + c,$$
$$n_I = \sin w, \qquad d_I = \sin 2w + b \sin w.$$

Substitution into (11.4.23) gives

$$|G(e^{jw})| = \left[\frac{(\cos w + a)^2 + (\sin w)^2}{(\cos 2w + b \cos w + c)^2 + (\sin 2w + b \sin w)^2} \right]^{1/2}, \tag{11.4.27}$$

which after some manipulation and the somewhat arbitrary but seemingly appropriate use of trigonometric identities yields

$$|G(e^{jw})| = \left[\frac{2a \cos w + (a^2 + 1)}{c \cos^2 w + 2b(c + 1) \cos w + (b^2 + c^2 - 2c + 1)} \right]^{1/2}. \tag{11.4.28}$$

Similarly,

$$\phi = \tan^{-1} \frac{-(\cos w + a)(\sin 2w + b \sin w) + \sin w(\cos 2w + b \cos w + c)}{(\cos w + a)(\cos 2w + b \cos w + c) + \sin w(\sin 2w + b \sin w)}$$

$$= \tan^{-1} \frac{a \sin 2w + (c - ab - 1) \sin w}{a \cos 2w + (c + ab + 1) \cos w + (b + ac)}. \tag{11.4.29}$$

To no one's surprise, there are several steps of manipulation between the two expressions in (11.4.29).

It is interesting to note that for

$$G(z) = z^k \frac{z + a}{z^2 + bz + c}, \tag{11.4.30}$$

the magnitude $|G(e^{jw})|$ is still given by (11.4.28). This is because for $z = e^{jw}$, $|z| = 1$ for all w, so multiplication by z^k does not change the magnitude of the response. This result is also logical in that multiplication of a function by an integer power of z corresponds to a translation in time of a finite number of periods, which will not affect the magnitude of the steady-state output. The phase angle will of course be affected unless by chance the length of the time period is an integer multiple of the period of the sinusoid.

The frequency response can be very helpful in designing systems to minimize the effects of sinusoids with known frequencies or with frequencies in certain ranges. A most fruitful area of application is in production and inventory control where the primary cost generators are changes in production and inventory levels and significant seasonal and other cyclic patterns of customer orders are anticipated. This problem is explored in one of the exercises.

Exercises

11.1   Derive an expression for the impulse response of the single-arc equivalent of the following networks, where r represents the input signal, c the output signal, and g_k the unit-impulse response of arc k, in terms of the g_k or, when necessary, their z transforms $G_k(z)$.

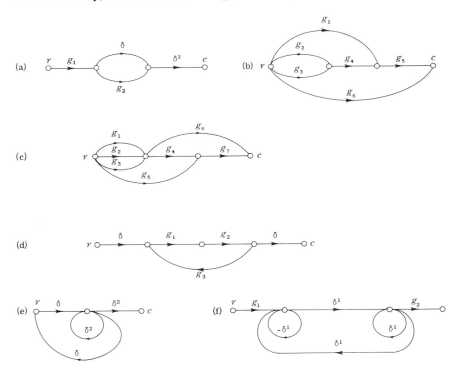

11.2   Derive a formula for the impulse response of the single-arc equivalent of the following networks, where r represents the input signal and c the output signal for the given arc unit-impulse responses:

a.

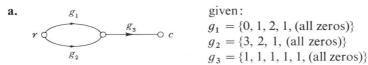

given:

$g_1 = \{0, 1, 2, 1, (\text{all zeros})\}$
$g_2 = \{3, 2, 1, (\text{all zeros})\}$
$g_3 = \{1, 1, 1, 1, 1, (\text{all zeros})\}$

b.   For the network of Exercise 11.1(a) given:

$g_1 = \{2, 1, (\text{all zeros})\}, \qquad g_2 = \{g_2(i) = (\tfrac{1}{2})^i,\ i = 0, 1, \ldots\}$

c.   For the network of Exercise 11.1(d) given:

$g_1 = \{2, 1, (\text{all zeros})\}, \qquad g_2 = \{1, 1, 1, (\text{all zeros})\}$
$g_3 = \{g_3(i) = (\tfrac{1}{3})^i,\ i = 0, 1, \ldots\}$

11.3

a.   Verify that the single-arc equivalent of the expanded network in Fig. 11.1.6a is equivalent to the single arc in Fig. 11.1.5b.

b.   Repeat part (a) for the expanded network in Fig. 11.1.6b.

11.4   Verify Eq. (11.1.28) for the "desired-level transfer function" of the discrete generalized process controller by reducing the flow graph of Fig. 11.1.10b to a single-arc equivalent.

11.5

a.   Reduce the graph in Fig. 11.1.13 of the control rule of the generalized nth-order process controller to an equivalent single arc.

b.   Substitute the transfer function of the single arc derived in part (a) for $G_D(z)$ in (11.1.24) to obtain the perturbation transfer function of the generalized process controller.

11.6   **(a)** and **(b)**.   The same as the corresponding part of 11.5 except for the graph in Fig. 11.1.4 of the control rule of the proportional-plus-difference-plus-summation controller.

11.7   The Fibonacci series of numbers is defined by the recursive relationship $F(n) = F(n - 1) + F(n - 2)$, where $F(n)$ is defined as the nth Fibonacci number and $F(0) = F(1) = 1$. Construct a signal flow graph whose unit-impulse response generates the Fibonacci series.

11.8   **(a)** through **(g)**.   Represent the system of Exercise 2.3 in flow graph form and find the transfer function of the single-arc equivalent which relates the perturbation $p(i)$ to system output $m(i)$ for the conditions given in parts (a) through (g), respectively, of 2.3. Compare answers to those of Exercise 5.7.

11.9   **(a)** through **(g)**.   The same as 11.8 except for the single-arc equivalent which relates the measurement error $\varepsilon(i)$ to process output $m(i)$. Compare answers to those of 5.7.

11.10 (a) through (g). The same as 11.8 except for the single-arc equivalent which relates the desired level of process operation D to process output $m(i)$. Compare answers to those of 5.7.

11.11 (a) through (g). The same as 11.8 except for the system of Exercise 2.4.

11.12 (a) through (g). The same as 11.9 except for the system of Exercise 2.4.

11.13 (a) through (g). The same as 11.10 except for the system of Exercise 2.4.

11.14 (a) through (f). Represent the system of Exercise 2.5 in flow graph form and find the transfer function of the single-arc equivalent which relates customer orders $p(i)$ to inventory level $m(i)$ for the conditions given in parts (a) through (f), respectively, of 2.5. Compare answers to those of Exercise 5.9.

11.15 (a) through (f). The same as 11.14 except for the single-arc equivalent which relates customer orders $p(i)$ to production level $a(i)$.

11.16 The same as Exercise 11.14(e) except instead of using single exponential smoothing to forecast customer orders, use the second-order forecaster $p(i, 0) = 2(1 - a)p(i - 1, 0) - (1 - a)p(i - 2, 0) + ap(i - 1)$, where $0 \le a \le 1$ and $p(0, 0) = p(1, 0) = 0$.

11.17 The same as Exercise 11.15(e) except using the second-order forecaster described in Exercise 11.16.

11.18 The same as 11.14(c) with $\tau_I = \tau_L = 0$ except instead of the rapid-response capability, g_P is represented by a distributed lag with probability of a lag of τ periods given by $f(\tau) = [1/(1 - A)]A^{-\tau}, A > 1$.

11.19 (a) through (g). Represent the simplified switching and delay model of the criminal justice system presented in Fig. 2.5.1 in flow graph form and find the transfer function of the single-arc equivalent which relates virgin arrests $V(i)$ to the dependent variable of the corresponding part of Exercise 2.7. Assume $\tau_I \ge \tau_{PC}$. Compare answers to those of 5.12.

11.20 Represent the water supply system of Exercise 2.9 in flow graph form and find the transfer function of the single-arc equivalent which relates demand to water intake from the river. Compare the answer to that of 5.13.

11.21 The same as 11.20 except for the single-arc equivalent which relates the leakage L_2 from the pure water reservoir to water intake from the river.

11.22   Represent the system derived in Exercise 2.10 in flow graph form and find the transfer functions of all single-arc equivalents relating each significant system input to the performance of each important system component.

11.23   For a proportional controller subject to feedback delay such that the system difference equation is given by (11.3.1), use the Jury test followed by numerical evaluation of the smallest positive root of the resulting polynomial to determine the stable range of the proportionality constant K for the following values of τ. Compare results with those of Table 11.3.3.

    a. $\tau = 2$ **b.** $\tau = 3$ **c.** $\tau = 4$

11.24   Use the Jury test to establish the stable range of control parameters for the following controllers subject to the delays listed:

    a. K_0 and K_1 for the generalized second-order process controller with feedback delay $\tau = 1$
    b. The same as part (a) except $\tau = 2$
    c. K and L for the proportional-plus-difference process controller with feedback delay $\tau = 1$
    d. The same as part (c) except $\tau = 2$
    e. K and N for the proportional-plus-summation process controller with feedback delay $\tau = 1$
    f. The same as part (e) except $\tau = 2$

11.25   For a proportional controller subject to feedback delay such that the system difference equation is given by (11.3.1), use the Routh–Hurwitz test followed by numerical evaluation of the smallest positive root of the resulting polynomial to determine the stable range of the proportionality constant K for the following values of τ. Use the results to extend Table 11.3.3.

    a. $\tau = 5$ **b.** $\tau = 6$ **c.** $\tau = 7$ **d.** $\tau = 8$

11.26   **(a)** through **(f)**.   The same as the corresponding part of Exercise 11.24 except using the Routh–Hurwitz test. If the appropriate part of 11.24 has been done, compare procedures and results.

11.27   For the production inventory control system described in Exercise 2.3 with g_P such that $a(i) = c(i)$ and g_D such that $c(i) = K_\rho \rho(i) + K_P p(i - 1)$, find the combinations of values of the constants K_ρ and K_P which lead to stable system operation for the following values of τ_I and τ_L:

    a. $\tau_I = \tau_L = 0$ (neither feedback nor feedforward delay)
    b. $\tau_I = 1, \tau_L = 0$ (feedback delay)
    c. $\tau_I = 0, \tau_L = 1$ (feedforward delay)
    d. $\tau_I = \tau_L = 1$ (feedback and feedforward delay)

11.28 Find the steady-state output c_i resulting from an input of $r(i) = \sin(wi + \theta)$ for a system with the following transfer functions:

a. $G(z) = \dfrac{3}{z + 0.1}$

c. $G(z) = \dfrac{z}{z^3 - 0.39z + 0.20}$

b. $G(z) = \dfrac{z}{z^2 - 0.1z - 0.2}$

d. $G(z) = \dfrac{z^2}{z^3 - 1.20z^2 + 0.57z - 0.10}$

11.29 (a) through (d). Find the maximum and minimum values of the amplitude frequency response and the values of w at which they occur for the system in the corresponding part of Exercise 11.28.

11.30 The production inventory control system described in Exercise 2.3 with $\tau_I = \tau_L = 0$ and g_P such that $a(i) = c(i)$ is subject to a pattern of customer orders which contains a sinusoidal component of angular frequency w radians per period.

    a. Determine the control rule g_D which will minimize the function $C = C_P A_P{}^2 + C_I A_I{}^2$, where A_P is the relative amplitude of the resulting steady-state sinusoidal variation in production level, A_I is the relative amplitude of the resulting steady-state sinusoidal variation in inventory level, and C_P and C_I are constants.

    b. Check the control rule derived in part (a) for stability.

    c. If the rule found leads to unstable operation, discuss how it might be modified to obtain a compromise between stability and minimization of C.

11.31 through 11.33 The same as 11.30 except for the following values of τ_I and τ_L:

    11.31 $\tau_I = 1, \tau_L = 0$
    11.32 $\tau_I = 0, \tau_L = 1$
    11.33 $\tau_I = \tau_L = 1$

11.34 For the generalized third-order process controller whose system difference equation is obtained by setting $n = 3$ in (2.3.34):

    a. Solve for $m(i)$ as a function of D, $\varepsilon(i)$, $p(i)$, the system parameters and appropriate initial conditions using variation of parameters. Be sure to consider all possibilities for types and multiplicities of the roots of the characteristic equation.

    b. The same as part (a) except using the z transform.

    c. Determine the combinations of K_0, K_1, and K_2 which result in stable system operation.

    d. Determine the combinations of K_0, K_1, and K_2 which result in each set of types and multiplicities of roots.

e. Derive expressions for the mean $E[m(i)]$ and variance $V[m(i)]$ of $m(i)$ given $\varepsilon(i)$ and $p(i)$ unbiased, identically, and independently distributed with variances σ_ε^2 and σ_p^2, and mutually independent.

f. Derive expressions for the amplitude and phase frequency responses.

11.35 **(a)** through **(f)**. The same as Exercise 11.34 except for the proportional-plus-difference-plus-summation controller whose system difference equation is given by (2.3.30) and in standard form for solution in (4.2.59). In this system the parameters involved are K, L, and N.

References

Azeltine, J. A. (1958). *Transform Method in Linear System Analysis*. McGraw-Hill, New York.

Beightler, C. S., Mitten, L. G., and Nemhauser, G. L. (1961). A short table of z-transforms and generating functions. *Oper. Res.* **9**, No. 4, 574–578.

Belcher, P. J. (1971). *Applications of Discrete Linear Control Theory to Automatic Control of Car-Following Headway*. Unpublished term paper submitted to A. Bishop.

Belkin, J., Blumstein, A., and Glass, W. (1973). Recidivism as a feedback process: An analytical model and empirical validation. *J. Crim. Justice* **1**, 7–26.

Bishop, A. B. (1960). *Discrete Random Feedback Models in Industrial Quality Control*. Eng. Experiment Station Bull. 183, pp. 70–75. Ohio State Univ., Columbus, Ohio.

Blumstein, A., and Larson, R. (1969). Models of a total criminal justice system. *Oper. Res.* **17**, No. 2.

Box, G. E. P., and Jenkins, G. M. (1970). *Time Series Analysis: Forecasting and Control*. Holden-Day, San Francisco, California.

Brockett, R. W. (1970). *Finite Dimensional Linear Systems*. Wiley, New York.

Brown, R. G. (1963). *Smoothing, Forecasting, and Prediction of Discrete Time Series*. Prentice-Hall, Englewood Cliffs, New Jersey.

Burington, R. S. (1957). *Handbook of Mathematical Tables and Formulas*. Handbook Publ., Sandusky, Ohio.

Cadzow, J. A. (1973). *Discrete-Time Systems*. Prentice-Hall, Englewood Cliffs, New Jersey.

Cadzow, J. A., and Martens, H. R. (1970). *Discrete-Time and Computer Control Systems*. Prentice-Hall, Englewood Cliffs, New Jersey.

Chang, S. S. (1961). *Synthesis of Optimal Control Systems*. McGraw-Hill, New York.

Churchman, C. W., Ackoff, R. L., and Arnoff, E. L. (1957). *Introduction to Operations Research*. Wiley, New York.

Cohn, A. (1922). Über die Anzahal der Wurzeln einer algebraichen Gleichung in einem Kreise. *Math. Z.* **14**, 110–148.

Conte, S. D. (1965). *Elementary Numerical Analysis—An Algorithmic Approach*. McGraw-Hill, New York.

Cosgriff, R. L. (1958). *Nonlinear Control Systems*. McGraw-Hill, New York.

Friedman, B. (1949). Note on approximating complex zeros of a polynomial. *Commun. Pure Appl. Math.* **2**, No. 2–3, 195–208.

Gardner, M. F., and Barnes, J. L. (1942). *Transients in Linear Systems*, Vol. 1. Wiley, New York.

Giffin, W. C. (1971). *Introduction to Operations Engineering*. Richard D. Irwin, Homewood, Illinois.

Gordon, R. L. (1969). A technique for control of traffic at critical intersections. *Transportation Sci.* **3**, No. 4.

Guillemin, E. A. (1949). *The Mathematics of Circuit Analysis*. Wiley, New York.

Hildebrand, F. B. (1952). *Methods of Applied Mathematics*. Prentice-Hall, New York.

Hitch, C. J., and McKean, R. N. (1965). *The Economics of Defense in the Nuclear Age*. Harvard Univ. Press, Cambridge, Massachusetts.

Howard, R. A. (1971a). *Dynamic Probabilistic Systems*, Vol. I, *Markov Models*. Wiley, New York.

Howard, R. A. (1971b). *Dynamic Probabilistic Systems*, Vol. II, *Semi-Markov and Decision Processes*. Wiley, New York.

Jolley, L. B. W. (1961). *Summation of Series*, 2nd rev. ed. Dover, New York.

Jury, E. I. (1962). A simplified stability criterion for linear discrete systems. *Proc. IRE* **50**, 1493–1500.

Jury, E. I., and Blanchard, J. (1961). A stability test for linear discrete systems in table form. *Proc. IRE* **49**, 1947–1948.

Kendall, M. G., ed. (1971). *Cost-Benefit Analysis*. Amer. Elsivier, New York.

Klir, G. J. (1972). *Trends in General Systems Theory*. Wiley, New York.

Laning, J. H., and Battin, R. H. (1956). *Random Processes in Automatic Control*. McGraw-Hill, New York.

LaSalle, J., and Lefschetz, S. (1961). *Stability by Luipunov's Direct Method, with Applications*. Academic Press, New York.

Lin, S.-N. (1941). Method of successive approximations of evaluating the real and complex roots of cubic and higher-order equations. *J. Math. Phys.* **20**, No. 3.

Luke, Y. L., and Ufford, D. (1951). On the roots of algebraic equations. *J. Math. Phys.* **30**, No. 2, 94–101.

Mangulis, V. (1965). *Handbook of Series for Scientists and Engineers*. Academic Press, New York.

Marden, M. (1966). *The Geometry of Polynomials*, 2nd ed. Amer. Math. Soc., Providence, Rhode Island.

Mason, S. J. (1953). Feedback theory—Some properties of signal flow graphs. *Proc. IRE* **41**, 1144–1156.

Mason, S. J. (1956). Feedback theory—Further properties of signal flow graphs. *Proc. IRE* **44**, 920–926.

Merriam, C. W. (1964). *Optimization Theory and the Design of Feedback Control Systems*. McGraw-Hill, New York.

Miller, K. S. (1960). *An Introduction to the Calculus of Finite Differences and Difference Equations*. Holt, New York.

Mitchell, L. C. (1964). *The Application of Discrete-Variable Servomechanism Theory to the Problems of Industrial Control*. Unpublished Master's thesis. Ohio State Univ., Columbus, Ohio.

Mood, A. M. (1950). *Introduction to the Theory of Statistics*. McGraw-Hill, New York.

Naddor, E. (1966). *Inventory Systems*. Wiley, New York.

Nielsen, K. L. (1956). *Methods in Numerical Analysis*. Macmillan, New York.

Quade, E. S., and Boucher, W. I. (1968). *Systems Analysis and Policy Planning: Applications in Defense*. Amer. Elsevier, New York.

Ragazinni, J. R., and Franklin, G. F. (1958). *Sampled-Data Control Systems*. McGraw-Hill, New York.

Ragazinni, J. R., and Zadeh, L. A. (1955). The analysis of sampled-data systems. *Trans. Amer. Inst. Elec. Eng. Part 2* **71**, 225–232.

Sargent, R. G. (1966). *A Discrete Linear Feedback Control Theory Inventory Model*. Unpublished Doctoral thesis. Univ. of Michigan, Ann Arbor, Michigan.

Scarf, H. E., Gilford, D. M., and Shelly, M. W. (1963). *Multistage Inventory Models and Techniques*. Stanford Univ. Press, Stanford, California.

Schur, I. (1917). Über Potenzreihen, die in Innern des Ernheitskreises beschrankt sind. *J. Reine Angew Math.* **147**, 205–232.

Schur, I. (1918). Über Polynome, die nur in Innern des Ernheitkreis verschwinden. *J. Reine Angew. Math.* **148**, 122–145.

Shinners, S. M. (1964). *Control System Design*, pp. 393–396. Wiley, New York.

Taylor, A. E. (1955). *Advanced Calculus*. Ginn, Boston, Massachusetts.

Tou, J. T. (1959). *Digital and Sampled Data Control Systems*. McGraw-Hill, New York.

Tou, J. T. (1963). *Optimum Design of Digital Control Systems*. Academic Press, New York.

Truxal, J. G. (1955). *Automatic Feedback Control System Synthesis*. McGraw-Hill, New York.

Truxal, J. G., ed. (1958). *Control Engineers' Handbook*. McGraw-Hill, New York.

Wagner, H. M. (1969). *Principles of Operations Research*. Prentice-Hall, Englewood Cliffs, New Jersey.

Widrow, B. (1956). A study of rough amplitude quantization by means of Nyquist sampling theory. *IRE Trans. Circuit Theory* **CT-3**, No. 4.

Wiener, N. (1948). *Cybernetics*. Wiley, New York.

Wiener, N. (1949). *The Extrapolation, Interpolation, and Smoothing of Stationary Time Series*. Wiley, New York.

Wilde, D. J., and Beightler, C. S. (1967). *Foundations of Optimization*. Prentice-Hall, Englewood Cliffs, New Jersey.

Index